The Present Situation in the Philosophy of Science

Proceedings of the ESF Research Networking Programme

THE PHILOSOPHY OF SCIENCE IN A EUROPEAN PERSPECTIVE

Volume 1

EUROPEAN SCIENCE FOUNDATION
SETTING SCIENCE AGENDAS FOR EUROPE

www.pse-esf.org

Friedrich Stadler
General Editor

Together with:
Dennis Dieks, Wenceslao J. González,
Stephan Hartmann, Thomas Uebel
and Marcel Weber

The Present Situation in the Philosophy of Science

Springer

Editor
Friedrich Stadler
Institute Vienna Circle
University of Vienna
Department of Philosophy and Department of Contemporary History
Campus der Universität Wien, Hof 1
A-1090 Vienna
Austria
Friedrich.Stadler@univie.ac.at

ISBN 978-90-481-9114-7 e-ISBN 978-90-481-9115-4
DOI 10.1007/978-90-481-9115-4
Springer Dordrecht Heidelberg London New York

Library of Congress Control Number: 2010926739

Cover design: eStudio Calamar S.L.

Printed on acid-free paper

Springer is part of Springer Science+Business Media (www.springer.com)

Table of Contents

Friedrich Stadler

Editorial: On the Present Situation in the Philosophy of Science

The University of Vienna saw witness to the Opening Conference of the ESF-Research Networking Programme "The Philosophy of Science in a European perspective" (PSE) which was organised by the Vienna Circle Institute[1] and took place on the 18–20 December at the Campus of the University of Vienna, 2008.[2] Its overarching aim was to set the background for a collaborative project organising, systematising, and ultimately forging an identity for, European philosophy of science by creating research structures and developing research networks across Europe to promote its development. As such under the general rubric of 'the present situation in the philosophy of science', the emphasis was on as a first step identifying traditions and research structures already present, and the directions in which this research was leading.

This volume presents the papers of the opening conference according to five pre-established groupings, each represented by speakers from a team:

Team E: Foundational and Methodological Debates (team leader Thomas Uebel)
Team A: Formal Methods and their Applications to the Philosophy of Science (team leader Stephan Hartmann)
Team B: Approaches to the Foundations of Science: the Place of the Life Sciences (team leader Marcel Weber)
Team C: The Present Situation of the Philosophy of the Cultural and Social Sciences: The "Naturalist Turn", the "Social Turn", and the Discussion on Scientific Realism (team leader Wenceslao J. González)
Team D: Philosophical Foundations of the Physical Sciences (team leader Dennis Dieks).

To start with a broad overview of the talks themselves, and the accompanying discussion, we can characterise the general themes that were pursued and issues that emerged within the frame of 'European philosophy of science'. One aspect that came readily to the fore was the strong historical aspects of the 'European perspective.' As the opening conference it was of course important for the philosophers

1 Cf. the activities of the Institut Wiener Kreis/Institute Vienna Circle: www.univie.ac.at/ivc.

2 See the detailed conference report: Friedrich Stadler, Donata Romizi, Miles MacLeod, "The Present Situation in the Philosophy of Science: Opening Conference of the ESF-Research-Networking Programme 'The Philosophy of Science in a European Perspective'", in: *Journal for General Philosophy of Science* (forthcoming). Published online August 13, 2009: DOI 10.1007/s10838-009-9088-y.

involved to define what could be meant by a European identity, and naturally they focused on the deep historical roots and their continuing traditions in modern philosophy of science. As such there was a strong attention to the historical relations and origins in European history of modern issues, and how in fact this context presented an ongoing influence on the modern practice.

The conference also revealed that this historical dimension is complemented by the evident broad scope of European philosophy of science which embodies not only a strong tradition of history and philosophy of science, history of philosophy of science, but also philosophy with respect to the cultural and social sciences as part of (not separate to) the discipline, combined with more traditional philosophical issues and approaches, such as the application of formal methods, the problem of realism, determinism and chance or the natural kinds debate. This consideration of general philosophical questions in science is married to a strong tradition of engaging naturalistically with the particular philosophical issues in individual sciences where there exists a prerogative of being closely schooled in the relevant scientific theory and research context. Additionally, one can refer to particular positions, like 'structural realism', as 'European', having their origin and their centre of pursuit, and indeed their historical links, in the context of European research.

Combining these elements is the quintessentially European self-reflection on the aims and values of philosophy of science in itself and the right methodology with which to do it. This is an ever-present theme, which traces its roots strongly also in the history of European philosophy. This was raised directly with respect to discussions on 'naturalism' but arose in the context of discussion over formal methods, natural kinds, and the relations between social and cultural sciences and the natural sciences.

The team leaders organised the 5 sections, elaborating generally on its theme. Each of the papers are complemented by prepared commentaries from one invited commentator, occasionally taking into account the general discussion.

The pre-history and research background of the programme running for 5 years up to 2013 is the beginning of a promising interdisciplinary networking and cooperation in the philosophy of science all over Europe with 17 participating countries and structured in 5 teams with more than 60 scholars coming from 22 countries – renowned scientists as well as younger gifted philosophers of science.[3] This collaborative enterprise is based on two previous ESF-research networks on "Historical and Contemporary Perspectives of Philosophy of Science" and on "The Philosophy of Physics". Together with the "European Philosophy of Science Association" (EPSA),[4] which was founded in 2006 this programme seems a promising forum to improve the cooperation and interaction in the flourishing philosophy of science. There is a long and powerful tradition in this research field and later on with the emergence of philosophy of science as a discipline since

3 www.pse-esf.org

4 www.epsa.ac.at

the beginning of the 20th century in the capitals of Berlin, Prague, Warsaw and in Vienna (with the Berlin Group around Reichenbach to the Vienna Circle around Schlick), but also in France, Italy and Great Britain. The forced migration of the movements of Logical Empiricism before World War II led to a radical transfer and cognitive transformation, which can be characterized as a turn "from *Wissenschaftslogik* (Logic of Science) to Philosophy of Science".[5] The institutional and academic discipline was a result of this transatlantic interaction and transition. In North America this move has led to an early institutionalization and professionalization of philosophy of science, as became manifest in 1934 with the founding of the Journal *Philosophy of Science* and later on the society *Philosophy of Science Association* (PSA). Only decades later, in the 1960s we can recognize a return of these currents back to Europe and a pleasing mutual communication between Europe and the USA. In the meantime there exists a lively cooperation with the North American community, as is documented partly with the PSA meetings, the International Society for History and Philosophy of Science (HOPOS), even if this was not always a symmetric interaction.

Therefore the recent developments and inceptions in Europe are seen as a welcomed scholarly counterpart and a collaborative research activity which is reviving the European tradition as well as fostering to days' increasing efforts and potentials in European's philosophy of science, but without aiming at an exclusive "Euro-centric" approach.

Already the heritage of the Vienna Circle was not theoretical uniformity but plurality, e.g., with the principle of tolerance and the acknowledgment of an integrated history and philosophy of science as a heuristic strategy completing the linguistic and semantic turns with a pragmatic-historical dimension. The reformulation and diversification of philosophy of science was thus pre-coined, even if not yet fully elaborated because of historical reasons.

Given this prehistory and intellectual context the ESF-Programme is designed as a further development based on earlier conceptions but also challenging some well known historically determined dogmas (like analytic/synthetic, theory/observation, context of discovery/context of justification, induction/deduction) incorporating recent European and global research results. Accordingly, the opening conference aimed at addressing the current situation of philosophy of science in Europe with reference to the main topics and recent results as a sort of description and critical account of the state of the art regarding the main foundational and methodological issues.

In the forthcoming years the five teams will focus in separate workshops related to annual topics from their specific perspective: "Explanation, prediction and confirmation" (2009), "Probability and statistics" (2010), and "The Sciences that

5 cf. Friedrich Stadler, "History of the Philosophy of Science. From *Wissenschaftslogik* (Logic of Science) to Philosophy of Science: Europe and America, 1930-1960", in: Theo Kuipers (Ed.), *Handbook of the Philosophy of Science – Focal Issues*. Amsterdam etc.: Elsevier 2007, pp. 577-658.

philosophy has neglected" (2011). A large closing conference will be held in Bologna on "New Directions in the Philosophy of Science". A strong interaction and cooperation between these groups with invited speakers in addition is intended and will lead to a series of 5 books as proceedings of the whole programme of which this volume is published as the first one.

In her closing remarks at the conference, Maria Carla Galavotti, the chair of the programme, summarized with the methodological lesson that a typical trait of European philosophy of science is attention to historical context and the use of history to identify trends in argumentation, and provide perspectives and interpretations on contemporary debates. Especially, she noted at least 3 important insights: (1) the importance of the historical research, (2) the roots of empiricism in Europe, and 3) the significance of pragmatism for the future investigations. The status quo of philosophy of science after the "historical turn" is characterized by plurality and specialization all over the world. The European perspective in philosophy of science is the inclusion of the historical roots of current debates and the focus on methodological problems that cross the various sub-disciplines.

This volume is a serious attempt to open up the subject of European philosophy of science to real thought, and provide the structural basis for the interdisciplinary development of its specialist fields, but also to provoke reflection on the idea of 'European philosophy of science'. This efforts should foster a contemporaneous reflection on what might be meant by philosophy of science in Europe and European philosophy of science, and how in fact awareness of it could assist philosophers interpret and motivate their research through a stronger collective identity.

Vienna, October 2009

Acknowledgment: Thanks go to the chair Maria Carla Galavotti and secretary Cristina Paoletti, the five team leaders and to all contributors, the members of the Steering Committee of the PSE-program for their cooperation. I am grateful to Donata Romizi and Miles MacLeod (University of Vienna, Doctoral Program "The Sciences in Historical Context") for their essential contribution to the basic conference report, and Robert Kaller (Institute Vienna Circle) for his layout and editorial work with this first volume in the series of the PSE-programme with Springer Publisher (Dordrecht).

University of Vienna
Department of Philosophy and Department of Contemporary History
Institute Vienna Circle
Universitätscampus, Hof 1
A-1090 Wien, Austria
Friedrich.Stadler@univie.ac.at

Part I
History of the Philosophy of Science

Thomas Uebel

Some Remarks on Current History of Analytical Philosophy of Science

For this first plenary conference of our network, the history of philosophy science team is presenting its "overview plus" of the current state of the discipline under fairly traditional headings. I should say right away therefore that the purpose of doing so is that of giving a fairly clear identification of each contribution's starting point. The categories chosen are meant to be neither exclusive nor exhaustive – nor, indeed, evaluative.

I

Let me begin by noting that the history of philosophy of science has made tremendous progress over the last two decades. That there now exists an international scholarly society (called "HOPOS") dedicated to work in this field with biennial conferences and a planned journal is even an institutional indicator of the progress made. But before we can ask "So where are we in our discipline?", we must ask not only "What made this development possible?" but also "What precisely is it that has developed here?" Both of the preliminary questions have quick answers, but there are considerable complexities hiding behind their superficial plausibility.

One of the things that made the growth of history of philosophy of science possible is the ever-increasing distance of philosophy of science from its beginnings. While such a distance inevitably encourages disciplinary self-reflection in a historical vein, there is in the present case an additional poignancy which derives from the fact that, at least as remembered, analytical philosophy of science originally professed to care little about history. So we get the apparent dialectic that, with growing age, a once a- or even self-consciously un-historical discipline acquires historical consciousness. One may remark that this is hardly surprising nor in itself newsworthy, but this is not the only development to be taken note of.

Here I'm thinking of a change in methodological attitude that late 20th century philosophy of science prided itself on, a change sometimes characterised as a naturalistic turn or even a turn to scientific practice: either way it involves the self-conscious rejection of *a priori* reflection about grand philosophical themes related to science and instead demands detailed knowledge of current scientific theories and experimental practices. So if history of philosophy of science wants to partake in this change – which I presume – its practitioners must engross themselves also in history of science and details of past scientific practices in particular scientific

F. Stadler (ed.), *The Present Situation in the Philosophy of Science*,
The Philosophy of Science in a European Perspective, DOI 10.1007/978-90-481-9115-4_1,

disciplines. And given the developments in history of science in turn, this typically involves, at a minimum, also an awareness, not only of developments internal to the scientific theorisings at issue, but also of the socio-political and cultural contexts of these theorisings.

A qualification is necessary here. The turn to details of first-order scientific theory and practice is demanded of history of philosophy of science to different degrees, namely, as appropriate to the cases at hand. In cases where the philosophies under consideration themselves concerned very general themes and were developed at arms length from scientific practice, extensive attention to details of the latter is less imperative than in cases where the philosopher in question was a practicing scientist. Parallel qualifications appear indicated concerning the consideration of the socio-political and cultural contexts of philosophies of science, but here the dynamic seems reversed: the greater their distance from scientific practice and the more general the topics and conclusion, the more such contextualising seems required to understand the presuppositions of these philosophies. Again it is difficult to generalise, but certain tendencies seem clear enough.

History of philosophy of science then, at least in its noblest intent, aspires to be an interdisciplinary undertaking to an even greater extent than current philosophy of science. It seeks to comprehend developments in the philosophy of science in relation both to the technical context of first-order scientific theorising and in relation the general socio-cultural context.

Note that I did not say "*the* development of philosophy of science", for so far such grand narratives have not been not on the agenda. This is not to say that ideas of historical trajectories do not inform work in the history of philosophy of science, but it seems to me that its practitioners are too aware of the pitfall of reaching for views of the development of the philosophy of science *sub specie aeternitatis*. Their work concerns more or less extended episodes and considers them from a particular philosophical viewpoint—be that problem or programme-based.

II

Turning now to a brief and somewhat rough catalogue of work done in recent history of philosophy of science, I will draw some minor morals bearing on the taxonomic misgivings I alluded to earlier. I will then go on to offer some thoughts on why work in history of philosophy of science may be turned to philosophical gain.

Sticking to my brief for the 20th century and the analytic tradition, it is possible categorise the work done in terms of philosophical movements and periods and in terms of the scientific disciplines with which it is concerned.[1] (That with some of the figures we reach back into the late 19th century seems inevitable: those with other foci are welcome to plunder the early and late 20th century in turn.)

Without any claim to completeness we find a certain concentration, *in terms of movements* (within or fading into the analytic tradition):

- Austro-German positivism (Mach, Petzold);
- French conventionalism (Poincaré, Duhem);
- British empiricism (Russell);
- logicism (Frege, Russell);
- formalism (Hilbert and his school);
- early logical empiricism (Vienna Circle and Berlin Society);
- Neo-Kantian philosophy of science (Cassirer);
- Lvov-Warsaw school of logic (Twardowski, Lukasiewicz, Tarski);
- orthodox logical empiricism (post-WWII North America);
- critical rationalism (Popper);
- post-WWII Austro-German philosophy of science (Stegmüller);
- the emergence of post-positivist philosophy of science (Feyerabend, Kuhn, Hanson);

in terms of disciplines:

- logic (predicate logic and model theory);
- mathematics (foundational issues);
- probability theory;
- space-time theories;
- quantum physics;
- chemistry;
- biology;
- psychology;
- social science.

Needless to say, these concentrations do cross-cut in various ways between movements and disciplines. For instance,

- work on Mach tends to focus on his philosophy of physics and psychology;
- work on Russell tends to focus on his philosophy of logic and arithmetic, of physics and of psychology;
- work on Poincaré and Duhem tends to focus on their philosophy of physics and the former's philosophy of arithmetic;
- work on early logical empricism predominantly focusses on its proponents' philosophies of logic and mathematics and of physics, esp. space-time, with little attention to its philosophies of social science;

1 Given the space constraints of this essay, even a sampling of the representative literature is out of the question. Readers may compare my list with the topics of relevant papers given at past HOPOS conferences (see http://www.hopos.org/conferences.html).

- work on Neo-Kantian philosophy of science tends to focus on its philosophy of physics, with its south-western wing's philosophy of social science coming into view;
- work on orthodox logical empiricism and critical rationalism on its proponents' philosophies of physics, now often quantum physics, and probability, with some attention to the former's philosophy of psychology;
- work on the emerging post-positivist philosophy of science typically focusses on its proponents' philosophies of physics.

Not surprisingly, of the empirical sciences, physics takes the lion's share of attention where the involvement of past philosophies of science with first-order theories is concerned. Work on past philosophy of biology and chemistry is catching up but would appear to escape, with only a few exceptions, association with the movements mentioned. Work on past philosophy of social science seems to continue to play a distinctly minor role.[2]

Considering even in briefest overview some of the variety and breadth of work in history of philosophy of science makes clear that it defies an informative summary. But we can ask: how does it bear on the self-image of analytical philosophy of science? First, we can see that the boundaries of the analytical tradition are by no means sharp. There is, for example, the fruitful exchange between Mach (who also stands on the borderline of 19th and 20th philosophy of science) and Duhem and, even more importantly, the lasting influence of the entire school of French conventionalism – Poincaré, Duhem, Abel Rey, less so LeRoy – on logical empiricism.[3] What the latter points to is that in philosophy of science whatever break occured between French philosophy of science and what we now call the analytic tradition, it occurred only well into the 20th century. (This break is notably later than the break which according to Dummett split analytical and continental philosophy of language between Husserl and the later Frege.)[4]

A second example of the boundaries of the analytical tradition being by no means sharp is presented by Neo-Kantian philosophy of science. Thus even though the southwestern wing of that movement (Windelband, Rickert) could be regarded as prototypically "continental", it turns out that Carnap's earliest major work was not insignificantly influenced by Rickert.[5] Of greatest importance here, however,

2 What's striking is the virtual absence of works on the history of philosophy of history in our sub-discipline: that there is no dearth of such work can be seen, e.g., by the contributions to the journal *History and Theory*.

3 See, e.g., Rudolf Haller, 'Der erste Wiener Kreis.' *Erkenntnis* 22 (1985) 341-358, trans. 'The First Vienna Circle' in T. Uebel (ed.). *Rediscovering the Forgotten Vienna Circle*. Dordrecht: Kluwer, 1991, pp. 95-108; Anastasios Brenner (ed.), *Interférences et transformations dans la philosophie francaise et autrichienne. Philosophia Scientiae* 3 (1998-99) Cahier 2.

4 Michael Dummett, "Origins of Analytical Philosophy", *Lingua e Stile* 23 (1988), 3-49, 171-210, repr. as a monograph Cambridge, mass.: Harvard University Press, 1993.

5 See Thomas Mormann, 'Carnap's Logical Empiricsm, Values and American Pragmatism." *Journal for General Philosophy of Science* 38 (2007), 127-146.

is that the heir of the Marburg wing of the movement, Cassirer, must be counted a most significant philosopher of physics in his own right whose work provided an indispensible reference point for the philosophy of space-time of early logical empiricism.[6] In this way, early analytical philosophy of science also taps into the Kantian tradition.

So history of philosophy of science helps to break down a certain insularity with which some practitioners of analytical philosophy of science look upon their tradition. None of this shows, however, that the term "analytical tradition" is without meaning, but it strongly suggests that attempts to define it in terms of necessary and sufficient conditions for membership are mistaken and it reminds us that, like all traditions, the analytic one can only be individuated in terms of historical lineage – such that connections, influences and overlaps as the ones just mentioned can only be expected.

But the dispelling of easy illusions about the past also has effects on the story that history of analytical philosophy of science tells about itself. Take, for example, my earlier claim that "as remembered, analytical philosophy of science originally professed to care little about history". Well, is that remembered correctly? When we look beyond the logical empiricist orthodoxy that institutionalised itself by the middle of the 20th century, it becomes obvious that early analytical philosophy of science itself was not as uniformly unhistorical as received wisdom has it.

What follows is that we historians of philosophy of science also must take care that in presenting our own case we do not trade on the very illusions that we (partly) make it our business to unmask. Moreover, our own enterprise is by no means as novel as it might appear. Thus historians of ancient and early modern philosophy would be right to remind us that a fair amount of their work can be also classified as history of philosophy of science, given that much of their subjects' philosophical thought distilled reflection on the science of their day.

III

I turn to the question whether the concern with historical facts exhausts the interest that history of philosophy of science possesses and what follows if it does not. As we shall see, this is closely related to the question why our work tends to be relatively local and shies away from the view from nowhere.

Let me approach the question via the commonplace that it's pretty much impossible to tell history "*wie es wirklich gewesen*", without any superaddition or distortion whatsoever. Like history in general, history of philosophy of science cannot do without what Arthur Danto called "narrative sentences" – sentences that

---

6 Ernst Cassirer, *Substanzbegriff und Funktionsbegriff*, Berlin: Bruno Cassirer, 1910, trans. in *Substance and Function and Einstein's Theory of Relativity*, 1923, repr. New York: Dover, 1953, pp. 3-346.

describe an event by reference to a later one – and these introduce the historian's perspective.[7] (Even a mere chronicle – as long as it is not the utopian "ideal" one – could be seen to introduce perspective due the inevitable selection of what it includes and excludes.)

Importantly, these narrative sentences do not impugn the factuality of the individual events that history discusses, nor that of any causal claims made about them, but the recognition of the inevitable perspectival nature of history has nonetheless given rise to worries about its objectivity. (This perspectival nature may express itself not only in conflicting evaluations of the events or developments under discussion, but also in the different temporal frameworks within which they are described as unfolding.) What validates the narratives historians lay out, given that there is no trans-historical standpoint we may take on them? The answer would seem to be that there simply is no validation available that singles out one and only one narrative as legitimate from all the different ones that deal with at least some of the same basic facts, so the challenge here is to hold on both to the factuality of history and the plasticity that comes with history being told from a perspective.[8]

To the idea that narrativity undermines truth it is rightly objected that the existence of distinct but compatible narratives does no such thing.[9] Obviously one event may figure in different stories. (The facts cited are either true of false and the selection of events as pertinent to the history told is subject to objective criteria of relevance etc.) But what of non-compatible narratives? Can we be sure that the narratives that we deem acceptable in principle are mutually compatible when we turn to history of philosophy (including philosophy of science)? Could we guarantee that a narrative of a given episode in the history of philosophy of science from a broadly Kantian perspective is compatible with one from a broadly empiricist perspective? I have no definite answer here, but I am sceptical. Consider the debate about the applicability of the Kantian framework to the epistemology of general relativity. Beyond agreement on the facts of who said and wrote what and when, there seems to be little compatibility between the narratives involving Schlick, Reichenbach and Cassirer in this respect as told, on the one hand, by a recent attempt to revive the research programme of transcendental idealism for the philosophy of physics and, on the other hand, a standard account of the rise

7 Arthur Danto, *Analytical Philosophy of History*. Cambridge: Cambridge University Press, 1965. Rev. and enlarged as *Narration and Knowledge*. New York: Columbia University Press, 1985, Ch. 8.

8 See the discussions in, e.g., Fred Ankersmith and Hans Kellner (eds.), *A New Philosophy of History*. Chicago: University of Chicago Press, 1995, or in Brian Fay, Philip Pomper and Richard Vann (eds.), *History and Theory. Contemporary Readings*. Oxford: Blackwell, 1998.

9 See Noel Carroll, "Interpretation, History and Narrative." *The Monist* 1990, repr. in Fay et al., op. cit., pp. 34-56, arguing against Hayden White, *Metahistory*. Baltimore: Johns Hopkins University Press, 1974.

of logical empiricism.[10] What's a genial cutting of a Gordian knot for the latter is a profound misunderstanding, an outright begging of the question for the former. Notably these are not mere value judgements but pertain to the question of what the problems are that were at issue.

So the challenge to the objectivity of history of philosophy of science does not appear to be as easily deflectable as in other historical inquiries. In response I'd like to argue that in our field history is not our only concern and that therefore worries about objectivity can be counterweighted appropriately, for the plasticity of history being told with an eye to the open future can in this case be suitably "disciplined". (This is not say that all history of philosophy of science must exemplify such perspective-taking, but that it is not illegitimate when this happens, especially if it remains suitably local.) As I understand it, history of philosophy of science, despite its strong emphasis on relevant aspects of the first-order science as practiced and on the cultural context of its theories, is still predominantly *philosophy* of science. Its *modus operandis* is historical, yet its aim is by no means exclusively to establish the mere facts of what was thought when and by whom (though they must be established as securely as possible as a matter of course). So my point is that since history of philosophy of science is philosophy of science by other means, the danger inherent in the perspectival nature of its historical methodology is neutralised by the nature of its subject matter. Let me explain further.

The absence of a presuppositionless starting point means that in philosophy uncontestable facts (beyond historical ones about who claimed what when) are far harder to come by than in the sciences. When we ascend to the meta-level, this becomes still more obvious: whether a philosophical research programme is progressive or regressive is much less clear cut than whether a scientific research agenda is. It seems we must accept the extreme contestability of its domain-specific facts as a fact about philosophy. Consider now whether the inevitable perspectival nature of *history* of philosophy of science introduces anything radically new, given that it is *philosophy* of science what the history is history of. My suggestion is that it does not. While history introduces a narrative perspective, philosophy already comes with its own potentially incompatible philosophical perspectives.

What does this mean for the narratives of the history of philosophy of science?[11] Does this mean that here two incompatible narratives can both be true – or that neither can be? Ordinary logic pushes us to the conclusion that, indeed, one of them would have to be false. If one is unwilling to urge the abandonment of bivalence (or the embrace of metaphilosophical relativism) then one is reduced to asking what, if anything, makes false one of these incompatible narratives which,

10 Compare Thomas Ryckman, *The Reign of Relativity. Philosophy in Physics 1915-1925*, Oxford: Oxford University Press, 1995, and, e.g., Hans Reichenbach, *The Rise of Scientific Philosophy*. Chicago: University of Chicago Press, 1951.

11 Similar considerations may well hold for history of philosophy generally, but I will keep my remarks limited to our subdiscipline.

after all, are supposed to respect the relevant historical facts. Clearly, it must be the philosophical presuppositions.

What follows for our historical narratives? Since philosophical disputes, given sufficient depth, tend to resist settlement, it follows that disputes about incompatible narratives in the history of philosophy of science may likewise tend to resist settlement. That is not unduly problematic, however, once we recognise that history of philosophy is part of the ongoing philosophical debate and contestation. So my suggestion is that, in the case of history of philosophy of science, the question of what fallibly validates the historian's narrative perspective finds its answer in the contestable philosophical perspective. The narrative is validated (to the extent that it is beyond the data it deals with) by the philosophical perspective it unravels. In other words, that different narratives in the history of philosophy of science which respect the relevant historical facts may still be incompatible need not be regarded as impugning the objectivity of history but as instead pointing to the philosophical nature of the history attempted.

So history of philosophy of science is not just like other history, nor is it just like other philosophy. This has consequences for the use of an epithet that historians are fond of labeling opponents with: "presentism". We can all agree that reading contemporary concerns into the past is a bad idea when done without sophistication and especially without any backing from the material at issue. But we must equally agree that already the phenomenon of Weberian value-relevance (that a theorist's values influence her choice of research programme) banishes the idea of wholly disinterested historical research. Yet the point I now want to press on the back of Danto's about narrativity is still stronger. It is that, once the historical facts of the case are established, history of philosophy of science can be as partisan as philosopy itself. Provided the proper precautions are taken, its search for contemporary relevance cannot be dismissed as "presentist", nor must its tendency be irenic (though it may delight in findings that may be considered ironic).

IV

Let's consider further then the legitimate partisanship of history of philosophy of science in the form of the exploration of "paths not taken". Let me lead into this by asking whether there are still more substantial illusions about its past than its easy separability from other traditions that inform the self-image of contemporary analytical philosophy of science and that are overturned by work on its history? I think there are and they can be summarised in terms of *bon mot* by Johann Nestroy that Wittgenstein adopted as the motto for his *Philosophical Investigations*, namely, that progress tends to look bigger than it is.

Here one may point to the common conceit to paint the logical empiricists – typically under the label "logical positivists" – as philosophers of astonishing naivity, in short, as negative poster boys for whatever new thesis a post-positivist philosopher is offering. In that capacity, logical empiricist philosophy is portrayed as pursuant of epistemological foundationalism and ontological reductionism, committed to a fact-fetishising instrumentalism that refuses to grant reality beyond the given and consequently as denying, with particularly disastrous effect in the social sciences, the critical force of investigating alternative possibilities – all this, moreover, from a perspective on scientific theorising that absolutises its formalisable aspects and abstracts from all socio-historical aspects of scientific practice. Needless to say, every one of these features is eminently criticisable, not only systematically but especially as attributed to logical empiricism.

Now there is no denying that some of the criticisms just surveyed do find a proper object among the theories and theorists that can be assembled under the umbrella of logical empiricism. For instance, the received view of scientific theories did give rise to all sorts of uncomfortable questions about the status of theoretical entities. Moroever, orthodox logical empiricism generally appears to have been caught up in a kind of formalisation frenzy, ignoring, under the guise of observing disciplinary strictures regarding the distinction between the contexts of discovery and justification, whatever resisted such treatment: it had no room for consideration of scientific practice and its social context and historical dimensions. But even in early logical empiricism we can find periods where the party-line, as far as it was observed, invites criticism of undue reductionism and where, once there was no longer a uniform party-line to speak of, we can find theorists who seemed to be pursuing the type of foundationalist programme that post-positivists reject so vigorously. Even so, however, there are also aspects of early logical empiricism that contradict the common stereotype most starkly and thoroughly.[12]

So the first factual lesson that history of philosophy of science has here is that early logical empiricism was by no means a monolithic movement and much more varied in philosophical outlook than the post-WWII orthodoxy. The second factual lesson is that much of the anti-foundationalism and anti-reductionism that post-positivism prides itself on can already be found in some versions of early logical empiricism. The third is that even the demand to pay attention to the historical and social dimension of scientific theorising held no surprise for proponents of this version of logical empiricism who sought to overcome the rigid dichotemy between the rational and the social. That this version thus anticipated some of the criticisms that later were raised against the orthodox logical empiricism of the '50s thus suggests, that it is largely immune to standard anti-positivist criticisms.

So what? you may be tempted to reply. Typically the version of logical empiricism just invoked was somewhat heterodox from the start and certainly was marginalised by the time the orthodoxy came to hold sway. The point of post-positivist

12 For a recent stock-taking of research on logical empiricism, see Alan Richardson and Thomas Uebel (eds.), *The Cambridge Companion to logical Empiricism*. Cambridge: Cambridge University Press. 2007.

pride rightly lies in that what was heterodox then is now common consensus: this progress may not be as radical as claimed, but surely that's progress enough.

I should stress that the historian's plaint and the unconvinced reaction to it are not unique to this case. Indeed, a somewhat parallel revisionist argument may be devised for transcendental idealism of the Marburg variety. Far from being stuck in the traditional Kantian mould, Cassirer developed the notion of the synthetic *a priori* and de-apodicticised it, rendering it relative to a historically given framework of theorising instead of conceiving of it as a constant presupposition of human reason as such (without, however, conventionalising it, as logical empiricists were wont to do).[13] Given how this conception of the *a priori* chimes with Kuhnian philosophy of science, again history of philosophy of science may lay claim to showing that a very topical and current philosophical thesis is not really terribly new. In this case it is the thesis that significant scientific theories tend to proceed against a background of substantive assumptions (often hidden by their adoption of new forms of mathematical representation) that they themselves are in no position to redeem.

So what? you may be tempted to reply again. The point of post-Kuhnian pride rightly lies in that what once was a maverick's view is now much more widely held: this progress too may not be as radical as claimed, but surely that's progress enough. This may be conceded. Yet conceding this does not undermine the significance of dispelling the illusion of the age to have made *radical* progress. For one thing, it counsels a certain modesty and that is no bad thing. For another, the recognition that not everything "logical positivist" or everything "transcendental idealist" belongs in the dustbin of history may also prompt the question whether that heterodox version of logical empiricism or that late development of transcendental idealism do not also hold other suggestions or insights that may be fruitfully brought to current debates.

This brings us to the exploration of "paths not taken". First, let me note that this is optional, not obligatory. Second, that pointing out that exploring paths not taken may hold profitable vistas need not represent messianic hubris leading us back to some "golden age". And third, that this does not mean crossing the line between history of philosophy of science and systematic philosophy of science.

Even history of philosophy of science that is philosophy of science by other means still uses those other means essentially. It does not explore philosophical topics *ab initio*, as it were, from first principles, but within frameworks or from assumptions that are historically given. To pursue paths not taken in this context means either to think through to their conclusion certain philosophical hypotheses more thoroughly than they were in the historical situation in which they arose or to vary certain elements in that original configuration of ideas and probe the plausibility of the target hypothesis under these strictly counterfactual circumstances.

13 See Michael Friedman, "Ernst Cassirer and the Philosophy of Science." In Gary Gutting (ed.), *Continental Philosophy of Science*. Oxford: Blackwell. 2005, pp. 71-83.

(Importantly, as I understand it here, pursuing paths not taken does not mean imagining the course of historical debates to have been different and imagining the consequences of different outcomes: it does not mean pursuing counterfactual history as such.) Yet the exploration of paths not taken may also have results for philosophy of science that are of significance for its self-understanding – in a way still different from the morals suggested earlier.

There's one case which I think raises the issues in a particularly stark way. (Let me also stress that this is an example from my own work and is not here presented as the only case of a path not taken that is worth exploring.) In the early 1930s there obtained a division in the Vienna Circle characterised somewhat mischieviously by Carnap, following Neurath, as "left wing" and "more conservative" or (misleadingly) "right wing".[14] The so-called left wing was comprised of Carnap, Neurath, Hahn (d. 1934) and Frank, the more conservative wing centred on Schlick and Waismann. Though the term "left wing" invoked a shared political outloook amongst its members, its main designation was philosophical. The left wing was characterised at first by its opposition to Wittgenstein's radical (namely complete) verificationism. Most importantly, it pioneered the thorough rejection of epistemological foundationalism in the philosophy of science. (Needless to say, enough topics of disagreement remained amongst them.)

One of the most striking of its shared philosophical theses was that traditional philosophy had outlived its usefulness and that what remained useful of philosophy was happily conceived as meta-theory of science, as a second-order inquiry of itself scientific nature, in other words, as science in the self-reflexive mode. What's particularly interesting is the form of the metatheory that we can ascribe to them. Importantly, it came in two forms. There was Carnap's "logic of science" and Neurath's "behavioristics of scholars", a naturalistic "pragmatics of science", in Frank's later terminology. The logic of science investigated scientific theories, their internal structure and their relation to their evidential base in purely logical terms (deductive and inductive). The pragmatics of science investigated scientific practice by means of the empirical sciences of science, the psychology and sociology as well as the history of science. So while the former investigated abstract relations of evidential support, the latter investigated concrete theory choice and change.

What was the relation between the two metatheories? Clearly, they are different both in methodology – one using formal *a priori*, the other material *a posteriori* reasoning – and in terms of their object. It is perhaps a standard view to regard them as standing, if not in outright opposition, at least in considerable tension with each other. Given a comparison of the exemplary clarity with which Carnap's inquiries proceeded with Neurath's decidely less clear explorations, it is perhaps no

14 See Rudolf Carnap, "Intellectual Autobiography", in P.A. Schilpp (ed.), *The Philosophy of Rudolf Carnap*, LaSalle, Ill.: Open Court, 1963, p. 57. Already in "Testability and Meaning", *Philosophy of Science* 3 (1936) 422, Carnap spoke of the "more conservative wing".

surprise that this view won the day, moreover, that the latter's "behavioristics of scholars" was not taken up in the burgeoning movement of logical empiricism.

Indeed, Carnap's claim that "once philosophy is purified of all unscientific elements, only the logic of science remains" is commonly read as saying that legitimate philosophy comprised only the logic of science, nothing else.[15] His logic of science is then easily assimilated to Reichenbach's "analysis of science" within the so-called context of justification, with all concern with history and sociology banished to the unphilosophical context of discovery. Such a reading overlooks, however, that Reichenbach allowed into his analysis of science not only the problems of logic, probability theory, but also "all the basic problems of traditional epistemology".[16] By contrast, Carnap stressed that to designate his logic of science as "theory of epistemology (or epistemology)" is "not quite unobjectionable, since it misleadingly suggests a resemblance between the problems of our logic of science and the problems of traditional epistemology".[17] Carnap's view of the role of logic of science as successor to philosophy and its relation to the pragmatics of science was quite different from Reichenbach's.

Now for Neurath throughout but explicitly so since the mid-'30s, the relation between the logic of science and the behaviouristics of scholars was one of coexistence and complementation.[18] Likewise, already in the early '30s Frank suggested that understanding science required adding the sociological dimension to a theory of science that was conducted mostly in terms of the (at the time syntactic) analysis of the symbol system it used.[19] This of course meant adding something like a behaviouristics of scholars to the logic of science. (In the 1950s Frank's work mainly concerned the pragmatics of science, but was widely ignored.) Both Frank and Neurath can thus be counted as supporters of a conception of the successor discipline to philosophy as consisting of both the logic of science and the pragmatics of science, that is, a bipartite meta-theory of science.

So what about Carnap? Carnap recognised as perfectly legitimate, "in addition to the logic of science ... also the empirical investigation of scientific activity, such as historical, sociological, and, above all, psychological inquiries" and grouped both of them together under the heading "theory of science".[20] This amounts to a

15 Rudolf Carnap. *Logische Syntax der Sprache*. Vienna: Springer, 1934. Revised and trans. *The Logical Syntax of Language*. London: Kegan, Paul, Trench Teubner & Cie, 1937. Repr. Chicago: Open Court, 2002, p. 279.

16 Hans Reichenbach. *Experience and Prediction*. Chicago: University of Chicago Press, 1938. Repr. Notre Dame: University of Notre Dame Press, 2006, p. 8.

17 Carnap, op. cit., p. 280.

18 See Otto Neurath. "Physikalismus und Erkenntnisforschung." *Theoria* **2**, 97-105, 234-7. Trans. "Physicalism and the Investigation of Knowledge" in Neurath, *Philosophical Papers 1913-1946* (ed. by R.S. Cohen and M. Neurath), Dordrecht: Reidel, 1983, pp. 159-67.

19 See Philipp Frank. *Das Kausalgesetz und seine Grenzen*. Vienna: Springer. Trans. *The Causal Law and its Limits*, Dordrecht: Kluwer, 1998, p. 14.

20 Carnap, op. cit., p. 279.

bipartite scientific metatheory which likewise divides into logical and empirical inquiries. Of course, Carnap, for his part, worked only on the logic of science. But he explicitly endorsed the bipartite nature of meta-theory by acknowledging the importance of the empirical aspects of the theory of science in his eulogy to Frank and noted in a comprehensive review of his own work that "unfortunately a division of labor is necessary, and therefore I am compelled to leave the detailed work in this direction to philosophically interested sociologists and sociologically trained philosophers".[21]

Carnap's view is commensurable with Neurath's and Frank's. This suggests that the "Left Vienna Circle" as such embraced the idea of a bipartite metatheory of science. (Needless to say, on occasion thorny questions of relative priority and general relevance can arise between the two parts of scientific metatheory: in fact, just those, with some personal issues added, did arise and led to the estrangement in later years between Carnap and Neurath, but this does not affect the principle.) Moreover, since Frank had recommended that it was precisely the recourse to the "more comprehensive study of the connections that exist between the activity of the invention of theories and the other normal human activities"[22] that obviated the need for metaphysical philosophy, one may state that that for the Left Vienna Circle, it was the bipartite meta-theory of science that was the successor of traditional philosophy.

So what? Well, it strikes me as not impossible that this conception of a bipartite meta-theory should prove useful in the further development of what nowadays would be called "deflationary" philosophy of science. So here's one "path not taken" that might, for colleagues with deflationist sympathies, merit further consideration.

Yet that is not all that the case may hold for us. For it might be argued that I went one step too far. It might be conceded that in principle the logic of science and the pragmatics of science are compatible and combinable, and yet contested that they were so put forward by the left wing of the Vienna Circle, indeed, that the very idea of that left wing as a philosophically coherent group and of Carnap, Frank and Hahn sharing a common programme is just too fanciful to be believed. This objection raises what I think is an important point not only about the philosophy of science here at issue, but also about philosophy of science in general.

Consider first that logical empiricists generally – though not universally – saw themselves engaged in the project of rendering philosophy of science itself sci-

21 See, respectively, Carnap, "A Few Words to Philipp Frank, for the Second Volume of the Boston Studies in the Philosophy of Science." In R.C. Cohen and M. Wartofsky (eds.). *In Honor of Philipp Frank. Boston Studies in the Philosophy of Science Vol. 2*, New York: Humanities Press 1965, pp. xi-xii. And Carnap, "Comments and Replies." In P.A. Schilpp (ed.), *The Philosophy of Rudolf Carnap*, LaSalle: Open Court, 1963, p. 868.

22 Frank, op. cit, p. 14.

entific.[23] According to the Circle's inofficial manifesto, "the scientific world conception" endeavours "to link and harmonise the achievements of individual investigators in their various fields of science. From this aim follows the emphasis on collective efforts, and also the emphasis on what can be grasped intersubjectively..."[24] Prominent here are two central features of philosophy as a scientific activity: co-operation and intersubjectivity, the former a method of inquiry, the latter a condition on its results. It is the former that I want to concentrate on here. (The latter feature proved crucial in bringing out the left wing's agreement on anti-foundationalism by 1932 which had not yet been achieved in 1929.)[25] Carnap, along with Hahn and Neurath, had been one of the signatories of the manifesto while Frank was the organiser of the conference at which it was first presented to the public. So when, as we saw, Carnap explicitly invoked the division of labour in 1963, he gave expression to a long-held view. This suggests that not only were the logic and the pragmatics of science compatible as each was conceived of from about 1932 onwards, but that they were recognised as such already then by the members of the Left Vienna Circle.

How then to explain the undeniable fact, which a detractor might focus upon, that Carnap, Frank and Neurath remarked upon this shared metaphilosophical conception so rarely in their publications? To claim that any compatibility in principle between theses they embraced was so patently obvious to them that no words needed to be wasted on it, would perhaps be a bit rash. (Particularly between Carnap and Neurath – though mainly during the protocol sentence debate that led to their shared anti-foundationalism – there are numerous cases where disagreement lurks even under confessed agreement.) All that's required though is that between explicit agreement and unnoticed complementarity lie a variety of possibilities, each of which suffice to underwrite my claim about the Left Vienna Circle's view on scientific metatheory.

Especially against their explicitly shared programme of rendering philosophy of science scientific, the documented convergence of their remarks on the nature of scientific metatheory as the successor to traditional philosophy can be taken to have been readily understood by Carnap, Frank and Neurath. It is an essential feature of the manifesto's programme that philosophy of science be a collective undertaking, requiring many minds and hands in different capacities and in different departments of unified science, as it were. The division of labour that Carnap

23 The significance of the project to render philosophy "scientific" has been explored by Alan Richardson in several papers; see, e.g., his "Scientific Philosophy as a Topic for History of Science", *Isis* 99 (2008) 88-96.

24 Rudolf Carnap, Hans Hahn and Otto Neurath. *Wissenschaftliche Weltauffassung - Der Wiener Kreis*. Vienna: Wolf, 1929. Trans. "The Scientific Conception of the World: The Vienna Circle". In Neurath, *Empiricism and Sociology* (ed. by M. Neurath and R.S. Cohen), Dordrecht: Reidel 1973, p. 306.

25 See my *Empiricism at the Crossroads. The Vienna Circle's Protocol Sentence Debate*. Chicago: Open Court, 2007.

remarked upon retrospectively in 1963 was already projected into forthcoming work in 1929, though not in the precise form that it was to take.

If it is correct that we may indeed attribute a shared perspective to the Left Vienna Circle along these lines, then there are consequences that in principle go far beyond the case at hand. History of philosophy of science may have to take some of the programmatic statements of the thinkers it studies more seriously than it often does, in particular their claims to be involved in a collective undertaking. There may be more to an individual's philosophy of science than what is apparent in that individual's work. The reason is that this individual may have seen him- or herself as working in a "community" which provided essential extensions or complementation to the work he or she performed and which was very much meant to be understood in this wider setting. For instance, to stay with Carnap, it may be gravely mistaken to criticise his philosophy of science as being purely formalist and unreflective of the practical and socio-historical dimensions of science. To be sure, the logic of science is, but considered "in the round" that is not all that his philosophy of science amounts to. What justifies us in rejecting such a criticism of Carnap and taking a wider view is precisely, of course, his documented commitment to philosophy of science as a collective enterprise: any one worker will only ever be able to work on some part or other, if lucky and particularly gifted perhaps on several, but never (or hardly ever) on all aspects of it.

V

I'm afraid I've talked a bit more about doing history of philosophy of science than I've done it, though I hope you will agree that what I've done supports what I said about doing it. In particular I hope to have shown that the history of analytic philosophy of science offers up an interesting suggestion of how philosophy of science may be thought to differ from other fields of philosophy. Of course, whether we all would wish to agree that philosophy of science is a significantly collective undertaking and that in this regard it differs from metaphysics, say, is yet another matter. Either way, however, by presenting us with such a case study, the history of analytical philosophy of science presents philosophy of science generally with yet another occasion for reflection on what it is and wants to be.

Philosophy
School of Social Science
University of Manchester
Manchester M13 9PL
United Kingdom
thomas.uebel@manchester.ac.uk

Thomas Mormann

History of Philosophy of Science as Philosophy of Science by Other Means? Comment on Thomas Uebel

1. Introduction

Adorno once remarked that the history of philosophy is the history of forgetting. Problems and ideas once examined fall out of sight and out of mind only to resurface later as novel and new. On the other hand, understanding history as historiography, the history of philosophy may be described as the organized and institutionalized attempt of overcoming the ever-growing threat of historical amnesia.

Even if „history of science is not just like other history" (Uebel), Adorno's dictum may well be true also for the (still rather brief) history of philosophy of science: problems and ideas once examined fall out of sight only to resurface later as novel and new. Examples are easily found: historical and sociological aspects of science that were treated by authors such as Duhem, Bachelard, Cassirer, Neurath or Fleck, fell into oblivion and were ignored by mainstream philosophy of science, only to be rediscovered decades later. In a different, perhaps even worse manner, the leading ideas of some historical currents, for instance classical logical empiricism of the Vienna Circle or Neokantian philosophy of science, suffered from partial historical amnesia: seriously distorted by later authors they sometimes served as their own caricatures. Thereby, to borrow Nestroy's well-known *bon mot* the progress in philosophy of science often tended to look bigger than it really was.

Adorno might have exaggerated the danger of historical amnesia that threatens philosophy, and certainly he did not spend a thought on the situation in philosophy of science. But his claim might serve as an antidote against the naive idea that the history of philosophy of science has been a history of permanent and unilateral progress in the sense that today we obviously possess a better philosophical understanding of the sciences than our forefathers. Rather, the more progress history of philosophy of science made in recent years the more it became clear how much remains to be done for achieving an adequate understanding of the still young discipline called philosophy of science.

This note will not deal with history of philosophy of science in general, it only has the modest aim to make some comments on Uebel's survey article on the history of *analytical* philosophy of science, in particular on his proposal to understand the task of history of philosophy of science as doing „philosophy of science by other means" conceiving it as an arsenal of abandoned or forgotten conceptual possibilities, or, as he put it, as the „exploration of paths not taken".

F. Stadler (ed.), *The Present Situation in the Philosophy of Science*,
The Philosophy of Science in a European Perspective, DOI 10.1007/978-90-481-9115-4_2,

The outline of this paper is as follows: In the next section I'd like to point out in what sense Uebel's topic – the history of *analytical* philosophy of science – may be considered as a particularly difficult but also particularly interesting topic of history of philosophy of science. The aim of section 2 is to show that the understanding of „history of philosophy of science as philosophy of science by other means" may be conceived as an expedient strategy of forestall a profusion of undesired meta(meta)disciplines which threaten the conceptual unity of an interdisciplinary research dealing with the history and philosophy of scientific culture. In section 3 I deal with some problems that may arise for Uebel's favorite example of a „path not taken", to wit, what he calls the „bipartite metatheory" of Carnap, Neurath, and Frank. In section 4 I argue that Uebel's proposal of a combined metatheory may be compared with Morris's earlier proposal of a synthesis of logical empiricism and American pragmatism in the 1930s. Morris's attempt failed, and it may be useful to inquire into the reasons of this failure to be in a better position to assess the prospects of Uebel's proposal.

2. The Many Issues on the Agenda of History of Philosophy of Science

Let me start with an observation concerning the programme of „History of Philosophy of Science" of this congress. From a logical point of view the collections of topics

- History of **Analytical** Philosophy of Science
- History of Philosophy of Science **in the French Tradition**
- History of the **19th Century** Philosophy of Science

may appear a bit mysterious. Comparing it with Borges's famous classification of „animals in a certain Chinese encyclopedia" intended to defy every attempt of explaining it rationally, may be exaggerated, but I think history of philosophy of science should make some efforts to explicate as clearly as possible the domain of possible issues that it wants to deal with. At least for some of the above mentioned items this seems easy enough. „History of Philosophy of Science in the French Tradition" and „History of the **19th Century** Philosophy of Science" may be characterized as „local" subdisciplines in a geographical and in temporal sense, respectively. Other examples for local subdisciplines in this sense easily come to mind: „History of Philosophy of Science in the Polish Tradition" or „in the German Tradition", or „History of the **18th or 17th Century** Philosophy of Science" and so on.

However, a „dimensional" classification of this kind by no means exhausts the possible topics of our discipline: In a recent newsletter the editors of the Society for History of Philosophy of Science (HOPOS) call for a series of „state of the art" essays dealing with „HOPOS figures" (sic) including Aristotle, Descartes,

Newton, Leibniz and many others. There is no reason to criticise such a personalized approach of history of philosophy, but it is certainly of a quite different kind than the one dealing with „local" topics of various type. Still another possibility is exemplified by Uebel's choice of „Analytical Philosophy of Science" as a topic of history of philosophy of science. Without arguing for it I think that Uebel's choice is „more philosophical" and more interesting, at least if we conceive history of philosophy of science as another way of doing philosophy of science.

Nevertheless, the topic of analytical philosophy of science is a somewhat delicate choice. Some philosophers, rooted in the analytical tradition, still believe that analytical philosophy of science is the only philosophy of science that is to be taken seriously. All other efforts undertaken by philosophers in the course of history to come to terms with science may simply be disqualified as metaphysical rubbish. A similar, slightly less pretentious stance is to consider all non-analytical philosophy of science as nothing but a historical precursor of the real thing, i.e. analytical philosophy of science. This way of interpreting the place of analytical philosophy of science is not a mere remote possibility, it has been common usage in many quarters. For instance, Kitcher and Salmon, in their anthology *On Scientific Explanation*, classify Duhem's *La Theorie Physique* explicitly as belonging to the "Modern Prehistory" of the subject (cf. Kitcher and Salmon 1989, *Chronological Bibliography*, 196). Uebel does not subscribe to these radical ways of determing the place of analytical philosophy of science in the history of the subject. Rather, he is at pains to point out that the boundaries between the analytical and other currents of philosophy of science are by no means sharp. Nevertheless, the place of analytical philosophy of science remains remains difficult to determine, and in any case it is somewhat special compared with the apparently easily located local topics mentioned above and the more traditional way of doing things by dealing with the philosophy of science of the great dead philosophers of the past. This becomes evident when one attempts to imagine a coherent "history of non-analytical philosophy of science" or, perhaps better, a "history of continental philosophy of science". If such a history existed, it would be a rather mixed bag.

From an "American perspective" the attitude of conceiving analytical philosophy of science as the culminating point of the history of philosophy of science may be tempting and appears perhaps almost natural, but this perspective is certainly not without presuppositions, and one may ask whether from a "European perspective" these should be taken for granted. It leads to certain difficulties even if we restrict our attention to local topics of the history of philosophy of science as, say, philosophy of science in the German or the Austrian tradition. For instance, take Schlick's empiriocriticism of *Allgemeine Erkenntnislehre*. It can hardly be classified as "analytical", but I doubt, whether characterizing it as "pre-analytical" or "proto-analytical" is really fair. A similar, even more important problem arises for Neokantian philosophy of science, in particular Cassirer's as Uebel correctly remarks. Analogous remarks hold for conventionalist philosophy of science in the French tradition and other currents of European philosophy of science.

Willy-nilly, then, the analytical perspective often tends to play down the proper value of other currents of philosophy of science. It tends to ignore the losses philosophy of science has suffered on its way toward its analytical realization. I have no quick recipe how to overcome these difficulties, the only thing I want to say is that the choice of the topic „history of analytical philosophy of science“ is not without presuppositions. Uebel seeks to avoid these difficulties. He rightly observes that Cassirer's Neokantian philosophy of science is important in its own right: it is problematic to conceive his account solely from the analytical perspective asking how it influenced Carnap on his way to analytical maturity.

Uebel attributes the growing importance of history of philosophy of science for philosophy of science to a *naturalistic turn* in philosophy that self-consciously rejects any a priori reflection about grand philosophical themes related to science. This is certainly correct. Moreover, I think Uebel is right in asserting that it is a trend to be heartily welcomed. A general *naturalistic* perspective seems to be less fond of producing intellectual fashions, necessarily accompanied by complementary blind spots, than aprioristic accounts. In this way naturalism may seem an antidote to the threat of amnesia mentioned in the beginning of this note.

3. History of Philosophy of Science as Philosophy of Science by Other Means

In the ongoing process of naturalization, which, of course, not only comprises the historical dimension, but also the sociological, the psychological and other ones, a lot of new meta-disciplines pop up:

- History of philosophy of science
- Psychology of philosophy of science
- Sociology of history of philosophy
- History of history of science

and so on. This may lead us to conceive "History of …", "Psychology of …" as sort of operators analogous to the modal operators such as „possibly“, „necessarily“, „obligatory“, and so on that are used in alethic or deontic modal logics. Iterations of such operators make sense formally, but become more and more opaque conceptually. Given a proposition p one may form expressions such as

$$\Box\Diamond\Diamond p \quad \text{or} \quad \Diamond\Box\Box\Diamond p$$

but probably very few people have an intuitive idea what these might mean. Analogously, the meanings of new metadisciplines such as „history of philosophy of sociology“, „philosophy of history of sociology“ tend to become obscure.[1] Uebel

1 Actually, things are more complicated than the operator analogy suggests: for instance, it is far from clear how the relation between the "history of science" and "philosophy of science" is to be conceived. One, not overly convincing, option is to understand the first as "descriptive" and the second as "normative". Another option is to claim an unspecified "complementarity" and "collaboration" of some kind between them.

puts forward an important thesis that can be used to cut off this undesired multiplication of possible meta-disciplines: “History of philosophy of science … is … predominantly *philosophy* of science. … History of philosophy of science is philosophy of science by other means …”

This thesis is, of course, not new. In various forms, it has been brought forward by many authors, usually not restricted to philosophy of science, but claimed to hold for philosophy in general. One might recall that already Windelband, more than one hundred years ago, in his influential *Lehrbuch der Geschichte der Philosophie* (Windelband 1889) pondered on the relation between philosophy and history of philosophy. According to him, history of philosophy should be considered as an integral part of philosophy, namely as its „organon“ (ibid., 567).[2] Nevertheless, he insisted that both disciplines should not be mixed up, philosophers should not forget the philosophical over the historical (ibid., iii).

It is a common place that traditionally the historical was a stronghold of continental philosophy, while the interest of analytical philosophy in history of philosophy was less fully developed. Recently, this state of affairs is changing as is evidenced, for instance, by the meanwhile well-developed history of the logical empiricism of the Vienna Circle and similar currents that are usually pursued from an analytical perspective.

Even if there is a rather general consensus among philosophers that history of philosophy should play a certain role for philosophy, it is far from clear what precisely this role is to be.[3] For instance, some contend that we have to know the history of philosophy (of science) so as not to avoid important alternatives to comtemporary proposals (cf. Curley 1986). This seems to be the stance of Uebel, Hardcastle, Richardson, and other philosophers of science. In the case of philosophy in general, some authors want to go further, claiming that philosophy is essentially an historicist endeavor (cf. Cohen 1986). How far this attitude may be applicable also to history of philosophy of science, remains to be investigated.

In any case, if we take into account something like Uebel’s thesis the profusion of meta-disciplines becomes less disturbing. Doing history of philosophy of science, or sociology of philosophy of science, just means doing philosophy of science in specific ways. Conceiving history of philosophy of science as one of the ways of doing philosophy of science, it is natural to ask why we should pursue

---

Further, as is exemplified by the logical empiricism and other currents of “scientific” philosophy, there are interesting relations between “history of philosophy of science” and “history of scientific philosophy” that render the agenda of “history of (philosophy of) science” rather complicated (cf. Richardson 2008).

2 Regrettably, Windelband nowhere explained exactly what he understood by “organon” here. There is no reason to expect that the role that he had in mind for history of philosophy resembled very much to that which Uebel is thinking of.

3 As it seems, there has not been made too much progress in this issue since the times of Windelband.

the historical way of doing philosophy of science, and what achievements we can expect from this endeavor.

Philosophers of science have dealt with this question for some time now. Roughly in line with Uebel's naturalistic turn, some years ago Hardcastle and Richardson spoke of a „historicist turn“ in philosophy of science that might help to overcome the crisis that plagues philosophy of science. By this they did not mean the turn inaugurated by Kuhn's *The Structure of Scientific Revolutions*:

> We refer to a more recent development in which philosophers have begun to recover the problems, solutions and motivations of earlier projects in the philosophy of science, paying attention to how the historical figures engaged in these projects understood them. ... Adapting what is perhaps the most famous sentence in the philosophy of science of the second half of the twentieth century, we can assert that the history of the philosophy of science is coming to be viewed as more than a repository for anecdote and chronology, and can, if we allow it, produce a decisive transformation in the *philosophy of science* we now possess.“ (Hardcastle and Richardson 2003, vii).

To be explicit, for Hardcastle and Richardson the „most famous sentence in the philosophy of science of the second half of the 20th century“ is Kuhn's dictum: "History, if viewed as a repository for more than anecdote or chronology, could produce a decisive transformation in the image of science by which we now are possessed." (Kuhn 1962, 1) In agreement with Uebel, then, Hardcastle and Richardson contend that history of philosophy of science is philosophy of science by other means. Moreover, they claim that philosophy of science urgently needs this new means. According to them, contemporary philosophy of science is entangled in a deep conceptual, almost existential crisis, and history of philosophy of science might help overcome it. Indeed, they invite us to tap the spiritual sources of the past:

> [I]t may well be time to return to the social spirit of philosophy of science of the 1930. Perhaps that is our best philosophical venture in a world of anxious social and technological Maybes.“ (ibid., xxvi)

Uebel does not speak of a crisis. Less alarmist he is to content to point out that the historical way of doing philosophy of science may help us to recover conceptual resources that we lost from sight. As he rightly contends, not „everything „logical positivist“ or everything „transcendental idealist“ belongs in the dustbin of history. In other words, he proposes to use history of philosophy of science as a source for exploring hitherto undeveloped or underestimated conceptual possibilities. In this respect I fully agree with him. History of philosophy of science might be a way of doing philosophy of science that overcomes the wide-spread *historical amnesia* reigning in many quarters of philosophy, as already Adorno lamented. On the other hand, history is certainly not a foolproof way of doing philosophy of science. It may lead us astray by inviting us to indulge in the idea of *an Golden Age of Philosophy of Science* when allegedly our discipline flourished in every respect. Due to historical contingencies, this narrative claims that later philosophy of science deviated from the right path and ended up in the morass where we presently find it. Uebel does *not* consider himself as an aficionado of the *Golden Age*.

4. The Bipartite Metatheory: A Path to be Taken?

Even if the conception of history of philosophy of science as a special means for doing philosophy of science is not the only way of pursuing it, for philosophers of science it is perhaps the most natural attitude. Uebel, apparently in line with Richardson, Hardcastle, Howard, and many others conceive the reflexive dimension that is opened up for philosophy of science by paying attention to its history as a means for making *proposals*. He invites us to employ the historicist option to do philosophy of science and to explore „paths not taken". More precisely, he proposes to reconsider a path that in the first half of the 20th century allegedly was persued by Carnap, Neurath, and Frank but later was abandoned by their successors. This path he dubs the path of a „bipartite metatheory of science". Uebel considers it as a promising strategy for contemporary philosophy of science. This „bipartite metatheory" has two components:

– a logical component in the sense of a Carnapian logic of science;
– an empirical part roughly in the sense of Neurath's „behavioristics of scholars" or Frank's „pragmatics of science".

Uebel contends that this bipartite theory should be considered as the common legacy of three great figures of the Vienna Circle, to wit, Carnap, Neurath, and Frank. I must confess, that I am skeptical about the prospects of pursuing further the "bipartite meta-theory", even if the great figures of the Golden Age of philosophy of science could be read as having subscribed to it as a desideratum, although in their later careers they did not undertake serious efforts to realize it. It seems not totally unreasonable, as Uebel admits, to doubt that such a theory as a more or less coherent conceptual enterprise has ever actually existed. To me, the mere coexistence and alleged complementation of Carnapian logic of science and Frank-Neurathian pragmatics of science, do not justify the claim that there was a kind of *theory* that comprised the logic and pragmatics as two sub-theories. After all, a theory should have a certain amount of theoretical unity, which the bipartite metatheory is clearly lacks. Uebel himself admits that the two halves of his envisaged bipartite theory neither share a common methodology nor a common object. In line with Neurath, for him it suffices that the relation between the logic of science and the pragmatics of science was one – or could have been one – of „coexistence and complementation".[4] In my opinion, Uebel is rather indulgent with the often vague programmatical announcements of the founding fathers of classical logical empiricism.[5] For instance, he emphasizes that it was an essential feature of the manifesto's programme that philosophy of science be a collective undertaking based on a well organized division of labour. By historical hindsight, however, the results of this envisaged "new way" of doing philosophy of science have been less than totally convincing. After all, up to now, the score of really collective work in philosophy of science has not been too impressive. In general, then, it is an important task for history of philosophy of science to identify the illusions and unfulfillable dreams that past philosophers cherished.

Let us come back to the issue of the bipartite metatheory. One may readily admit that it is, of course, always possible to weaken the requirements a theory has to satisfy in such a way that any iuxtaposition of more or less unrelated theoretical endeavors counts as a theory. The question is, whether it is useful to conceive such a thing as a theory. For the time being, the expression "bipartite metatheory" seems to be me more a name of a problem rather than a framework for a truly comprehensive philosophy of science. Be this as it may, my negative assessment should not be considered as an objection to Uebel's main thesis, namely, that history of philosophy of science is an expedient means for exploring conceptual possibilities for contemporary philosophy of science.

Having expressed my doubts about the feasibility of Uebel's proposal of exploring further the path the bipartite metatheory may be interpreted as sort of an obligation to make myself a proposal of an interesting path not taken that would be worth to be explored in the present situation. I think, however, that it would be more appropriate to the context of history of philosophy of science to point out that Uebel's proposal of a synthesis of logical and pragmatical currents of philosophy of science had an interesting precursor some sixty or seventy years ago. This attempt of synthesis failed for reasons that we do not fully understand up to now. Thus, it may be justified to rescue that unifying attempt from oblivion – not the least of these reasons the one that this example perhaps could shed some light on the feasibility of Uebel's proposal.

4 I consider this condition as too weak as though it could distinguish clearly the scientific complementation and collaboration of partisans of the project of a bipartite meta-theory from dubious endeavors such as the recent fashionable attempt of certain philosophers to construe "religion" and "reason" as "complementing and cooperating elements" of the modern condition.

5 He shares this attitude with many scholars engaged in the history of early logical empiricism. This attitude might be understandable in view of the fact that the "true" history of logical empiricism and related currents has been unduly neglected for a long time, and is still neglected in some quarters of analytic philosophy perhaps even now. Nevertheless, I see certain dangers in this attitude.

5. A Comprehensive Metatheory that Failed: Morris's Scientific Empiricism

Already in the thirties of the last century Morris had urged the logical empiricists of the Vienna Circle to think over their narrow concept of scientific philosophy as syntax of the language of science. Against this overly narrow conception of philosophy Morris argued for a pragmatist scientific philosophy that comprised four different stages: Painting with a broad brush Morris identified four realms of scientific philosophizing labeling them with the names of Carnap, Peirce, Dewey, and Whitehead (cf. Morris 1937, 8ff.)[6]

- Philosophy as logic of science (Carnap)
- Philosophy as clarification of meaning (Peirce)
- Philosophy as empirical axiology (Dewey)
- Philosophy as empirical cosmology (Whitehead)

In this schema, Carnap's purely theoretical account of scientific philosophy as syntax of the language of science figured as the first and most restricted level of a comprehensive scientific philosophy which whould take into account not only the logical but other dimensions of of a scientific culture as well (cf. Morris 1937, 8ff). Morris readily admitted that moving from "Carnap" to "Whitehead" amounted to lowering the standards of exactness and certainty (ibid., 19). But he was convinced that scientific philosophy had to pay this price, if it wanted to be relevant for life in a comprehensive manner that took into account theory *and* practice of human existence. Moreover, he gave a compelling naturalist reason why it might be unscientific or even unreasonable to insist on Carnapian standards of exactness throughout:

> Science reveals no absolute break between theory and practice, and there is no clear reason why the situation should be different in philosophy. Meaning at the level of philosophical generality has its pragmatic dimension just a have the meanings at other levels. ... It would be a signal instance of ethical irresponsibility ... to turn the world over to the exclusive control of dreamers, adventurers, men of action, and technicians. (ibid., 20).

Carnap never showed much sympathy with the pragmatist unification programme even if he did not militate against it explicitly. Rather, he tried to eschew it in some way or other. In his reply to Morris's proposal (in the Schilpp volume) at the end of the day he had not more to offer than the bland assertion: „I am inclined to agree

6 Uebel does not mention pragmatist philosophy of science in his list of „movements (within or fading into the analytical tradition)", probably because European philosophers have not contributed much to a philosophy of science from a pragmatist point of view. Be this as it may, history of pragmatist philosophy of science would certainly an important issue on the agenda of history of philosophy of science.

with Morris that the difference between my view and that of the pragmatists is not as large as it might appear at first glance" (Carnap 1963, 862). Notwithstanding this conciliary assessment he stuck to his ethical non-cognitivism clinging to the existence of „pure optatives" and refusing the the pragmatist „mean-end continuum".

Despite Carnap's half-hearted conciliation the conceptual differences between Carnapian logical empiricism and American pragmatism of Dewey, Lewis, and Morris (to say nothing about Peirce and Whitehead) essentially remained as they were. Probably they can be attributed to the quite different conceptions of science underlying this currents of scientific philosophy. For Carnap, science was a system of theoretical knowledge – a set of consistent and rationally justifiable statements (Carnap 1935, 32). For Dewey, to take him as the most outspoken representative of a genuine pragmatist philosophy of science, science was rather a process or activity. Science, according to him, was not knowledge, but a process for solving problems. This entailed that Dewey and the other pragmatists contended that valuation was essential to the production of scientific knowledge, whereas Carnap insisted on a radical separation between knowledge and valuation. The gap between these fundamentally different philosophical perspectives on science was never really overcome as is shown by the difficult and finally rather unsatisfying coexistence between empiricist and pragmatist currents of scientific philosophy in the second half of the 20th century.

The failure of constructing a comprehensive „scientific empiricism" in Morris's sense should be taken into account when we seek to assess the chances of Uebel's „bipartite metatheory", even if the parallelism between Uebel's and Morris's proposals is limited. At first look, it might be tempting to associate Neurath's „behaviorism of scholars" and Frank's „pragmatic of science" at least *grosso modo* with the part that Morris reserved for the pragmatists in his sketchy program of a comprehensive philosophy of science. But I am not sure how far this goes. If Neurath's and Frank's accounts could serve as a pragmatic (or pragmatist?) complementation of Carnapian logic of science somehow analogous to classical American pragmatism that Morris had envisaged some decades ago then Uebel's bipartite metatheory would be confronted with similar difficulties that led to the abandonment of Morris's program.[7] On the other hand, if it would turn out that the complementation envisaged by Uebel were of a quite different kind than that which Morris envisaged, interesting problems concerning the relation between genuine American pragmatist philosophy of science and the Viennese ersatz pragmatism of Neurath and Frank would arise. In any case, there are still a lot of issues on the agenda of history of analytical philosophy of science that deserve to be studied in the future.

7 Actually, I think that Neurath's and Frank's "pragmatism" or "pragmatic" would make a rather poor substitute of the real thing, but this is not an issue to be discussed here.

References

Carnap, R., 1935, *Philosophy and Logical Syntax*, London, Kegan, Trench, and Trubner.

Carnap, R., 1963, "Reply to Morris", in P.A. Schilpp (ed.) *The Philosophy of Rudolf Carnap*, La Salle, Open Court.

Cohen, L., 1986, "Doing Philosophy is Doing its History", Synthese 67, 51 – 55.

Curley, E., 1986, "Dialogues with the Dead", Synthese 67, 33 – 49.

Giere, R., 1973, "History and Philosophy of Science: Intimate Relationship or Marriage of Convenience", The British Journal for the Philosophy of Science 24, 282 – 297.

Kitcher, P., Salmon, W., 1989, *Scientific Explanation*, Minnesota Studies in the Philosophy of Science XIII, Minneapolis, University of Minnesota Press.

Kuhn, T.S., 1962, *The Structure of Scientific Revolutions*, Chicago, Chicago University Press.

Morris, C., 1937, *Logical Positivism, Pragmatism, and Scientific Empiricism*, Actualités scientifiques et industrielles 449, Paris, Hermann.

Morris, C., 1963, "Pragmatism and Logical Empiricism", in P.A. Schilpp (ed.), *The Philosophy of Rudolf Carnap*, LaSalle, Open Court, 87 – 98.

Richardson, A. W., Hardcastle, G.L., 2003, "Logical Empiricism in North America", in G.L. Hardcastle and A.W. Richardson (eds.), *Logical Empiricism in North America*, Minnesota Studies in the Philosophy of Science XVIII, University of Minnesota Press, Minneapolis, London, vii – xxix.

Richardson, A.W., 2002. "Engineering Philosophy of Science: American Pragmatism and Logical Empiricism in the 1930s", Philosophy of Science 69, S36 – S47.

Richardson, A.W., 2008, "Scientific Philosophy as a Topic for History of Science", Isis 99, 88 – 96.

Uebel, T., 2007, *Empiricism at the Crossroads*, LaSalle and Chicago, Open Court.

Windelband, W., 1892(1912, 1993[18]), *Lehrbuch der Geschichte der Philosophie*, J.C.B. Mohr (Paul Siebeck), Tübingen.

Department of Logic and Philosophy of Science
University of the Basque Country UPV/EHU
Avenida de Tolosa 70
200.80 Donostia San Sebastian
Spain
ylxmomot@sf.ehu.es

Cristina Chimisso

Aspects of Current History of Philosophy of Science in the French Tradition

French philosophy of science and 'mainstream' philosophy of science

When Thomas Uebel invited me to write a paper on the current situation of history of French philosophy of science, I must admit that I found the task a little daunting. I do not think that it is possible to do justice to the diverse research programmes that scholars in different countries are developing, or to present them as a coherent whole. I would like, however, to make some remarks on the state of this particular field of study, with two provisos: one is that my perspective is somewhat centred in my experience in Great Britain, although it is not limited to it, the other is that I do not aim at an overall presentation of the current state of the study of history of French philosophy of science. Inevitably, my remarks will mainly refer to that part of French philosophy of science that is the object of my own research. I am confident, however, that Anastasios Brenner in his commentary will correct my necessarily partial presentation.

Especially from the point of view of somebody working in the English-speaking world, French philosophy of science appears to be an area of study with clearer boundaries than other national traditions. There seems to be a general understanding that French philosophy of science is different from 'mainstream' philosophy of science: this difference has been made official, as it were, in reference works and Encyclopaedias. In this, the *Routledge Encyclopedia of Philosophy* is paradigmatic: it has two entries, one for 'Philosophy of Science', and another, contributed by Gary Gutting, for 'French philosophy of science'.

French philosophy of science is not perceived as autonomous only by English-speaking philosophers. Indeed, the same distinction as that of the *Routledge Encyclopedia of Philosophy* has been proposed by French-language scholars. Dominique Lecourt, for instance, in his overview of the philosophy of the sciences, has presented this discipline first as a largely Austrian and Anglo-American affair (although Auguste Comte is present as a founder father), and then has introduced the 'French tradition of philosophical reflection on the sciences' as autonomous from the tradition of logical positivism and its legacy. Lecourt has explained that the distinctive identity of this tradition mainly rests on its constant link between

F. Stadler (ed.), *The Present Situation in the Philosophy of Science*,
The Philosophy of Science in a European Perspective, DOI 10.1007/978-90-481-9115-4_3,

history and philosophy of science, and on the rejection of empiricism and of a 'certain logical formalism'.[1]

In fact, there is little difference between Gutting's and Lecourt's choices of illustrious names for the pantheon of French philosophy of science, and both place Gaston Bachelard and Georges Canguilhem at the centre of their presentations. Unsurprisingly for a philosopher who has promoted Bachelard's ideas arguably like no other, Lecourt has declared the former to be the 'emblematic figure' of the 'French tradition of philosophy of science', and has presented Georges Canguilhem as developing Bachelard's philosophy. In Lecourt's account, François Dagognet is the direct inheritor of this tradition, which for him has also produced thinkers who do not sit completely comfortably under the heading of philosophy of science: Michel Foucault and Louis Althusser.[2] Like Lecourt, Gutting has dedicated in-depth analyses not only to Bachelard and Canguilhem, but also to Michel Foucault and Michel Serres.[3] The presence of Foucault is particularly important for an English-speaking readership, who is much more likely to be familiar with his writings than with those of either Bachelard or Canguilhem, to this day not all translated into English.[4] Indeed, some readers would have heard of them because of Foucault, not least due to Gutting himself: his book on Foucault opens with a chapter dedicated to these two philosophers.[5]

Both Lecourt and Gutting provide backgrounds for the major philosophers of the 'French tradition' in philosophy of science: the former presents as the 'fathers' of this tradition Condorcet, Augustin Cournot and Auguste Comte, but mainly focuses on Pierre Duhem, Henri Poincaré, Emile Meyerson, Abel Rey, Léon Brunschvicg and Alexandre Koyré. Gutting introduces the main part of his article by sketching a history of French philosophy of science in which the main characters are Descartes, the Enlightenment, Auguste Comte, Pierre Duhem, Emile Meyerson and Henri Poincaré. However, the centrality of Bachelard and

1 Dominique Lecourt, *La philosophie des sciences.* Paris: Presses Universitaires de France 2001, p. 90.

2 Ibid, pp. 113-4.

3 Gary Gutting, "French philosophy of science," in Craig (Ed), *Routledge Encyclopedia of Philosophy*. London: Routledge, Retrieved March 09, 2009, from http://www.rep.routledge.com/article/Q038 1998.

4 Canguilhem's works that have not as yet been translated into English include even Georges Canguilhem, *La formation du concept de réflexe aux XVIIe et XVIIIe siècles.* Paris: Presses Universitaires de France 1955; Bachelard's books not available in English include: Gaston Bachelard, *L'activité rationaliste de la physique contemporaine.* Paris: Presses Universitaires de France 1951; Gaston Bachelard, *Le matérialisme rationnel.* Paris: Presses Universitaires de France 1972 [1953]; Gaston Bachelard, *Le rationalisme appliqué.* Paris: Presses Universitaires de France 1986 [1949]; Gaston Bachelard, *Le pluralisme cohérent de la chimie moderne.* Paris: Vrin 1973 [1932]; Gaston Bachelard, *L'intuition de l'instant.* Paris: Stock 1992 [1931].

5 Gary Gutting, *Michel Foucault's Archaeology of Scientific Reason.* Cambridge: Cambridge University Press 1989.

Canguilhem is hard to miss in these as in other presentations of French philosophy of science. Indeed, Bachelard and Canguilhem's 'historical epistemology' has come to be synonymous with French philosophy of science.

The trademark of these philosophers, and what arguably most sharply distinguishes them from their Anglo-American counterparts, is the importance of history for their philosophy – and indeed of philosophy for their history. In fact, Lecourt also calls French philosophy of science 'philosophical history of the sciences'.[6] As a consequence, often its practitioners have been variously designated as philosophers or historians: Pierre Duhem and Alexandre Koyré have been considered sometimes historians and sometimes philosophers; Abel Rey published both philosophical and historical works;[7] Hélène Metzger may be chiefly known for her works on seventeenth and eighteenth-century chemistry, but she also wrote a philosophical book on scientific concepts, and many historiographical papers;[8] Léon Brunschvicg considered himself a philosopher, but this did not stop him being regarded by some, including George Sarton, as a historian of science, such was his interest in history of science and his attention to the historical detail.[9] Indeed, Bachelard's and Canguilhem's doctrines have not only been called 'historical epistemology', but also 'epistemological history', notably by Michel Foucault;[10] Dominique Lecourt has distinguished between Bachelard's 'historical epistemology' and Canguilhem's 'epistemological history', a distinction further developed

6 Lecourt, *La philosophie des sciences,* p. 90.

7 Abel Rey's philosophical works comprise monographs (e.g. Abel Rey, *Le retour éternel et la philosophie de la physique.* Paris: Flammarion 1927), textbooks (Abel Rey, *Eléments de philosophie scientifique et morale.* Paris: Cornely 1903) and articles (e.g. Abel Rey, "Sur le positivisme absolu", in *Revue philosophique* 34, no. 68 1909, Abel Rey, "Vers le positivisme absolu", in *Revue philosophique* 34, no. 67 1909, Abel Rey, "Pour le réalisme de la science et de la raison", in *Revue de métaphysique et de morale* 19, no. 4 1911); his historical works include a five-volume history of ancient science: Abel Rey, *La science dans l'antiquité, vol.1: La science orientale avant le grecs.* Paris: La Renaissance du livre 1930; Abel Rey, *La science dans l'antiquité, vol. 2: La jeunesse de la science greque.* Paris: La Renaissance du livre 1933; Abel Rey, *La science dans l'antiquité, vol. 3: La maturité de la pensée scientifique en Grèce.* Paris: La Renaissance du livre 1939, Abel Rey, *La science dans l'antiquité, vol. 4: L'apogée de la science technique greque: Les sciences de la nature et de l'homme, les mathématiques d'Hippocrate à Platon.* Paris: La Renaissance du livre 1946, Abel Rey, *La science dans l'antiquité, vol. 5: L'apogée de la science technique greque: L'essor de la mathématique.* Paris: La Renaissance du livre 1948.

8 Hélène Metzger, *Les concepts scientifiques.* Paris: Alcan 1926; her historiographical papers have been re-published in: Hélène Metzger, *La méthode philosophique en histoire des sciences. Textes 1914-1939, réunis par Gad Freudenthal.* Paris: Fayard 1987.

9 See Brunschvicg, letter to Sarton of 2 February 1923 (Houghton Library, Sarton Papers, bMS Am 1803/1803.1).

10 Michel Foucault, *The Archaelogy of Knowledge.* London: Tavistock 1972 [1969], p. 190.

by Jean Gayon and Hans-Jörg Rheinberger.[11] Along with history, another research focus of extreme importance for many leading French philosophers of science has been the study of the mind.

Understanding science and understanding the mind in history

The integration of history and philosophy did not start with Bachelard or Canguilhem, although these two philosophers have often been presented, or perceived, as initiators of a tradition, especially in English-language criticism.[12] From the point of view of pure intellectual history, the philosophical tradition of interpreting historical development was obviously not new to French philosophy. More specifically, there existed a tradition that was aimed at sketching the history of the mind, which included such classic models of the progress of the mind in history as Condorcet's, Cournot's and Comte's. Twentieth-century projects of studying the mind in history, which were central to French philosophy of science, certainly did not ignore those illustrious models (Bachelard even recalled the law of the three stages at the beginning of *La formation de l'esprit scientifique*,[13] without needing to mention Comte). However, twentieth-century scholars at the centre of the reflection on science took history more seriously, in some cases extremely more seriously. Moreover, most of them opposed positivistic and mechanicistic views of historical progress. In this respect, Hélène Metzger's remarks on Comte are emblematic: she argued that his numerous examples from the history of science inevitably 'prove' his law of the three stages of the development of the mind, but only because he postulates this law as an 'inviolable dogma'.[14] For her, in such philosophical representations of history as Comte's, historical events are chosen and interpreted to illustrate a theory, rather than being the basis for the theory. Not only the historian Metzger, but also philosophers, notably Brunschvicg, insisted that the mind should rather be studied *a posteriori*. Twentieth-century scholars did carefully consider previous models of the history of the mind, but often in a polemical way.

11 Dominique Lecourt, *Pour une critique de l'épistémologie*. Paris: Maspero 1972 ; Jean Gayon, "The Concept of Individuality in Canguilhem's Philosophy of Biology", in *Journal of the History of Biology* 31 1998.; Hans-Jörg Rheinberger, "Reassessing the Historical Epistemology of Georges Canguilhem," in Gutting (Ed), *Continental Philosophy of Science*. Oxford: Blackwell 2005.

12 The reception of Bachelard and Canguilhem in English-language criticism presented them as the beginning of a tradition that continued with Althusser and Foucault, as I shall discuss below in this article. An example of this is Gutting's book on Foucault cited above (note 5).

13 Gaston Bachelard, *La formation de l'esprit scientifique: contribution à une psychanalyse de la connaissance objective*. Paris: Vrin 1993 [1938], p. 8.

14 Hélène Metzger, "Tribunal de l'histoire et théorie de la connaissance scientifique," in Gad Freudenthal (Ed), *La méthode philosophique en histoire des sciences, textes 1914–1939*. Paris: Fayard 1987[1935], p. 27.

As I have discussed at length in my book *Writing the History of the Mind*,[15] there is a much more historically-situated story to tell in order to understand the twentieth-century projects that are seen as the core of French philosophy of science, and that made possible the development of historical epistemology. These projects developed in an intellectual and institutional context that made a meaningful dialogue between history and philosophy possible. The importance that history of philosophy came to acquire in the first half of the twentieth century, especially at the Sorbonne, created the ideal environment for the development of historical epistemology. The analysis of professorships, courses, doctoral dissertations and publications, as well as the views of contemporaries, all demonstrate the strong development of history of philosophy in higher education in the first half of the twentieth century, in particular after the First World War.[16] Not only did history of philosophy come to be regarded as an important subject, but it was also regarded as a philosophical subject. Many scholars, far from considering history as irrelevant to philosophy, believed that history was, in a fortunate expression, 'the laboratory of philosophy', that is to say the discipline that provides the empirical data to philosophy, and that allows philosophical doctrines to be tested. This is particularly true of the specific research programmes, which were elaborated during the Third Republic, whose questions Bachelard and Canguilhem inherited. The historians of philosophy Lucien Lévy-Bruhl and Léon Brunschvicg were central to these research programmes.

Lucien Lévy-Bruhl and Léon Brunschvicg, who between them occupied the Sorbonne chair of history of modern philosophy for almost the whole first half of the twentieth century,[17] both aimed to study the mind. They both believed that the mind was not fixed, but rather changed in different times and places. As a consequence, it was not possible for them to study it *a priori*, without recourse to

15 Cristina Chimisso, *Writing the History of the Mind : Philosophy and Science in France, 1900 to 1960s*. Aldershot: Ashgate 2008.

16 For an analysis of this development in the early twentieth century, see Ibid., Ch.1 and Jean-Louis Fabiani, *Les philosophes de la République*. Paris: Editions de Minuit 1988. Nineteenth-century history was rather different (about it see John I. Brooks, *The Eclectic Legacy: Academic Philosophy and the Human Sciences in Nineteenth-Century France*. Newark and London: University of Delaware Press and Associated University Press 1998), and it is important not to confuse the two periods.

17 Lévy-Bruhl was appointed to the Sorbonne chair of history of modern philosophy in 1908 and retired in 1927 (his chair was of 'histoire de la philosophie moderne', which includes what in English is generally called history of early modern philosophy. There was, however, disagreement about the exact chronological limits of 'philosophie moderne'). Brunschvicg replaced Lévy-Bruhl in 1927 and retired in 1940. Lévy-Bruhl had been appointed *maître de conférences* at the Sorbonne in 1899, and was active after his retirement. Brunschvicg was made professor Emeritus at his retirement, but had to go into hiding, where he died in 1944, as the Germans were occupying France. For the details of their appointments, see Albert Guigue, *La Faculté des Lettres de l'Université de Paris depuis sa fondation (17 mars 1808) jusqu'au 1er Janvier 1935*. Paris: Alcan 1935.

empirical research. Indeed, they intended to study the mind *a posteriori*, that is to say through documents that would show the way it works. They however elected to employ different sources. Brunschvicg chose to analyse intellectual history, in particular history of philosophy and history of science, whereas Lévy-Bruhl turned to the study of ethnologists' reports on the way of thinking of peoples in Papua New Guinea, Africa and South America. However, he had already developed his research aims and methods in his works on history of philosophy and ethics. Moreover, as he explicitly explained, even his work on primitive mentality was not meant to be a contribution to ethnology but rather to the study of human nature; his aim was to investigate the truth of Hume's and Comte's claims that human nature is universal.[18]

The importance of Brunschvicg's and Lévy-Bruhl's doctrines for the French tradition in philosophy of science cannot be overstated. Brunschvicg was Bachelard's mentor and supervisor on one of his doctoral dissertations,[19] and supported him in his career. When Bachelard's dissertations were published, Brunschvicg, who had just been appointed to the prestigious chair of history of modern philosophy at the Sorbonne, immediately reviewed them in one of the two major philosophy journal, the *Revue de métaphysique et de morale*. In one of his reviews, he saluted his former student as a 'thinker of the first order';[20] his validation could not fail to produce a profound impression on the philosophical establishment. More importantly, many aspects of Brunschvicg's philosophy were the starting point of Bachelard's philosophy, including the aim to understand the mind by examining intellectual history, the view that the mind changes through history, and the idea that the objects of knowledge are not mind-independent. It is hardly surprising that Bachelard's philosophy has been presented as an original development of Brunschvicg's.[21]

18 Jean Duvignaud, *Le langage perdu. Essai sur la différence anthropologique*. Paris: Presses Universitaires de France 1973, p. 126; Lucien Lévy-Bruhl and al., "La mentalité primitive. Séance du 15 février 1923." in *Bulletin de la Société française de Philosophie* 23 1923.

19 Gaston Bachelard, *Etude sur l'évolution d'un problème de physique: la propagation thermique dans les solides*. Paris: Vrin 1973 [1927].

20 Léon Brunschvicg, "*Etude sur l'évolution d'un problème de physique. La propagation thermique dans les solides*, par Gaston Bachelard", in *Revue philosophique* 54 1929, p. 94; Léon Brunschvicg, "*Essai sur la connaissance approchée*, par Gaston Bachelard", in *Revue philosophique* 54 1929.

21 Jean Wahl, *Tableau de la philosophie française*. Paris: Gallimard 1962, p. 114; Gary Gutting, *French Philosophy in the Twentieth Century*. Cambridge: Cambridge University Press 2001, pp. 85-6; François Dagognet, "M. Brunschvicg et Bachelard", in *Revue de métaphysique et de morale* 70 1965, pp. 43-54; Gary Gutting, "Introduction: What is Continental Philosophy of Science?," in Gutting (Ed), *Continental Philosophy of Science*. Oxford: Blackwell 2005, p. 14.; Jacques Gagey, *Gaston Bachelard ou la conversion a l'imaginaire*. Paris: Rivière 1969, pp 30, 54; Carlo Vinti, *Il soggetto qualunque: Gaston Bachelard fenomenologo della soggettività epistemica*. Napoli: Edizioni scientifiche italiane 1997, pp. 168, 427–52; Teresa Castelão-Lawless, "Gas-

Lévy-Bruhl's impact on a variety of disciplines was remarkable; here it is sufficient to notice the important role that the reception of his work played in the doctrines of philosophers and historians of science. Just to mention a few, Hélène Metzger developed her concepts of mental *a priori* and expansive thought, as well as her theory of active analogy, with direct reference to Lévy-Bruhl's theory of primitive mentality;[22] Abel Rey extensively cited Lévy-Bruhl in his discussion of the *outillage mental*, or mental tool, in the first article of the *Encyclopédie française*;[23] both Léon Brunschvicg and Gaston Bachelard referred to Lévy-Bruhl's theory, accepting some aspects and rejecting others, in order to define their own views of past intellectual history and their conceptions of the mind.[24]

The use of history in order to answer philosophical questions, and in particular questions about the functioning of the mind, was what characterized the work not only of philosophers, but also of historians – first of all the historians of *mentalités*, including Lucien Febvre, and the historians of science, including Alexandre Koyré. However, the present disciplinary distinctions did not hold in the inter-war period in France. New disciplines, including ethnology,[25] sociology, experimental psychology and general history of the sciences had strong links with philosophy, from which they originated. Their practitioners, to different degrees, did aim to differentiate their disciplines from philosophy, but they kept their institutional and intellectual links with it very well alive. Philosophers, sociologists, ethnologists, psychologists, historians of science and others discussed the mind, history, society and science together at the Société française de philosophie, the Centre de synthèse, and international conferences, and shared students and projects. This does not mean that they necessarily agreed with one other's perspectives and methods, but even when they disagreed, they did so reflectively, referring to the other scholars' approaches.

---

ton Bachelard et le milieu scientifique et intellectuel français," in Pascal Nouvel (Ed), *Actualité et postérités de Gaston Bachelard*. Paris: Presses Universitaires de France 1997, pp. 101-15.

22 Hélène Metzger, „Lucien Lévy-Bruhl, *L'âme primitive*", in *Isis* 9 1927, p. 486, n. 1.; Hélène Metzger, "L'*a priori* dans la doctrine scientifique et l'histoire des sciences", in *Archeion* 18 1936, p. 37.; Hélène Metzger, "La philosophie de Lévy-Bruhl et l'histoire des sciences", in *Archeion* 12 1930; Hélène Metzger, *Attraction universelle et religion naturelle chez quelques commentateurs anglais de Newton. Première partie. Introduction Philosophique*. Paris: Hermann 1938.

23 Abel Rey, "L'évolution de la pensée: De la pensée primitive à la pensée actuelle," in Febvre (Ed), *Encyclopédie française*. Paris: 1937.

24 Léon Brunschvicg, *Les étapes de la philosophie mathématique*. Paris: Alcan 1912; Ch. 1; Léon Brunschvicg, "Nouvelles études sur l'anime primitive", in *Revue des deux mondes* 52 1932; Gaston Bachelard, *La psychanalyse du feu*. Paris: Gallimard 1949 [1938].

25 For the sake of simplicity, I translate *ethnologie* with 'ethnology'; however, *ethnologie* is more correctly translated as 'cultural anthropology'.

It would be far too long here to discuss the disciplinary, institutional and personal networks of French academia, but it may be interesting to recall Abel Rey's career as an example of the disciplinary fluidity that was standard at that time in France. Abel Rey was Bachelard's other supervisor, and was also his predecessor in the Sorbonne chair of history and philosophy of the sciences. Before his Sorbonne appointment, he had been professor of philosophy at Dijon where he founded the laboratory of experimental psychology. At the Sorbonne he founded the Institut d'histoire des sciences et techniques, and, outside academia, closely collaborated with the historians of the Centre de synthèse. Lucien Febvre, who generally speaking did not particularly like philosophers, had nevertheless a close collaboration with Rey; indeed he entrusted the latter with the first volume of the *Encyclopédie française,* dedicated to the *outillage mental.* Once again, the study of the mind through history was what linked many of these scholars. French intellectual and institutional history created a fertile soil for the development of a distinctive tradition in philosophy of science, that produced historical epistemology.

Has historical epistemology stolen the show? Other aspects of French philosophy of science

The centrality accorded by Dominique Lecourt and Gary Gutting to historical epistemology in their presentation of French philosophy of science is shared by the large majority of representations of this tradition, both inside and outside France. There is little doubt that historical epistemology has been the dominant image of French philosophy of science. Indeed, for a long time in France Bachelard's philosophy has represented a sort of orthodoxy. As Claude Debru has put it, the central concept of Bachelard's philosophy of science became the 'catechism' of the philosophy of science 'made in France'.[26] This image is now being challenged as partial. Moreover, historians of philosophy of science are presenting works that show that it is not correct to see a complete separation, indeed an opposition, between French philosophy of science on the one hand, and logical positivism, its legacy, and current mainstream philosophy of science, on the other. An excellent example of these attempts is Anastasios Brenner's *Les origines françaises de la philosophie des sciences.*[27] Brenner explicitly points out two assumptions that are often made concerning French philosophy of science: one is that its starting point is the philosophy of Gaston Bachelard, and the second is that the French tradition in philosophy of science is autonomous and irremediably different from mainstream philosophy of science, logical positivism and post-positivism. He has shown that

26 Claude Debru, *Georges Canguilhem, science et non-science.* Paris: Editions rue d'Ulm/Presses de l'Ecole normale supérieure 2004, p. 67.

27 Anastasios Brenner, *Les origines françaises de la philosophie des sciences.* Paris: Presses Universitaires de France 2003.

French conventionalism, especially in Poincaré's and Duhem's versions, but also in those of Edouard Le Roy and Gaston Milhaud, played an important role in the formation of current philosophy of science. In so doing, Brenner aims to revise the widespread view that current (one could add analytical-oriented) philosophy of science has its roots only in Austria and in logical positivism.

It is very welcome that scholars have been working towards showing that the French tradition in philosophy of science has been far richer, and more complex than the standard image would allow. For example, the philosophy of one of Bachelard's critical targets, Emile Meyerson – the 'forgotten philosopher' as both Jean Largeault and Eva Teklès-Klein have called him[28] – is being brought to the attention of scholars again, for instance in the project of Bernadette Bensaude-Vincent and Frédéric Fruteau de Laclos.[29] Some scholars have also turned their attention to Meyerson outside France: among the latter group they have been intellectual historians and historians of philosophy, such as Mario Biagioli and Michael Heidelberger respectively, but also philosophers of science in the analytical tradition, such as Elie Zahar and Peter Lipton.[30]

The epistemology of other scholars, for instance of Hélène Metzger, has been re-discovered; Gad Freudenthal has given a tremendous impulse to the study of her work; many other critics have analyzed her philosophical work, including two members of our Team E, Michael Heidelberger and I, and many others, such as Ian Golinski, John Christie, Ilana Löwy, Michel Blay, Christine Blondel, Pietro Redondi, Lucia Tosi and Bernadette Bensaude-Vincent.[31] Metzger had not been forgotten, but she had been mainly remembered as a historian of chemistry.

28 Jean Largeault, "Emile Meyerson, philosophe oublié", in *Revue philosophique*, no. 3 1992; Eva Telkès-Klein, "Emile Meyerson: A Great Forgotten Figure", in *Iyyun* 52 2003; see also Eva Telkès-Klein, "Emile Meyerson, d'après sa correspondence. Une première ébauche", in *Revue de synthèse* 5e série 2004.

29 See Bernadette Bensaude-Vincent, "Chemistry in the French Tradition of Philosophy of Science: Duhem, Meyerson, Metzger and Bachelard", in *Studies in History and Philosophy of Science* 36 2005; Frédéric Fruteau de Laclos, 'La philosophie de l'intellect d'Emile Meyerson. De l'épistémologie à la psychologie' (thesis: Université de Paris-X Nanterre 2004). Fruteau de Laclos' dissertation has now been partly turned into a book: Frédéric Fruteau de Laclos, *L'épistémologie d'Emile Meyerson. Une anthropologie de la connaissance* Paris: Vrin 2009. See also Anastasios Brenner, "Le statut de l'épistémologie selon Meyerson", in *Archives de philosophie* 70, no. 3 2007. I understand that a number of publications on Meyerson, including of volume of unpublished primary sources, is coming out in 2009, and a conference is being organized to mark the event.

30 Elie Zahar, "Meyerson's 'Relativistic Deduction': Einstein Versus Hegel", in *The British Journal for the History of Science* 38 1987; P. Lipton, "Explanation in the Sciences – Meyerson, E", in *Annals of Science* 51, no. 2 1994.

31 See for instance the articles in Gad Freudenthal (Ed.), *Etudes sur/Studies on Hélène Metzger*. Leiden: Brill 1990: J.R.R. Christie, "Narrative and Rhetoric in Hélène Metzger's Historiography of Eighteenth Century Chemistry"; Jan Golinski, "Hélène Metzger et l'historiographie de la chimie du XVIIIe siècle; Bernadette Bensaude-

Because of the established tradition of studying the history of philosophy, the study of the history of philosophy of science has also been lively, and we can say mainstream in France. Moreover, although in France the reception of French philosophy of science has gone well beyond philosophy – it would be enough to recall how central many concepts of historical epistemology have been to the sociology of Pierre Bourdieu – the study of its history is a perfectly standard field of study for philosophers. A number of distinguished philosophers have worked on it, including Dominique Lecourt, François Dagognet, Jean Gayon, Etienne Balibar, Bernadette Bensaude-Vincent, Jean-Francois Braunstein, Michel Fichant, Anastasios Brenner, Claude Debru, Guillaume Le Blanc and Didier Gil, just to mention a few, in no particular order.[32] In Italy too several philosophers have worked on

---

Vincent, "Un essai de vulgaritation: *La chimie* dans l'*Histoire du monde*"; Gad Freudenthal, "Epistémologie des sciences de la nature et herméneutique de l'histoire des sciences selon Hélène Metzger"; Gad Freudenthal, "Hélène Metzger: Eléments de biographie"; Gad Freudenthal, "Epistémologie des sciences de la nature et herméneutique de l'histoire des sciences selon Hélène Metzger"; Ilana Löwy, "Constructivist epistemologies:Metzger and Fleck"; Michel Blay, "Léon Bloch et Hélène Metzger: La quête de la pensée newtonienne"; Christine Blondel, "Hélène Metzger et la cristallographie: de la pratique d'une science à son histoire"; Martin Carrier, "Some aspects of Hélène Metzger's philosophy of science"; Pietro Redondi, "Henri Berr, Hélène Metzger et Alexandre Koyré: la religion d'Henri Berr"; in Agnès Biard, Dominique Bourel and Eric Brian (Eds.), *Henri Berr et la culture du XXe siècle. Histoire, science et philosophie*. Paris: Albin Michel/Centre international de synthèse 1997; Lucia Tosi, "Hélene Metzger y la historia de la química", in *Saber y Tiempo* 9 2000; Cristina Chimisso, "Hélène Metzger: the history of science between the study of mentalities and total history", in *Studies in History and Philosophy of Science Part A* 32, no. 2 200; Cristina Chimisso and Gad Freudenthal, "A Mind of Her Own: Hélène Metzger to Emile Meyerson, 1933", in *Isis* 94, no. 3 2003; Bensaude-Vincent, "Chemistry in the French Tradition of Philosophy of Science: Duhem, Meyerson, Metzger and Bachelard", in *Studies in History and Philosophy of Science* 36 2005; Gad Freudenthal, "Hélène Metzger (1888–1944)", in Bitbol and Gayon (Ed.), *L'épistémologie française, 1830–1970*. Paris: Presses Universitaires de France 2006.

32 Some of these scholars have extensively published on the history of French philosophy of the sciences, and I will not attempt to give a full list of their publications here. Elsewhere in this article I cite some of the works on this subject by Lecourt, Debru, Bensaude-Vincent and Brenner; I have also cited an article by Dagognet on Brunschvicg and Bachelard, but I would like to add here at least two other works: François Dagognet, *Gaston Bachelard: sa vie, son œuvre*. Paris: Presses Universitaires de France 1965, and François Dagognet, *Georges Canguilhem: Philosophe de la vie*. Le Plessis-Robinson: Institut Synthélabo 1997. Balibar is better known for his works on Althusser, but he has also published on Canguilhem: see Etienne Balibar, "Science et vérité dans la philosophie de Georges Canguilhem," in E. Balibar et al. (Ed.), *Georges Canguilhem: Philosophe, historien des sciences. Actes du colloque (6-7-8 décembre 1990)*. Paris: Albin Michel 1993. See also Guillaume Le Blanc, *Canguilhem et les normes*. Paris: Presses Universitaires de France 1998, Guillaime Le Blanc, *La vie humaine: anthropologies et biologie chez Georges Canguilhem*. Paris: Presses Universitaires de France 2002. Didier Gil, *Bachelard et la culture scientifique*. Paris: Presses Universitaires de

the history of French philosophy of science, including Francesca Bonicalzi, Pietro Redondi, Gaspare Polizzi and Carlo Vinti.[33]

The present discussion about the identity of French philosophy of science is possible thanks to traditions that value the study of the history of philosophy, including as a tool to stimulate further philosophical research. In this context, this discussion has both historical and philosophical meanings.

The curious fate of French philosophy of science in the English-speaking lands

The fortunes of French philosophy of science in English-speaking countries have been uneven to say the least. In the 1970s, Anglophone readerships did show a considerable interest in Bachelard's and Canguilhem's ideas, which they received mainly by reading Dominique Lecourt's works.[34] Through Lecourt, Bachelard's

France 1993; Michel Fichant, "L'épistémologie en France," in Chatelet (Ed), *Histoire de la philosophie: le 20ᵉ siècle*. Paris: Hachette 1973; J.-F. Braunstein, "Canguilhem avant Canguilhem", in *Revue d'histoire des sciences* 53, no. 1 2000, and Jean-François Braunstein, "Abel Rey et les débuts de l'Institut d'histoire des science et techniques (1932–1940)," in M. Bitbol and J. Gayon (Ed.), *L'épistémologie française, 1830–1970*. Paris: Presses Universitaires de France 2006. Articles on Canguilhem by Braunstein, Lecourt, Debru and Delaporte (and Ian Hacking and Arild Utaker) are collected in Jean-François Braunstein, (Ed.), *Canguilhem: Histoire des sciences et politique du vivant*. Paris: Presses Universitaires de France 2007; in fact, some of these authors have published on French philosophy of science in other edited volumes, including François Bing, Jean-François Braunstein and Elisabeth Roudinesco, (Eds.), *Actualité de Georges Canguilhem. Le normale et le pathologique*. Paris: Synthélabo 1998. and Jean Gayon and Jean-Jacques Wunenburger, (Eds.), *Bachelard dans le monde*. Paris: Presses Universitaires de France 2000.

33 Francesca Bonicalzi, *Leggere Bachelard: le ragioni del sapere*. Milano: Jaca Book 2007; Gaspare Polizzi, *Forme di sapere e ipotesi di traduzione; materiali per una storia dell'epistemologia francese*. Milano: Angeli 1984; Gaspare Polizzi, *Tra Bachelard e Serres: aspetti dell'epistemologia francese del Novecento*. Messina: Armando Siciliano 2003; Pietro Redondi, *Epistemologia e storia della scienza. Le svolte teoretiche da Duhem a Bachelard*. Milano: Feltrinelli 1978, Pietro Redondi, "Science moderne et histoire des mentalités. La rencontre de Lucien Febvre, Robert Lenoble et Alexandre Koyré", in *Revue de synthèse* 104 1983; Pietro Redondi and P.V. Pillai, (Eds.), *The History of Science: The French Debate*. London: Sangam Books 1989; Pietro Redondi, "Henri Berr, Hélène Metzger et Alexandre Koyré: la religion d'Henri Berr," in Agnès Biard, Dominique Bourel, Eric Brian, (Eds). *Henri Berr et la culture du XXᵉ siècle. Histoire, science et philosophie*, Paris, Albin Michel / Centre international de synthèse 1997; Sabyasachi Bhattacharya and Pietro Redondi, (Eds.), *Techniques to Technology: A French Historiography of Technology*. London: Sangam Books 1990; Carlo Vinti, *Il soggetto qualunque: Gaston Bachelard fenomenologo della soggettività epistemica*.

34 Dominique Lecourt, *Marxism and Epistemology. Bachelard, Canguilhem and Foucault*. London: NLB 1975. This book comprises the translation of two French books: Lecourt,

and Canguilhem's philosophies were read in relation to Althusser; with the lessening of the interest in Althusser, the attention to Bachelard's and Canguilhem's philosophies, and I daresay to French philosophy of science, also waned. In the 1980s, Mary Tiles made an exemplary effort to render Bachelard comprehensible and acceptable to Anglo-American philosophers of science.[35] Despite the quality of her work, her effort had a limited effect, apart from sporadic articles, like for instance one by Mary Tijattas who attempted to reconcile Bachelard's ideas with a view of scientific realism common in the analytical tradition, and one by Dan McArthur, who nine years later responded to her,[36] or recent attempts to bring together analytic philosophy of science with 'Continental' philosophy of science, such as Christopher Norris'.[37] Whereas in France Bachelard's philosophy may have been a sort of 'orthodoxy', in English-language countries it has been regarded as a niche interest.

The little attention that 'mainstream' philosophers of science have paid to the French tradition is largely due to its image, dominated by historical epistemology, which suggests an intimate integration of history and philosophy. In truth, this image on the whole is not misleading. Although not all philosophers of science in this tradition have put history at the core of their doctrines, it is undeniable that history has played a major role in French philosophy of science, and not only in historical epistemology, but also in other doctrines. Even critics who aim to show the rich tradition in philosophy of science outside historical epistemology, nevertheless stress the importance of history for other philosophers of science, as Brenner does in relation to Duhem.[38]

This centrality of history has been an obstacle for the reception of French philosophy of science by analytical philosophers, who have been by and large little interested in history. I do not mean to ignore the impact that French philosophy of science has had in English-language philosophy of science. The use that philosophers of science writing in English have made of works in the French tradition has even prompted critics like Denis Vernant to include under the 'historical epistemology' heading not only Koyré, Bachelard and Canguilhem, but also Paul Feyer-

Pour une critique de l'épistémologie, Dominique Lecourt, *L'épistémologie historique de Gaston Bachelard.* Paris: Vrin 1969.

35 Mary Tiles, *Bachelard: Science and Objectivity.* Cambridge: Cambridge University Press 1984.

36 M. Tijiattas, "Bachelard and Scientific Realism", in *Philosophical Forum* 22 1991; Dan McArthur, "Why Bachelard is not a Scientific Realist", in *Philosophical Forum* 33, no. 2 2002.

37 Christopher Norris, *Minding the Gap: Epistemology and Philosophy of Science in the Two Traditions.* Amherst, Mass.: University of Massachusetts Press 2000, Christopher Norris, *Epistemology.* London: Continuum 2005.

38 Anastasios Brenner, "The French Connection: Conventionalism and the Vienna Circle," in Michael Heidelberger and Friedrich Stadler (Eds.), *History of Philosophy of Science: New Trends and Perspectives*. Dordrecht–Boston–London: Kluwer Academic Publishers 2002.

abend and Thomas Kuhn.[39] It is well-known that in the preface to his *The Structure of Scientific Revolutions*, Kuhn acknowledged the importance that the writings of Alexandre Koyré, Emile Meyerson and Hélène Metzger had for the development of his own view of science.[40] It is also true that from Kuhn onwards there has been a reception, although selective, of French philosophy of science into Anglo-American philosophy of science, and a keener attention to history, such as in the works of Ian Hacking and Gerald Holton.[41] However, the reception of French philosophy of science has not become mainstream. If any proof were necessary, it would suffice to see the little space that the French tradition in philosophy of science finds in mainstream English-language publications dedicated to the philosophy of science. To mention an example, in the journal *Philosophy of Science* in the years between 1934 and 2008 I could only find one full article dedicated to Bachelard, Teresa Castelão-Lawless' piece on phenomenotechnique,[42] and none about Canguilhem. Duhem fares better, relatively speaking, as he is mentioned in eleven articles, although often within the expression 'Duhem's problem', with little or no direct reference to Duhem himself. This of course does not mean that there are no academics who are members of philosophy departments in English-speaking countries and who at the same time work on French philosophy of science and its history: they do exist, and indeed I am one of them. However, there is no escaping the fact that we are in a small minority.

In fact, the reception and use of French philosophy of science in the English-speaking world seems to be stronger outside philosophy of science. For instance, after Canguilhem's death in 1995, several volumes and journals' special issues on this philosopher were published. In English, it was *Economy and Society*, a social sciences journal, that dedicated a double issue to him.[43] The guest editors, Nikolas Rose and Thomas Osborne, are sociologists, as are others among the English-language contributors to the volume, namely Monica Greco, Lorna Weir and

39 D. Vernant, "Epistémologie," in S. Auroux (Ed.), *Encyclopédie philosophique universelle. Vol. 2 : Les notions philosophiques*. Paris: Presses Universitaires de France 1990. On this theme, see also A. Brenner, "Which historical epistemology? Kuhn, Feyerabend, Hacking and Bachelard's school", in *Revue de métaphysique et de morale*, no. 1 2006.

40 Thomas S. Kuhn, *The Structure of Scientific Revolutions*. Chicago: The University of Chicago Press 1996 [1962], pp. vii-viii.

41 See for instance Gerald Holton, *Thematic Origins of Scientific Thought: Kepler to Einstein*. Cambrige, Massachussets: Harvard University Press 1973; Gerald Holton, "Einstein and the Cultural Roots of Modern Science", in *Science in Culture* 127, no. 1 1998; Ian Hacking, *The Emergence of Probability*. Cambridge: Cambridge University Press 1975; Ian Hacking, *The Taming of Chance*. Cambridge: Cambridge University Press 1990; Ian Hacking, *Historical Ontology*. Cambridge, Mass.: Harvard University Press 2002.

42 Teresa Castelão-Lawless, "Phenomenotechnique in Historical Perspective: Its Origins and Implications for Philosophy of Science", in *Philosophy of Science* 62 1995.

43 *Economy and Society* 27, 2/3 (1998).

Mike Gane; Paul Rabinow, who also contributed an article, is a cultural anthropologist.[44] The only contributor based in an English-language philosophy department appeared to be Ian Hacking, who later moved to the Collège de France. By contrast, the articles by Francophone authors, although comprising a paper by Pierre Bourdieu, were mainly by philosophers, including Dominique Lecourt, Alain Badiou and François Delaporte.[45] At least in Britain, the few scholars who work on French philosophy of science have a variety of backgrounds, including French, as in the case of Mary McAllester.[46] Social scientists have employed in particular the work of Michel Foucault, who, for the scholar of history of philoso-

44 As is often the case in English-speaking scholarship, here the interest in Canguilhem is linked to a previous interest in the work of Foucault; indeed several of the English-speaker contributors to the *Economy and Society*'s special issue on Canguilhem have extensively worked on Foucault; see for instance: Hubert L. Dreyfus and Paul Rabinow, *Michel Foucault: Beyond structuralism and Hermeneutics*. Brighton: Harvester 1982; Michel Foucault, *The Birth of the Clinic: An Archaeology of Medical Perception*. trans. A.M. Sheridan, London: Routledge 2003 [1963]; Michel Foucault, *The Essential Works of Michel Foucault, 1954–1984. Vol. 2, Aesthetics*, edited by Paul Rabinow. London: Penguin 2000; Michel Foucault, *The Essential Works of Michel Foucault, 1954–1984. Vol. 3, Power*, edited by Paul Rabinow. London: Penguin 2002; Paul Rabinow, (Ed.), *The Foucault Reader*. London: Penguin, 1991 [1986]; Andrew Barry, Thomas Osborne and Nikolas Rose, (Eds.), *Foucault and Political Reason: Liberalism, Neo-liberalism and Rationalities of Government*. Chicago and London: Chicago University Press and UCL Press 1995. Rabinow also wrote the introduction of the English-language anthology of Canguilhem's work: Paul Rabinow, "Introduction: A Vital Rationalist," in F. Delaporte (Ed), *A Vital Rationalist: Selected Writings from Georges Canguilhem*. New York: Zone Books 1994.

45 See T. Osborne and N. Rose, "Introduction", in *Economy and Society* 27, no. 2-3 1998; and the articles in the same double issue: Alain Badiou, "Is there a Theory of the Subject in Canguilhem?"; Pierre Bourdieu, "Georges Canguilhem: An Obituary Notice"; Georges Canguilhem, "The Decline of the Idea of Progress"; François Delaporte, "Foucault, Epistemology and History"; Monica Greco, "Between Social and Organic Norms: Reading Canguilhem and 'Somatisation'"; Ian Hacking, "Canguilhem amid the Cyborgs"; Dominique Lecourt, "Georges Canguilhem on the Question of the Individual"; David Macey, "The Honour of Georges Canguilhem"; Paul Rabinow, "French Enlightenment: Truth and Life"; Nikolas Rose, "Life, Reason and History: Reading Georges Canguilhem"; M. Gane, "Canguilhem and the Problem of Pathology"; C. Gordon, "Canguilhem: Life, Health and Death"; L. Weir, "Cultural Intertexts and Scientific Rationality: The Case of Pregnancy Ultrasound".

46 Mary McAllester has worked on Bachelard's philosophy as a whole, rather than only on his philosophy of science (see for instance Mary McAllester Jones, *Gaston Bachelard, Subversive Humanist: Texts and Readings*. Madison, Wisc.: University of Wisconsin Press 1991; Mary McAllester Jones, "Bachelard's Metaphors of the Self", in *French Studies* 54, no. 1 2000; Mary McAllester Jones, "The Redemptive Instant – Bachelard on the Epistemological and Existential Value of Surprise", in *Philosophy Today* 47, no. 5 2003.); she has also worked on Canguilhem: Mary McAllester Jones, "Georges Canguilhem on science and culture: learning biology's lessons", in *French Cultural Studies* 11, no. 31, 2000.

phy of science, is the inheritor of the tradition of historical epistemology. Within the social sciences, Foucault's work has generally not been considered philosophy, but rather 'theory', that is the theoretical part of the sociologists' work. Foucault has also been vindicated for science studies, as Martin Kusch has done.[47] The large majority of 'Continental philosophers' do not focus on philosophy of science. However, there are exceptions; for instance David Webb, who mainly works on Michel Foucault and Michel Serres, and also on Bachelard and Cavaillès.[48] I shall not go on with this list; my general point is that French philosophy of science has been received across several disciplines, but at the same time has played a minor role within philosophy of science.

In addition to the difficulties of the encounter between traditions that have regarded science from different perspectives, another obstacle has been the little attention that analytical philosophers, who are still very dominant in the English-speaking world, on the whole pay to the history of their own discipline. Not only history of French philosophy of science, but history of philosophy of science in general appears to be a minority interest; indeed there is a rather weak presence of history of philosophy, let alone of history of philosophy of science, in the universities of English-speaking countries, especially in the UK. This presence is particularly weak within philosophy. As I have discussed, history of philosophy, and intellectual history in general, played a crucial role in the philosophical debate in France, indeed in the debate across many disciplines. History of philosophy, not just history of philosophy of science, has not been as integral to philosophy in the English-speaking world as it has been in France, and, in a different way, in Italy. At least in Britain, it is not unusual for philosophy departments not to have even a single historian of philosophy. Moreover, history of philosophy has largely been history of early modern philosophy, as even a cursory glance at the publications in journals dedicated to the history of philosophy would show. This narrow focus has left out the bulk of philosophy of science. However, the situation is changing. The *British Journal for the History of Philosophy* has recently appointed two associated editors, one for ancient philosophy, and one for history of philosophy from Kant onwards, because it has now been recognized that history of philosophy does include these two periods. The creation, in the mid-1990s, of HOPOS, the learned society specifically dedicated to the study of the history of the philosophy of science, has also created an interesting international context for this subdiscipline.

I probably presented a rather divided image of the current situation in the history of French philosophy of science, or at least of that part of this field with which

47 Martin Kusch, *Foucault's Strata and Fields: An Investigation into Archaeological and Genealogical Science Studies*. Dordrecht: Kluwer 1991.

48 See for instance David Webb, "Microphysics – from Bachelard and Serres to Foucault", in *Angelaki-Journal of the Theoretical Humanities* 10, no. 2 2005; David Webb, "The Complexity of the Instant: Bachelard, Levinas, Lucretius," in R. Durie (Ed.), *Time and the Instant: Essays in the Physics and the Philosophy of Time*. Manchester: Clinamen Press 2000.

I am most familiar. I am afraid I have also presented a somewhat bleak image of the state of health of this subdiscipline in the English-speaking world. Many of us historians of philosophy of science think that the study of the history of philosophy of science should be part of philosophy of science, or at least very relevant to it, but this is not a universally shared view, to say the least. However, although I think that the history of French philosophy of science does not receive the attention it deserves within philosophy of science, I also think that the present situation is open to change, and indeed changing, and that programmes and initiatives like the present one will have a significant impact.

Department of Philosophy
Arts Faculty
The Open University
Walton Hall
Milton Keynes MK7 6AA
United Kingdom
c.chimisso@open.ac.uk

Anastasios Brenner

Reflections on Chimisso: French Philosophy of Science and the historical Method

Introduction

Over the past several years there has emerged a collective and conscious effort aiming to understand the history of philosophy of science. This has led to the renewed examination of the Vienna Circle and logical positivism, which is considered as one of the main sources of philosophy of science in the English-speaking world. Yet there have also been attempts to explore the development of other schools of thought. A study of the French tradition raises several questions, in particular the reception of this tradition and its salient recourse to a historical approach. Attention has turned from the well-known Bachelardian school to earlier philosophers.

Cristina Chimisso has provided us with a broad and stimulating picture of French philosophy of science. She can draw on her recent book *Writing the history of the mind: philosophy and science in France, 1900 to 1960s*. Chimisso takes us back prior to those doctrines that continue to pervade current views, that is postpositivism in English-speaking lands and historical espistemology in French-speaking countries. The 1960s mark a shift, and what lies before is now part of history. Chimisso's interests are not merely antiquarian; she leads us to philosophical issues. In particular, she draws our attention to Lucien Lévy-Bruhl and Léon Brunschvicg, who, although prominent in their time, have long been neglected. Chimisso thereby points to works that are beginning to receive interest again. She includes several other thinkers who played a role in the development of philosophy of science, including Henri Berr, Abel Rey, Hélène Metzger and Alexandre Koyré. A whole community makes its re-appearance. By taking us back before World War Two, Chimisso directs us to a time of intense philosophical debate. Philosophy of science as carried out at this time appears however quite different from what we practice today under the same heading. This has the effect of making us sensitive to the historical dimension: we may measure the distance covered, evaluate the persistent core of our discipline and scrutinized the background with respect to which new methods and theses arose.

How to justify historical study? I believe that returning to the primary sources already provides an answer. The picture of earlier philosophy of science as it was handed down to us by way of retrospective testimony or in the general surveys or introductions to philosophy of science does not correspond to the historical record. Chimisso's study, I believe, translates a new sensitivity: the need to push further as

F. Stadler (ed.), *The Present Situation in the Philosophy of Science*,
The Philosophy of Science in a European Perspective, DOI 10.1007/978-90-481-9115-4_4,

regards our methods, our situation as observers and the evolution of the constitutive notions of our discipline. Such a sensitivity has been termed variously: history of philosophy of science, historical semantics and meta-epistemology.

My aim here is to reflect upon Chimisso's results and to bring in further material. How did Lévy-Bruhl and Brunschvicg contribute to philosophy of science? What were their relations with other scholars working in the field? How to understand their markedly historical approach with respect to the application of logic to philosophy that came to dominate English-language philosophy? I wish also to inquire into the nature of historical method as put to philosophical use as well as the difference between the philosophical traditions.

1. Lévy-Bruhl, Brunschvicg and the *a posteriori* exploration of the mind

Chimisso devotes herself to philosophical reflection on science produced during the first half of the 20th century. One could of course extend her inquiry further back in time to the founding fathers of the French tradition. A complete history would certainly include Auguste Comte. His *Cours de philosophy positive* provided an impressive picture of the entire spectrum of the sciences and initiated several major topics of this new field of studies, such as the classification of scientific disciplines, the role of hypotheses and the empirical criteria of meaning.[1] Comte set the agenda in several respects for philosophy of science in France. Positivism, in one form or another, dominated here the philosophical scene until World War One, and even later thinkers who had relinquished positivism continued to pay tribute to him, most notably Canguilhem. First and foremost is Comte's decision to favor a historical approach over a logical one. Philosophy of science, he continually asserts, must be grounded on history of science. This trend was to characterize French philosophy of science generally. As an attempt to direct philosophical reflection toward science and to make scientific knowledge a model, positivism, in its various forms has been intimately bound up with a large portion of philosophy of science either as a source of inspiration or as a target for criticism: from Comtian positivism to logical positivism and even to postpositivism. It is thus important to come to grips with the significance and role of this doctrine. There were other significant figures of the time: André-Marie Ampère, Antoine-Augustin Cournot, Claude Bernard and Charles Renouvier. They all made significant contributions to the philosophy of science and helped to shape the early stage of the field.

My space is however strictly measured, and I shall keep to Chimisso's main focus. Lévy-Bruhl came of age in 1875 and Brunschvicg a decade later. They both died around the time of the second World War. Their active life spans what I shall characterize as the second stage in the development of philosophy of science. The

1 This six-volume work was published between 1830 and 1842.

Franco-Prussian War in 1870 not only signaled a change of political system – the end of the Second Empire and the beginning of the Third Republic – but led a whole generation to reflect on French science and to seek to emulate the German university system. We may mark out here a fifty-year period running until the end of the first World War in 1918. It is characterized by the early institutionalization of the discipline. Thereafter followed the interwar period, which represents a new phase, that of expansion of the discipline and development of a reflection on the latest scientific discoveries. I shall thus take the story back to the formative years of Lévy-Bruhl and the factors that explain the new departures of the early 20th century. Bachelard and Koyré, who started their carriers during the interwar period, will be considered here only in so far as they were influenced by the theories of their predecessors; their work has indeed received a good deal of attention.

The importance of Lévy-Bruhl and Brunschvicg in the constitution of philosophy of science in France is due to several facts. The former paved the way for the latter: Lévy-Bruhl had started teaching at the Sorbonne in 1902 and was elected to the chair of history of modern philosophy in 1908; he gave a new direction to the discipline, studying philosophers of the past in relation to the context of their epoch in its various aspects, with particular emphasis on the scientific background. To be sure, Émile Boutroux had already initiated a change with respect to the literary approach characteristic of the school of Victor Cousin, which have been influential until then.[2] But the arrival at the Sorbonne of Lévy-Bruhl followed by Brunschvicg and Milhaud, all of whom insisted on bringing science to bear on philosophy, marked a decisive shift.

Lévy-Bruhl is responsible for having forged the modern notion of mentality.[3] This notion was to play a central role not only in philosophy but also in history; the French historical school in the 20th can be characterized in the main by its recourse to mentalities. This provides the interpretative thread of Chimisso's study, centered on the "history of the mind". Shunning logic, French philosophy of science made extensive use in its investigations of the social sciences (sociology, anthropology and psychology), often combined with history. Anti-psychologism did not have a strong hold in France, excepting phenomenologists. This leads to several differences with respect to philosophy of science in German-speaking countries. Chimisso singles out several endeavors that are closely related methodologically:

> The underlying assumptions that united these projects were that the mind could not be studied *a priori*, and that ways of thinking were different in different civilizations. As a consequence, history was as a rule an essential component of research. Past philosophy and past science were expected to reveal worldviews and mental processes that differed from current ones.[4]

2 Boutroux replaced Paul Janet, a disciple of Victor Cousin, in 1888.

3 This is not unrelated to Auguste Comte's notion of mind or *esprit*, which corresponds to the three states of humanity: theological, metaphysical and positive.

4 Chimisso, 2008, p. 3. Cf. p. 73, 168.

Indeed, Lévy-Bruhl and Brunschvicg elaborated an *a posteriori* method of exploration of the mind, based on the historical documents that it yielded in its aim to understand the world.

Let us now turn to Brunschvicg. He had been teaching at the Sorbonne since 1905, and in 1927 he replaced Lévy-Bruhl in the chair of history of modern philosophy. As he came to elaborate his philosophical position, he acknowledged his debt to his predecessor.[5] There are many connections between the two thinkers; they were associated in many networks, and together they represent a strong line of development. Brunschvicg was to exert an ascendancy over French philosophy, establishing a particular brand of rationalism and idealism as well as forming many students. In particular he was Bachelard's doctoral supervisor, and Chimisso stresses the many similarities of their philosophies.[6]

2. The moment 1900, scientific revolutions and philosophical reflection

I have suggested that several factors explain "the moment 1900".[7] We should not forget that Lévy-Bruhl belongs to the same generation as a number of other important figures for the philosophy of science: Poincaré, Duhem, Milhaud and Meyerson. It seems that Lévy-Bruhl's shift from a rather traditional history of philosophy to a new approach occurred at a time when he was working on his book on Comte.[8] The originality of this book is to depart from the hagiographic writings of Comte's disciples and to provide a more distanced reading, by setting his doctrine more precisely within its historical context. One should not forget, however, a synchronous attempt by Milhaud to evaluate Comte's legacy: *Le positivisme et le progrès de l'esprit: études critiques sur Auguste Comte.*[9] Milhaud likewise was proposing a critical evaluation of this thinker. Both Lévy-Bruhl and Milhaud nevertheless retained something of the attitude that Comte had initiated. Furthermore, Lévy-Bruhl's ethnology or anthropology is not wholly unrelated to Comte's sociology, which aims to develop a "positive" study of humankind drawing largely on history. In a sense Comte's positivism gave rise to several parallel developments: sociology, anthropology and history of science. These were to replace metaphysics. Paul Tannery's history of science was one such outcome. Brunschvicg could call on both anthropology and history of science.

---

5 See "L'idée de la vérité mathématique", in Brunschvicg, 1958, vol. 3.

6 See Chimisso, 2008, p. 141.

7 I am referring here to the conceptualisation given by Frédéric Worms, *Le moment 1900 en philosophie*.

8 Lévy-Bruhl, *La philosophie d'Auguste Comte*, 1900.

9 This work was published in 1902, but Milhaud already criticizes Comte in his first book, 1893, p. 205.

It is worth to point out the underlying controversies; these helped to shape the movement we are interested in. One may note that all our authors developed their methods in opposition to the school of Victor Cousin. Milhaud brings this out clearly in speaking of Paul Tannery's contribution to the history of philosophy:

> You know what academic philosophy was like for a long time in France, I mean that kind of naïve and banal catechism which the school of Cousin had resulted in; and you know to what extent rhetoric, to which was given free rein, had inevitably divorced philosophy from science.[10]

One should not omit the scientific factors coming into play. A succession of revolutions in science had taken place that called for a reworking of the picture of knowledge, in succession: non-Euclidian geometry, the theory of evolution, thermodynamics and electromagnetism. One of the leading figures of the time was Henri Poincaré. His research in mathematics convinced him that non-Euclidian geometry was not a mere fiction but a fruitful conceptual construction. Meditating on the nature of geometrical hypotheses, Poincaré advanced the idea that they are conventions.

Pierre Duhem formulated a similar idea with respect to physics. Hypotheses are not directly derived from experience; they are founded on the free choice of the theorist. Experimental refutation is more complex than it was generally believed. This led to the holist thesis, which Neurath, followed by Quine, was to take up in the context of a logical analysis of science. These striking results were seized upon by several philosophers and scientists. Édouard Le Roy perceived here the rise of an intellectual movement that he labeled "a new positivism". Gaston Milhaud went so far as to speak of logical positivism or *positivisme logique* as early as 1905.[11] This reformulation of positivism attracted the attention of young Austrian scholars who were to found the Vienna Circle and provides us with a noteworthy connection between the philosophical traditions of France and Austria.

Le Roy emphasized the novelty of these reflections on science; he was one of the first to make use of the term *épistémologie* or epistemology. The term designates in French usage philosophy of science rather than theory of knowledge. What was being proposed was an investigation precisely centered on scientific activity. This carried an implicit criticism of earlier philosophy of science, as practiced by Comte, and signaled a shift in the discipline.

In connection with these debates over the nature of scientific theories early attempts were made to introduce philosophy of science into the university curriculum. In 1892 a chair of "General history of science" was instituted at the Collège de France. In 1909, a chair of "History of philosophy in its relation to science" was created for Milhaud at the Sorbonne. He thus came to work in the same university

10 Milhaud, 1911, p. 2. Similar criticism is voiced by Brunschvicg in the second edition of his thesis. A point also made by Bouglé as quoted by Chimisso, 2008, p. 73.

11 Milhaud, 1927, p. 55, reproducing an article published in 1905.

as Lévy-Bruhl and Brunschvicg. Milhaud's chair was to play a pivotal role in the future of the field, being held successively by Abel Rey, Bachelard and Canguilhem.[12]

Taking up Poincaré's ideas, Abel Rey was careful to emphasize the tendency toward realism. He was in particular struck by the recent discoveries of atomic theory, and was led to elaborate a historical approach employing techniques developed in the social sciences. His thesis, a synthetic presentation of the turn-of-the-century debates, was seized upon by the logical positivists. Abel Rey was furthermore included among Neurath's collaborators to the *Encyclopedia of Unified Science*. However, this promising connection between French conventionalism and Austrian positivism was cut short[13].

Bachelard, who succeeded to Rey in 1940, can be credited with having forcefully directed philosophical attention to the latest scientific theories. Along with Alexandre Koyré, he was convinced that the succession of revolutions that had shaken science since the discovery of non-Euclidian geometry called for a "philosophical revolution". Borrowing a phrase from Reichenbach, Bachelard spoke of a "conflict of generations", and he was quickly led to spell out the inadequacies of the philosophical conceptions of his predecessors. Thus was brought to a close a particular phase in the development of philosophy of science. However the historian may question this portrayal and seek deeper links and transmissions.

3. On historical method

What characterizes a large portion of French philosophy of science is the importance allotted to history. This is apparent in the early formulation of the discipline by Comte as well as its later institutional establishment. Of course, a historical approach can be pursued in many ways. One direction consists in grounding philosophy of science upon the history of science. In the absence of empirical testing, history of science provides a means of assessing philosophical conceptions of science; it provides the analogue of the laboratory[14]. This is particularly clear in Duhem. His *Aim and Structure of Physical Theory* furnished an analysis of the stages involved in the construction of a scientific theory. But this "logical analysis"[15], as he termed it, was to be followed by a historical study, and the numerous volumes he devoted to the evolution of science since Antiquity bear witness to this preoccupation. Such a method was followed by many of his contemporaries, for example Meyerson. Postpositivists were later to call on this tradition in their effort

12 Concerning the filiation between Tannery, Milhaud and Rey, see Brenner, 2005.

13 Contingent historical factors enter here.

14 This metaphor used by Brunschvicg is quoted by Chimisso, 2008, p. 73. Cf. p. 168.

15 Duhem, 1906, p. XV. Cf. Duhem, 1913, p. 115.

to reassert the importance of history, and this was one of the trends of the French tradition that received the most sustained interest abroad.

"History of philosophy in its relation to the sciences", to use the title of the chair created for Milhaud, constitutes another significant line of research. In introducing philosophy of science within the university curriculum, Milhaud was careful to link this speciality with the history of philosophy, which occupied an important role in France. Brunschvicg and Rey fit well into this program[16]. Such an interdisciplinary approach allowed for various collaborations and many topics of inquiry. It characterizes the institutional situation in France and marks a difference with respect for example to philosophy in Great Britain.

In a sense Bachelard and Koyré built on these antecedent efforts, the former in the direction of a historical philosophy of science and the latter in the sense of a philosophical history of science. But they gave a new twist to this approach. Both had misgivings over earlier conceptions of scientific growth as a continuous process. They set about to elaborate what has been named a "historical epistemology". Study of past science still retained its importance. But it was to be placed within a clearly discontinuist conception, inspired by the recent discoveries in science. Scientific revolutions are accompanied by breaks between common knowledge and scientific knowledge. Bachelard especially made explicit the position from which the philosopher observes the past: reading is necessarily retrospective or *récurrent*.

One may call here on Ian Hacking, who throws light on this issue. He brings out clearly the difference between Bachelard and Foucault, in other words the evolution undergone by historical method. He himself takes Foucault's historical epistemology or historical ontology a step further and gives expression to a whole trend of research being done today. Although educated in the analytic tradition, Hacking does not hesitate to call on French history and philosophy of science. Foucault, enlarging on Bachelard's perspective, had made a broader and more systematic use of history, which he in due course named "archeology of knowledge" or "historical ontology". Hacking takes up this approach, applying it more specifically to philosophy of science. In particular he gives a concrete meaning to the attempt to relate discourse to its context of formulation. And Hacking offers a careful analysis of the sites of production of experimental science: the laboratories, the observatories and the research centers.

He claims that it is quite possible to recover thereby the concerns of analytic philosophy. Historical ontology is just another way of pursuing analysis: the conceptual usages are referred chronologically to their site of enunciation. This is how he presents his program:

16 Although Rey obtained the change of this chair to "History and philosophy of science", he nevertheless admitted to pursuing the path opened by Milhaud. For more, see Brenner, 2005.

> Historical ontology is about the ways in which the possibilities for choice, and for being, arise in history. It is not to be practiced in terms of grand abstractions, but in terms of the explicit formulations in which we can constitute ourselves, formulations whose trajectories can be plotted as clearly as those of trauma or child development, or, at one remove, that can be traced more obscurely by larger organizing concepts such as objectivity or even facts themselves[17].

One can then submit the constitutive notions of science to a historical analysis, recording the discursive formulations and mapping out their development.

Conclusion

The period 1870–1920 that I have singled out for examination is very different from the founding years of philosophy of science; many new objects of inquiry arose, and the analysis of scientific knowledge provided was rich, original and fruitful. It is worthwhile to return to this epoch in order to sharpen our tools and to enlarge our list of problems. Furthermore, a complete picture of philosophy of science requires us to understand the transformation that brought about the conceptions of the mid-twentieth century. I believe that one way to move ahead is to be clear as to the objects, methods and aims of our inquiry. It is essential that we plot the trajectories of the tools of our trade.

In the past twenty years several French philosophers of science of the period prior to the Second World War have become the object of a more thorough and systematic investigation: first Duhem and Poincaré, then Meyerson and Metzger. Chimisso has convincingly argued in favor of adding to our list Lévy-Bruhl and Brunschvicg. We now have a whole series of philosophers, among whom the connections are numerous. Historical research has not only focused on individuals; work is currently been directed toward the content of journals as well as societies and institutions. Networks of relations among scholars are being extensively explored. In consequence, our picture of the field and our understanding of the nature of philosophy of science is being deeply modified.

Bibliography

Bachelard Gaston (1934), *Le nouvel esprit scientifique*, Paris, PUF, 1971.

— (1972), *L'engagement rationaliste*, Paris, PUF.

Bitbol Michel and Jean Gayon (2006) (eds), *L'épistémologie française: 1830-1970*, Paris, PUF.

Brunschvicg Léon (1958), *Écrits philosophiques*, Paris, PUF.

17 Hacking, 2002, p. 23.

Brenner Anastasios (2003), *Les origines françaises de la philosophie des sciences*, Paris, PUF.
— (2005), "Réconcilier les sciences et les lettres: le rôle de l'histoire des sciences selon Paul Tannery, Gaston Milhaud et Abel Rey", *Revue d'histoire des sciences*, vol. 58, p. 433-454.
Brenner Anastasios and Jean Gayon (2009) (eds), *French Studies in the Philosophy of Science*, Vienna, Springer.
Canguilhem Georges (1966). *Le normal et le pathologique*, Paris, PUF.
— (1968), *Études d'histoire et de philosophie des sciences*, Paris, Vrin, 1970.
Chimisso Cristina (2008), *Writing the History of the Mind: Philosophy and Science in France, 1900 to 1960s*, Aldershot, Ashgate.
Comte Auguste (1830–1842), *Cours de philosophie positive*, M. Serres, F. Dagognet A. Sinaceur and J.P. Enthoven (eds), new ed. revised, 2 vols, Paris, Hermann, 1998.
Duhem Pierre (1906), *La Théorie physique, son objet, sa structure*, Paris, Vrin, 1981.
— (1913), *Notice sur les titres et travaux scientifiques*, Bordeaux, Gounouilhou.
Hacking (2002), *Historical Ontology*, Cambridge, Mass., Harvard University Press.
Koyré Alexandre (1966), *Études d'histoire de la pensée scientifique*, Paris, Gallimard, 1973.
Lecourt Dominique (2008), *Georges Canguilhem*, Paris, PUF.
Le Roy Édouard (1901), "Un positivisme nouveau", *Revue de métaphysique et de morale*, 9, p. 138-153.
Lévy-Bruhl Lucien (1900), *La philosophie d'Auguste Comte*, Paris, Alcan.
Meyerson Émile (2009), *Lettres françaises*, B. Bensaude-Vincent and Eva Telkes Klein (eds), Paris, CNRS Éditions.
Milhaud Gaston (1893), *Leçons sur les origins de la science grecque*, Paris, Alcan.
— (1902), *Le positivisme et le progrès de l'esprit: Études critiques sur Auguste Comte*, Paris, Alcan.
— (1911), *Nouvelles etudes sur l'histoire de la pensée scientifique*, Paris, Alcan.
— (1927), *La philosophie de Charles Renouvier*, Paris, Vrin.
Poincaré Henri (1902), *La Science et l'hypothèse*, Paris, Flammarion, 1968.
Wagner Pierre (2002) (ed.), *Les philosophes et la science*, Paris, Gallimard, 2002.
Worms Frédéric (2005) (ed.), *Le moment 1900 en philosophie*, Lille, Septentrion.

Université Paul Valéry-Montpellier III
Route de Mende
34199 Montpellier Cedex, France
anastasios.brenner@wanadoo.fr

Michael Heidelberger

Aspects of Current History of 19th Century Philosophy of Science

"Through history, philosophy seeks,
in its past, its eternal present."
(Bréhier 1940, 44)

The attitude of philosophy towards its own history and historiography has been subject to change over the ages. Periods of deep interest in the history of philosophy and periods where this interest fades into the background alternate in the course of time. One reason for this variation might be a natural life-cycle in the succession of generations: the young must be iconoclasts and destroy their heritage to a certain extent in order to develop a new and original one. When they grow old, however, they sooner or later reach the point where they start wondering whether their promises have been kept and whether their own achievements really fulfil what they intended at the start – and a new interest in the recent past sets in again.

I think that, if appearances are not deceptive, we are witnessing today a renewed interest in the history of the philosophy of science, and especially of the 19th century. This interest might be to some extent the result of the life-cycle just described, but there seem to be additional and deeper motives for it. In the following, I would like to ask what these deeper reasons might be, where we stand today in relation to 19th century philosophy of science, and how we can fruitfully develop this interest further and gain better insight into the work of our forebears and, with this, into our own situation.

The interest in a bygone period of philosophical thought has first of all an emotional dimension: On the one hand one can see the past with some wistfulness as an age that already realized what one wishes for one's own present and future, and what one laments as being lost, maybe even forever. In this sense the philosophical past can appear as a golden age and serve as a projection screen of hopes and ideals for the future, but also as a dreamland in which one can seek refuge from the cruel present. On the other hand, the past can, of course, also appear as a barren age whose limitations and mistakes can and must be overcome by a new philosophical spirit that is immune to the failures of the past. There is also the third position that, without taking sides, so to speak, uses the former age as a mirror, in which the present can be reflected and reappraised. This can lead to a better understanding of one's own situation and to a strengthening of one's own identity.

F. Stadler (ed.), *The Present Situation in the Philosophy of Science*,
The Philosophy of Science in a European Perspective, DOI 10.1007/978-90-481-9115-4_5,

I think that our attitude today towards the philosophy of science of the 19th century is fed by all three attitudes: We can admire the sublime innocence and simplicity with which many 19th century thinkers laid the foundations for countless philosophical views in the philosophy of science whose origins have often been forgotten and buried under their elaboration by later generations. We can, of course, also often see in 19th century thought "bewitchments of our intellect," as Wittgenstein wrote, that still exert a deleterious influence on us today. Both these attitudes can help us better understand our own standpoint in the present. This is all the more possible as agreement among philosophers of science on their subject matter has definitely diminished since the 1960s. Herbert Feigl could still claim in 1964 that in the field of philosophy of science "there is perhaps a larger measure of agreement […] than in any other area of philosophy" (Feigl 1964, 467). The strictures of logical empiricism have been lowered since then and have led to a more large-minded attitude as a result of which the variety of individual approaches and procedures has definitely increased.

For a long time, 19th century philosophy of science appeared to many as a mere outgrowth of a bygone age that either unduly praised science as the redemption from all evil or, equally unduly, condemned it as godless materialism and the origin of modern society's faults. In any case, the philosophy of science of the 19th century carried with it an overtone of *Weltanschauung* that made it often indigestible for the philosophical outlook of most of the 20th century. The negative image of the 19th century allowed for only a handful of exceptions, mostly "forerunners" like e.g. the towering philosopher-scientists Mach or Helmholtz, or some isolated geniuses like Poincaré or Einstein, who became household gods of logical empiricism, the major school of philosophy of science in the 20th century.

The image of the 19th century as an obsolete and outdated age with which philosophy of science has little or nothing to do anymore was also due to the devastating experience of the First World War. In this vein, Hans Reichenbach wrote in the first issue of *Erkenntnis* of the "breakdown of traditional emotional worlds" (*Gefühlswelten*), of "disenchantment" (*Entzauberung*) and "desoulment" (*Entseelung*). The great disillusionment (*Ernüchterung*) of the time is for Reichenbach, as he wrote, not only "a dominant feature of scientific research, but it is also the main feature of our daily existence, the category under which we have to see our present" (Reichenbach 1930/31, 70). Another reason for leaving the 19th century behind at the time can be seen in the revolutionary developments in physics: with relativity theory and quantum mechanics the old mechanist worldview of the 19th century can be thrown into the dustbin of history and with it the associated philosophy of science.

In the meantime, however, it seems that, if I am not mistaken, a more positive attitude has set in towards the 19th century. 19th century philosophy of science is increasingly seen in a much more positive light or at least felt as a sounding-board that still resonates in today's philosophical work. The reasons for this development are manifold. A first reason lies in the fact that in the wake of post-positivist philosophy of science, i.e. the philosophical development after Kuhn, Feyerabend, Lakatos and others, philosophers increasingly left their armchairs and tried to do justice to concrete history of science in their work. As a result of this, philosophers of science have developed an interest in the history of their own field. One of the first steps was to have a closer look at the "official" forerunners of enlightened 20th century philosophy of science as they are codified in the 1929 manifesto of the Vienna Circle, for example. This closer look revealed that these "heroes" were much less an exception of their time and much more embedded into the contemporary mainstream of thought than is widely assumed. In my own work, I have found out and shown that Ernst Mach, one of the founding fathers of our field, is deeply rooted in the psychophysics of G. T. Fechner and the intricacies of the mind-body problem as it posed itself at the time of the newly rising field of experimental psychology (Heidelberger 2004, chs. 4 and 6).

This trend towards more historical context is sometimes correlated with a growing awareness that some of the hotly debated issues in today's philosophy of science are taking on a more *weltanschaulich* dimension as well that is getting closer to standpoints of the 19th century in some respects. With the anti-metaphysics of Carnap and Neurath and others it seemed for a while as if genuine philosophical controversies had come to an end. There cannot really be any difference, e.g., between idealism and realism, Carnap held, other than a metaphysical one. And if metaphysics is nonsense, the difference between idealism and realism disappears. Yet the development of philosophy of science has shown that this hope was much too rash. Even the most hardboiled follower of logical empiricism has to admit that there is at least a difference between an empiricist and a realist attitude towards empirical theories. There are many more substantial philosophical problems that are not just pseudo-problems and are not to be solved just by additional empirical information or logical tricks.

The newly found appreciation of substantive philosophical problems in the philosophy of science is complemented by a trend in society at large to connect science with more than just technical expertise. In an extreme scenario, there is the possibility that today's controversies about science risk conjoining with the conflict of Muslim culture with the West some day. A glimpse of this can already be felt in the disputes about intelligent design. As a result of this development, philosophy of science finds it has a deeper affinity with the way philosophical problems were debated in the 19th century than with the deflationary attitude of much of 20th century philosophy.

Take, for example, the second part of Friedrich Albert Lange's *History of Materialism* from the 1870s, and read what he has to say about the "philosophical materialism since the time of Kant" (Lange 1873-75). You get a "déjà vu" except that the materialist prophets of today no longer teach at German or Swiss universities but, rather, at some remote Australian or Texan one. Or flip the same book open to the chapter on "brain and soul" and you will almost get the impression that you are reading a present-day philosopher of science rebutting the claim of some neuroscientist that Libet's experiments have done away with freedom of the will once and for all. Another example is Lange's chapter on "Darwinism and Teleology" where you find material that could easily be used today as arguments against the belief in intelligent design.

Also, Lange's passionate outlook on the newly emerging scientific psychology of his time has much more similarity with our discipline's enthusiasm with cognitive science than with the methodological games philosophy of science played with logical or empirical behaviourism in the early 20th century. Lange, however, is not the only 19th century author who can speak to us directly without much hermeneutics involved. I think that there is hardly a better way to introduce a student to the enigmas of realism and antirealism in philosophy of science than to read with her or him a chapter of Vaihinger's *Philosophy of the 'As If'* or to compare some suitable paragraphs of the *Mechanics* of Mach, the antirealist, with a corresponding text from a talk of Helmholtz, the arch-realist.

It sometimes looks as if the two World Wars and their aftermath distracted philosophy of science from its real topics and that the decline of Logical Empiricism – at least in the perspective proposed by George Reisch – and the more liberal attitude towards metaphysics and other schools of thought has brought us back again to the point from where Logical Empiricism started its journey. Historians sometimes call the 20th century the "short century" and compare it to the preceding "long" one because the deep changes that took place in the 1900s only span from 1914 to 1989, whereas the 19th century as an epoch must be considered from 1789 to 1914. It seems to me that the 20th century was short also in the second sense that the number of durable revolutionary achievements in philosophy of science turns out to be less than one originally thought.

Even the call to revert to Kant that could be heard in philosophy from the 1830s onward and that dominated so much of the 19th century, also in philosophy of science, has its serious present-day counterpart. Michael Friedman tirelessly tries to convince us that the relativism that arose in the wake of Thomas Kuhn's work can be overcome by making use of Reichenbach's distinction between two meanings of the Kantian a priori, necessary and unrevisable on the one hand, and constitutive of the concept of the scientific object and its knowledge on the other. Einstein's revolution has shown, says Friedman, that the first meaning must be relativized in order to allow for historical change whereas the second meaning can and must be retained. Questions arising inside a scientific paradigm can be likened to Carnap's "internal questions" and the relative a priori, whereas external questions that concern the framework itself correspond to Kuhn's scientific revolutions and to the constitutive a priori. In this sense, Friedman is able to argue that the parting of the ways of so-called "analytical" and "continental" philosophy (of science) was not an inescapable fate but more the result of the contingencies of the twisted 20th century. In the end, Carnap appears as the central figure of 20th century philosophy of science and philosophy *tout court* because his principles fit best into the neo-neo-Kantianism proposed by Friedman (Friedman 2001).

The call to Kant is not the only nostalgic return to the 19th century that can be noted in the philosophy of science today, although it is perhaps the most notable one. We are also witnesses to a revival of an American and other brands of pragmatism that seems, however, a little less homogeneous than the neo-Kantian one. And we can finally note a growing interest in so-called "continental philosophy of science" generally. This is no longer just a fringe interest in dubious and frivolous ideas, as the word "continental" originally was supposed to signify, but in genuine and viable alternatives to logical empiricism.

The use of the term "continental" has led to a curious development in this respect. From the early days of Frege, Russell and Wittgenstein, analytical philosophy has inherited a strong anti-naturalist tendency. Since the philosophy of science grown out of logical empiricism also partially sympathized with a philosophical programme of naturalisation, but was and is at the same time part of the analytic, and thus anti-naturalist, movement, a tension developed, from the time of Neurath and Carnap onward, between a more logical and a more naturalist outlook in philosophy of science. As a result, someone like Ernst Mach, who is justly called the forefather of logical empiricism, is discussed today under the label of "continental philosophy of science" – i.e. of a movement that is supposed to be in opposition to analytical philosophy (Babich 1994, Norris 1999). A logical consequence would be to categorize Quine in the same way: Although he was one of the foremost logicians of the 20th century, his "epistemology naturalized" made him a critic of logical empiricism and thus of "analytic" philosophy of science. In addition, his critique was developed with the help of at least two typical "continental" philosophers: Duhem and Meyerson (cf. Laugier 2009). The resulting obliteration of categories has a salutary effect: One can admit that philosophy of science has never been intrinsically or essentially analytic. This in turn allows us to give up more easily the ahistorical posture entertained by "analytical" philosophy and admitting our affinity to 19th century philosophy of science.

A similar remark can be made about the French philosophy of science tradition with Bachelard, Canguilhem and others. They seem even more to constitute a paradigm of a "continental" approach than Mach does. But this view neglects the fact that without the conventionalism of Henri Poincaré, which deeply informed, and still informs, French *épistémologie*, logical empiricism would not have developed in the way it did. Curiously enough, the French philosophy of science tradition seems for many to be rooted much more in the 19th century than Carnap or Neurath and other logical empiricists, although both sides lived and worked in the early 20th century.

The return of the 19th century in today's philosophy of science is not only visible in publications, but also, I think, in teaching, at least in my own experience. If you really want to enthuse beginning students for the subject, it is very efficient to take some scientist-philosopher of the late 19th century as a starting point – thinkers like Mach, Helmholtz, Boltzmann, Darwin, Hertz, Poincaré, Duhem, Einstein or even Ostwald – or a scientifically enlightened philosopher like F. A. Lange,

Alois Riehl or Hans Vaihinger. Even Schopenhauer and Nietzsche and, yes, Henri Bergson, can do the same job to some extent. Incidentally, most of these authors served as inspirations for our logical empiricist heroes when they attended the *Gymnasium*. That is already enough reason not to forget about them today.

I find it hard to seduce raw recruits directly to Carnap or Quine or Neurath, although they should, of course, get excited about them at a later, more advanced, stage. There might be a slight difference between teaching philosophy of science to beginning philosophers and teaching it to beginning science students. The latter do not need as much 19th century flavour because they can much more identify with 20th century science than most "pure" philosophy students, but they see it as a challenge to be mastered and to be connected right away with a philosophical outlook. But even among some physics students of today one can find a feeling of *fin de siècle* as it was present at the end of the 19th century: Most of the fundamental problems of physics seem to have been solved by now, and those that are left might be unsolvable, so that the only job left is to expand the realm of application of fundamental scientific insights and not so much to find new revolutionary beginnings.

Perhaps you have become suspicious by now of my enthusiasm for the 19th century and tend to regard it as mere sentimentalism that shies away from addressing the challenges of the science of the 21st century. I think, however, that if we want to understand where we stand today as philosophers of science, and as *European* philosophers of science at that, we definitely have to come to terms with our own history. Alan Richardson has recently put it in a succinct way:

> Most philosophers of science engage with logical empiricism only in so far as they are concerned to claim that they have gone beyond it. [...] Yet the question of what contemporary philosophy of science owes to logical empiricism and how it has advanced beyond it can be adequately answered only with [...] a history [of logical empiricism as a project in twentieth century philosophy]. Only through such a history can philosophers fully understand both their sense of what is philosophical in their own project and how they ought to engage in philosophical inquiry" (Richardson 2008, 96).

I can very much agree with this, but I think that as European philosophers of science trying to find our own way today, we cannot limit ourselves and be content with just a history of logical empiricism. For North American philosophy of science and perhaps for those parts of the international community that are exclusively influenced by an Anglo-Saxon outlook, this might be enough for some time. But even from this outlook it is necessary to consider also the history of American pragmatism as well: someone like Quine cannot be understood without taking account of a pragmatist influence. From a genuinely European perspective, however, there is – and should be – a special interest in the development of philosophy of science that goes beyond the formative period of logical empiricism. European philosophy of science comprises more: There are many different and rich traditions of reflecting on science that already existed before the rise of logical empiricism and still await reappraisal. The history of the philosophy of science to be reconsidered and understood in order to develop a *European* identity for philosophy of science definitely extends beyond the formation period of logical empiricism and beyond the official forerunners of this movement further into the past. So I would counter the allegation of being sentimentalist about the past against those who disdain the history of their field with the charge of perhaps blitheful but mindless ignorance of one's own identity.

In conclusion, let me reflect a little on the consequences that result from this defence of the history of 19th century philosophy of science for our future work. For me, it has become evident by now that we should start to undertake *comparative* studies in the history of the philosophy of science. I think that little work has been done so far in this direction, regardless of how we define the precise meaning of "comparative". It can mean that we start studying the different transfers of ideas from one country or culture to another during the 19th century and beyond. It can also mean comparing the philosophical outlook on the different sciences from one country or society to another. And it can likewise imply the study of the difference or similarity among concepts that played – and still play – a role in the philosophy of science in general. This should not be misunderstood as a plea for national(ist) histories. On the contrary: it should and can make visible the subtle but effective interdependence of a philosophical spirit in relation to science that has after all outlived many wars and hostilities among the different European societies and has in the end led to cross-cultural movements such as logical empiricism.

Even if it might in the end be farfetched to assume a common European spirit in all the different approaches that have been developed in our history, it would be wrong just to forget these different attitudes. Recently, I have read a French introduction to the philosophy of science that draws a veil of silence over the peculiar tradition of French *épistémologie* (Barberousse 2000). It otherwise gives a quite respectable overview of the field as it stands today in the Anglo-Saxon world. It is true that parts of the French academic scene still have to catch up with the international development and are sometimes too wrapped up in (often exclusively their own) history, thereby neglecting systematic approaches. But this should not be a reason to hush up one's own roots. And even if one feels alienated from one's own tradition, it is still worth the effort to learn from the failures of the tradition – if there really are any. But in order to learn from something, one has at least to take note of it.

References

Babich, Babette E. (1994), "Philosophies of science: Mach, Duhem, Bachelard", in *Twentieth-Century Continental Philosophy*, ed. by Richard Kearney.

(Routledge History of Philosophy, Volume VIII) London: Routledge, 144-183.

Barberousse, Anouk, Max Kistler and Pascal Ludwig (2000), *La philosophie des sciences au XXe siècle*. Paris: Flammarion.

Bréhier, Émile (1940), *La philosophie et son passé*. Paris: Presses universitaires de France.

Feigl, Herbert (1964), "Philosophy of Science", in *Philosophy*, ed. by Roderick M. Chisholm et al. Englewood Cliffs, NJ: Prentice-Hall, 465-540.

Friedman, Michael (2001), *Dynamics of Reason: The 1999 Kant Lectures at Stanford University*. Stanford: CSLI Publications.

Heidelberger, Michael (2004), *Nature from Within: Gustav Theodor Fechner's Psychophysical Worldview*. Pittsburgh: The University of Pittsburgh Press.

Lange, Friedrich Albert (1873-75), *Geschichte des Materialismus und Kritik seiner Bedeutung in der Gegenwart*. 2nd rev. and expanded ed. 2 vols., Iserlohn: Baedeker. Vol. 1: 1873; vol. 2: 1875. [Reprint in 2 vols., ed. Alfred Schmidt. Frankfurt am Main: Suhrkamp 1974.] Transl. of the 2nd German ed. as *The History of Materialism and Criticism of its Present Importance* by Ernest Chester Thomas. 3 vols., London: The English and Foreign Philosophical Library 1877-79. Reprint of the 3rd ed. of 1892 in one vol. with introd. by Bertrand Russell. London: Kegan, Paul 1925. Also New York: Humanities 1950, and London: Routledge 2000. 1st German ed., Iserlohn: Baedeker 1866.]

Laugier, Sandra (2009), "Science and Realism: The Legacy of Duhem and Meyerson in Contemporary American Philosophy of Science", in *French Studies in the Philosophy of Science*, ed. by Anastasios Brenner and Jean Gayon. (Boston Studies in the Philosophy of Science, 276) Dordrecht: Springer, 91-112.

Norris, Christopher (1999), "Continental Philosophy of Science", in *The Edinburgh Encyclopedia of Continental Philosophy*, ed. by Simon Glendinning. London: Routledge, 402-415.

Reichenbach, Hans (1930-31), "Die philosophische Bedeutung der modernen Physik". *Erkenntnis 1*: 49-71.

Richardson, Alan (2008), "Scientific Philosophy as a Topic for History of Science". *Isis 99* (1): 88-96.

Universität Tübingen
Philosophisches Seminar
Bursagasse 1.
D-72070 Tübingen
Germany
michael.heidelberger@uni-tuebingen.de

Massimo Ferrari

Well, and Pragmatism?

Comment on Michael Heidelberger's Paper

Michael Heidelberger suggests that "we are witness today of a renewed interest in the history of philosophy of science, especially of the 19th century". This statement involves a broad historical perspective and asks not only that we take a deeper look into the philosophy of science of the age of Helmholtz and Mach, of Poincaré and Hertz, but also that we consider them and other leading figures of the time in the context – as Heidelberger correctly says – of national traditions (such as that of France or even Italy) and within a more articulated historical background.[1] I agree in particular with the proposal for "comparative studies in the history of philosophy of science". The transfer of ideas from one country to another during the 19th century represents a crucial historical issue for our research and may contribute to a new interpretation of the history of philosophy of science, not only during the "long century", but also in regard to the "short" one, i.d. looking forward – broadly speaking – to our present debates and our philosophical agendas.

Heidelberger offers very stimulating considerations about the "nostalgic return" to Kant that can be noted today and suggests that philosophers such Friedrich Albert Lange can be read in a different perspective. His great work on the *History of Materialism* was actually a *Standardwerk* for almost two generations of philosophers of science; and as Wilhelm Dilthey wrote in 1877, Lange's *opus magnum* was destined to remain a book marking a turn point in the philosophical debate of late 19th century.[2] But the *History of Materialism* was a very important reference for Carnap, Reichenbach and Schlick still in the age of early Logical Empiricism. Though it may seem strange, we can read in the third issue of *Erkenntnis* an enthusiastic portrait of Lange, one of the few philosophers who was able – according to the editors of his correspondence with Anton Dorn – to engage in a dialogue with the natural sciences and to acquire in this sense the great «merit» (quite similar to the Marburg School of Neo-Kantianism) to have rediscovered "Kant as a natural scientist", in opposition to the metaphysical interpretation of

1 Among recent publications on this issue I would like to mention the stimulating collections of essays Jean-Claude Pont/Laurent Freland/Flavia Padovani/Lilia Slavinskaia (Eds.), *Pour comprendre le XIXe. Histoire et philosophie des sciences à la fin du siècle*. Firenze: Olschki 2007 and Michael Heidelberger/Friedrich Stadler (Eds.), *History of Philosophy of Science. New Trends and Perspectives*, Dordrecht–Boston–London: Kluwer 2002.

2 Wilhelm Dilthey, *Gesammelte Schriften*, vol. XVII, *Zur Geistesgeschichte des 19. Jahrhunderts*, ed. by U. Hermann, Göttingen: Vandenhoeck & Ruprecht 1974, p. 101.

F. Stadler (ed.), *The Present Situation in the Philosophy of Science*,
The Philosophy of Science in a European Perspective, DOI 10.1007/978-90-481-9115-4_6,

Kant's philosophy endorsed by "professional philophers". It will be useful to remember, that such a praise of Lange was drawn from no less than Rudolf Carnap and Hans Reichenbach.[3]

Otherwise it is well known that Neokantianism, first of all the Marburg Neokantianism of Cohen, Natorp and Cassirer, but also the Neokantianism of Alois Riehl or, to some extent, of Hans Vaihinger, had a great influence on the philosophy of science which starts its journey at the Vienna Station. Historical and systematic reconstructions – in the case of Carnap it will be enough to remind you of the contributions of Alan Richardson and Andre W. Carus – show in a very exciting way how the received view and the current geneaologies of Logical Empiricism must be corrected within the framework of another story.[4] That story starts from scientific Neokantianism and reformulates some crucial aspects of this tradition in a really revolutionary new perspective. But the breakdown of the older world in every revolution is more complicated than a mere farewell to the previous age or, in this case, to previous conceptual tools. In my opinion, Neokantianism was in this context not only a philosophical stream, but a branch of late 19th century philosophy of science that, first of all in the German speaking world, was very influential on, and at some length was elaborated from, the leading scientists in their own work.[5] There was a time, to put it differently, in which a great physicist as Heinrich Hertz was able to read *Kant's Critique of Pure Reason* or his *First Metaphysical Principles of Natural Science* after long hours of hard laboratory work, obviously not as a "moral holiday" from his scientific engagement.[6] If our historical and philosophical task is the contextualization of philosophy of science as well as of epistemological frameworks elaborated in the late 19th century, it seems unavoidable to elucidate the assimilitation within philosophy of science of the Kantian and Neo-Kantian heritage, or – to quote Michael Friedman's statement – to describe

3 I refer to „Dokumente über Naturwissenschaft und Philosophie. Briefwechsel zwischen Friedrich Albert Lange und Anton Dohrn", in: *Erkenntnis* 3, 1932/33, pp. 262-300 (quotation from pp. 262-263).

4 Alan W. Richardson, *Carnap's Construction of the World. The "Aufbau" and the Emergence of Logical Empiricism*, Cambridge: Cambridge University Press 1998 and Andre W. Carus, *Carnap and Twentieth-Century Thought. Explication as Enlightenment*, Cambridge: Cambridge University Press 2007.

5 For an excellent overview on this topics see Michael Friedman/Alfred Nordmann (Eds.), *The Kantian Legacy in Nineteenth-Century Science*, Cambridge (Massachusetts)–London: The MIT Press 2006. See also Massimo Ferrari, "Il Kant degli scienziati: immagini della filosofia kantiana nel tardo Ottocento tedesco", in: Giusppe Micheli (Ed.), *Momenti della ricezione di Kant nell'Ottocento*, Milano: Franco Angeli 2006, pp. 183-201. For the philosophy elaborated by scientists more generally, see the noteworthy book by Erhard Scheibe, *Die Philosophie der Physiker*, München: Beck 2007.

6 Heinrich Hertz, *Erinnerungen, Briefe, Tagebücher*, ed. by Mathilde Hertz and Charles Susskind, San Francisco: Physik Verlag 1977, p. 190.

"how the original Kantian position was successively transformed by a long tradition of scientific thinkers leading all the way up to the present day"[7].

Michael Heidelberger reminds us that we are now also witness "of a revival of American pragmatism", although "less homogeneous than the neo-Kantian camp". This is a very interesting point and I would like to develop some reflections about the transfer of American Pragmatism from Harvard to Europe and, particularly, to Italy on the one side and to Vienna on the other. I think that the list of scientist-philosophers or philosophers of the late 19th century that Heidelberg cites as fruitful teaching material (Mach, Helmholtz, Poincaré, Duhem, Lange or even Bergson) can be enriched by the name of a leading figure of Pragmatism: William James.

The standard view of the topic "Pragmatism and European Philosophy of Science" is well known. According to it, the emigration of Logical Empiricism from Germany, Austria and Central Europe between the wars and the intricate process of its alteration in the "new world" created a context in which European philosophy of science was contaminated by North American ways of thinking, especially the tradition of pragmatism. This standard view has indeed overlooked two aspects. On the one side, recent scholarship has showed that the transfer of Logical Empiricism in the U.S.A involved an increasing professionalization of philosophy of science and, at the same time, the lost of the typical political and cultural engagement of its heyday in Vienna.[8] On the other side, and this is much more important for our present perspective, a relationship between European philosophy of science and Pragmatism was established long *before* the intellectual emigration from Europe between the World Wars. Especially James' pragmatistic insights – certainly more James' version of Pragmatism than Peirce's – travelled from America to Europe at the very beginning of 20th century in precisely the opposite direction of the later, more well-known journey from Weimar Germany and 'red' Vienna to American departments of philosophy.

In other words, there is another version of the story of the relationship between Pragmatism and Logical Empiricism which starts at the end of 19th century and whose direction is – paraphrasing the title of Gerald Holton's contribution on "the Americanization of the *Wissenschaftliche Weltauffassung*" – from Harvard Square to the Vienna Circle.[9] A brief account, particularly, of the reception of William James' pragmatism within European philosophy of science would un-

7 Michael Friedman, "History and Philosophy of Science in a New Key", in: *Isis* 99, 2008, p. 133.

8 See the illuminating reconstruction offered by George A. Reisch, *How the Cold War Transformed Philosophy of Science. To the Icy Slopes of Logic*, New York: Cambridge University Press 2005. Important contributions on this topic can be found also in Gary L. Hardcastle/Alan W. Richardson (Eds.), *Logical Empiricism in North America*, Minneapolis-London: University of Minnesota Press 2003

9 Gerald Holton, "From the Vienna Circle to Harvard Square: The Americanization of a European World Conception", in: Friedrich Stadler (Ed.), *Scientific Philosophy: Origins and Developments*, Dordrecht–Boston–London: Kluwer 1993, pp. 47-73.

doubtedly deal with Mach and his entourage in Vienna. As Holton points out, James' "philosophy of Pragmatism, developed in the first instance as a way out of a personal struggle that has been called James' 'Kant crisis', overlapped with Machian empiricist position in many ways, for example, in finding the meaning of ideas in the sensations that may be expected from their realization".[10] To be sure, James was well acquainted with Mach's works which he had read carefully, making annotations, marginalia, queries and so on; and, particularly, James was deeply interested not only in Mach's *Analyse der Empfindungen*, but also in his book on *Mechanik*, especially Mach's famous discussion of Newton's views on time, space and causality[11]. For his part, Mach was indeed a convinced supporter of James' work on *The Principles of Psychology*, but on the other hand his disagreement with Pragmatism as philosophical orientation was quite clear: an interesting proof of his critical evaluation may be found in a letter to Anton Thomsen from January 1911.[12]

Nevertheless, the connection James-Mach suggests first of all another connection which has to do with both the American thinker and the Viennese scientist. We mean the Italian philosopher of science and language Giovanni Vailati, a former collaborator of Giuseppe Peano's *Formulario mathematico* and a convinced supporter of Mach's historical and epistemological work, who was also engaged, at the very beginning of the century, to endorse a "logical pragmatism" quite different from the "magic pragmatism" of his friend Giovanni Papini. Vailati had a great admiration for Peirce and his pragmatic rule of meaning (i.e. the rule formulated by Peirce in his seminal essay "How to Make Our Ideas Clear"), but he also was aware immediately of the *epistemological* relevance of the *Jamesian* pragmatism. In his reviews both of *The Will to Believe* and some years later of James' famous *Pragmatism*, Vailati emphazises James' great merit of having offered a certain rehabilitation of «the *constructive* and *anticipating* activities of human understanding». According to Vailati, James was right to criticise as the common understanding scientific and philosophical truth has underestimated this aspect and consequently has endorsed an image of mental activity which is limited to a mere classification and, so to speak, a recording of empirical data. In Vailati's opinion, James is in this respect perfectly in agreement with the recent "logic of science", namely with the analyses developed by Mach, Clifford and others of the methods, history and principles of modern science. On the other hand, Vailati underlines the epistemological importance of James' critical assessment of positivism as well as of the sometimes «narrow-minded» philosophy nourished by the

10 Holton, "From the Vienna Circle to Harvard Square", loc. cit., p. 50.

11 Holton, "From the Vienna Circle to Harvard Square", loc. cit., p. 51.

12 *Ernst Mach als Außenseiter. Machs Briefwechsel über Philosophie und Relativitätstheorie mit Persönlichkeiten seiner Zeit*, ed. by J. Blackmore and K. Hentschel, Wien: Braumüller 1985, p. 86 („Der Schwerpunkt seiner Arbeit liegt gewiß in seiner ausgezeichneten Psychologie. Mit seinem Pragmatismus kann ich mich nicht ganz befreunden").

scientists. According to Vailati, James is perfectly right in emphasizing the crucial role in the scientific inquiry of audacious formulations of hypotheses;[13] similarly, he points out that James has recognized better than any other philosopher of science the function of belief for the scientific method.[14] Broadly speaking, Vailati appreciates the pragmatic view according to which scientific knowledge is always the result of a *mental construction*, whereas the empirical, factual basis seems to be not as foundational and unavoidable as the (positivistic) standard view tends to suggest.[15]

The great merit of Vailati seems to have been to have understood, quite unlike his contemporaries, that James was elaborating a version of Pragmatism that was in no way to be thought of as a mere voluntaristic or even "irrationalistic" philosophy. And we may recognize that Vailati's suggestions are correct. In his book on *Pragmatism*, indeed, James offers a short but very illuminating account of contemporary philosophy of science. Mach, Duhem and Poincaré – says James – are "teachers", according to which "no hypothesis is truer than any other in the sense of being a more literal copy of reality. They are all but ways of talking on our part, to be compared solely from the point of view of their *use*."[16] Moreover, James gives an holistic account of what means the acquisition and growth of truth within the historical process of knowledge which seems undoubtedly 'up to date' to a reader well acquainted with the following philosophy of science from Neurath to Quine[17]. James says, for instance:

> [A] new idea is [...] adopted as the true one. It preserves the older stock of truths with a minimum of modification, stretching them just enough to make them admit the novelty, but conceiving that in ways as familiar as the case leaves possible [...] New truth is always a go-between, a smoother-over of transitions. It marries old opinion to new fact so as ever to show a minimum of jolt, a maximum of continuity. We hold a theory true just in proportion to its success in solving this "problem of maxima and minima"[18].

James is fully convinced that an anti-foundationalist account of knowledge is required when we want to take into account that our thinking develops in quite a different way from that offered by traditional philosophy since Descartes:

13 Giovanni Vailati, *Scritti*, Firenze: Seeber & Barth 1911, p. 270.

14 Regarding Vailati's position within European philosophy of science between 19th and 20th century I would like to refer to my book *Non solo idealismo. Filosofi e filosofie in Italia tra Ottocento e Novecento*, Firenze: Le Lettere 2006, pp. 141-164.

15 Vailati, *Scritti*, op. cit., p. 283.

16 William James, *Pragmatism. A New Name for Some Old Ways of Thinking*, Cleveland and New York: Meridians Books 1963, p. 125.

17 On James and Quine see I. Nevo, "James, Quine, and Analytic Pragmatism", in: R. Hollinger/D. Depew (Eds.), *Pragmatism. From Progressivism to Postmodernism*, Westport (Connecticut)–London: Prager 1995, pp. 153-161

18 James, *Pragmatism*, op. cit., pp. 50-51.

To begin with, our knowledge grows in *spots*. The spots may be large or small, but the knowledge never grows all over: some knowledge always remains what it was [...] Our minds thus grow in spots; and like grease-spots, the spots spread. But we let them spread as little as possible: we keep unaltered as much of our old knowledge, as many of our old prejudices and beliefs, as we can. We patch and tinker more than we renew. The novelty soaks in; it stains the ancient mass; but it is also tinged by what absorbs it. Our past apperceives and co-operates; and in the new equilibrium in which each step forward in the process of learning terminates, it happens relatively seldom that the new fact is added *raw*. More usually it is embedded cooked, as one might say, or stewed down in the sauce of the old. New truths thus are resultants of new experiences and of old truths combined and mutually modifying one another[19].

We may consequently affirm that Vailati was right in emphazising the epistemological core of James' Pragmatism: this makes him an excellent exception in the philosophical landscape at the beginning of 20th century in Europe. But there is another meaningful historical circumstance that supports the relevance of Vailati in this context. In September 1908 Vailati was in Heidelberg in occasion of the Third International Congress of Philosophy. The European quarrel about pragmatism started just there, in the section of the Congress devoted to the discussion of Ferdinand Schiller's talk about the pragmatic theory of truth. The critical reaction of the German philosophical establishment towards the "yankee" philosophy just arrived in Europe was extremely unfavourable and the debate following Schiller's lecture was, according to the congress report, very lively.[20] It is noteworthy, however, that the only participants to the Congress being in agreement with the pragmatic method in philosophy were Vailati and a philosophical outsider from Vienna, Wilhelm Jerusalem. In the same year as the Congress in Heidelberg Jerusalem published a very good German translation of James' *Pragmatism* and wrote a highly interesting preface to it. First of all Jerusalem expressed the hope that James' contribution could be welcome in Germany and be able to renew its philosophic spirit. In the second place he underlined that Pragmatism was not a system, but a method, which finds its centre of gravity in the refusal of a priori, a sacred place for German philosophers. Finally, Jerusalem claimed that the pragmatist view of truth – which is here by no means associated with the "yankee" spirit of dollar pursuit – ought to be integrated into the historical investigations of the growth of knowledge and into his "sociology of knowledge" – which studies truth as a "social condensation" – thus achieving a convergence of Pragmatism and sociology.[21] In the same year of

19 James, *Pragmatism*, op. cit., pp. 112-113.

20 Theodor Elsenhans (Ed.), *Bericht über den III. Internationalen Kongress für Philosophie zu Heidelberg*, Heidelberg: Winter 1909, pp. 711-740.

21 Wilhelm Jerusalem, „Vorwort des Übersetzers", in: William James, *Der Pragmatismus. Ein neuer Name für alte Denkmethoden*, übersetzt von W. Jerusalem, Leipzig: Klinkhardt 1908 pp. V, VIII-IX . We must also remember his paper „Soziologie des Erkennens" published in May 1909 in *Die Zukunft* (and available also in Wilhelm Jerusalem, *Gedanken und Denker. Gesammelte Aufsätze. Neue Folge*, Wien und Leipzig: Braumüller 1925, pp. 140-153).

1908, Jerusalem took up such an alternative view of Pragmatism supporting it in a paper, which represents, so to speak, the missed road of German reception of Pragmatism. He confirmed his struggle against apriorism and presented Jamesian Pragmatism as the irreplaceable ally in order to offer an alternative solution to Kant's theory of knowledge. Furthermore, he strongly insisted - on the basis of their common view of biologic roots of human mind - on James' and Mach's affinities, thus drawing an ideal axis between Vienna and United States, a move which appeared to aim at avoiding the encumbering defensive wall of German *Geist* [22].

A closer account of Jerusalem's contribution to the discussion about the philosophy of pragmatism as well as about its theory of truth and knowledge in the German speaking culture at the beginning ot 20th century goes beyond the limits of the present comment. It must nevertheless be emphazised that Jerusalem represented the essential connection between American Pragmatism and the future Viennese Logical Empiricism, not only due to his mediation between James and German speaking culture, but more specifically due to his relationship with Otto Neurath, a crucial figure in the history of the Vienna Circle. If their personal connections are still to be documented in detail, it is not hard to suppose that Jerusalem – who was active in Vienna not only in the strictly academic environment, but also in wider intellectual circles, in the press and in cultural associations well represented in the Austrian capital during Neurath's early years – was well-known also to the future promoter of the "left Vienna Circle".[23] It was not by accident that Neurath mentions Jerusalem not only in a late work of 1935, *Le développement du Cercle de Vienne et l'avenir de l'empirisme logique*, where he placed him in the main stream of anti-Kantianism typical of both Austrian philosophy and the Vienna Circle, but particularly in a brief text that followed shortly afterwards. There he depicts Jerusalem as the "pioneer (*Vorkämpfer*) of a pragmatist conception", underlying his membership of the characteristic stream of Habsburg thought and especially of Vienna University tradition.[24] Thanks to Jerusalem's mediation, therefore, a connection seems to have taken place between Pragmatism and Logical Empiricism. While well-known in its general outlines, it would be better described in Neurath's case by the light of a certain ideal filiation James-Jerusalem-Neurath, as

22 See Wilhelm Jerusalem, „Der Pragmatismus. Eine neue philosophische Methode", in: *Deutsche Literaturzeitung*, 29, 25. Januar 1908, coll. 197-206 (republished in: *Gedanken und Denker*, op. cit., pp. 130-139). On Jerusalem and Pragmatism see Ludwig Nagl, „Wilhelm Jerusalems Rezeption des Pragmatismus", in: Michael Benedikt/Reinhold Knoll/Cornelius Zehetner (Eds.), *Verdrängter Humanismus – verzögerte Aufklärung*, vol. V, *Im Schatten der Totalitarismen*, Wien: Fakultas Verlags-und Buchhandels AG 2005, pp. 344-353.

23 See the documentation available in Thomas Uebel, *Vernunftkritik und Wissenschaft: Otto Neurath und der erste Wiener Kreis*, Wien–New York: Springer 2000, esp. pp. 164-167, 292-295.

24 Otto Neurath, *Der Logische Empirismus und der Wiener Kreis*, in: Otto Neurath, *Gesammelte philosophische und methodologische Schriften*, ed. by R. Haller and H. Rutte, Wien: Hölder-Pichler-Tempsky 1981, vol. II, p. 742.

long as the convergence of his anti-fondationalist epistemology and the outcomes of American Pragmatism from Peirce to Dewey is recognised.[25] It short, it would not be implausible to claim that many issues characterising Neurath's philosophy (mainly in the 1930s) are at least in agreement with both James' Pragmatism and its "enlargements" proposed by Jerusalem *sub specie* the sociology of knowledge. There was a place also for James and for the one who has brought him to light in the German-speaking philosophical culture at the beginning of twentieth century on Neurath's famous boat, to use his metaphor for inquiry and knowledge as always travelling through the sea of history unable to assume a *tabula rasa* or build on a certain foundation once and for all.[26]

All this has obviously to do with an "image" of James quite different from the image that was widely dominant in early 20th century. He was in no way the philosopher supporting the *yankee* way of thinking deplored by his most prominent German colleagues at the time of the International Congress of Heidelberg. James was rather a philosopher of late 19th century who was perfectly aware of his commitment to recent philosophy of science. In his essay *Humanism and Truth* (1904) James pointed out how deeply the pragmatistic way of thinking was connected with the increasing transormations in exact and natural science during the last decades.

As I understand the pragmatist way of seeing things, it owes its being to the break-down which the last fifty years have brought about in the older notions of scientific truth. "God geometrizes", is used to be said; and it was believed that Euclid's elements literally reproduced his geometrizing. There is an eternal and unchangeable 'reason'; and its voice was supposed to reverbeate in *Barbara* and *Celarent*. So also of the "laws of nature", physical and chemical, so of natural history classification – all were supposed to be exact and exclusive duplicates of pre-human archetypes buried in the structure of things, to which the spark of divinity hidden in our intellect enables us to penetrate. The anatomy of the world is logical, and its logic is that of a university professor, it was thought. Up to about 1850 almost everyone believed that sciences expressed truths that were exact copies of a definite code of non-human realities. But the enormously rapid multiplication of theories in these latter days has well-night upset the notion of any one of them being a more literally objective kind of things than another. There are so many geometries, so many logics, so many

25 On Neurath and Pragmatism see Thomas Mormann, „Neuraths anticartesische Konzeption von Sprache und Wissenschaft", in: Elisabeth Nemeth/Richard Heinrich (Eds.), *Otto Neurath: Rationalität, Planung, Vielfalt*, Wien–Berlin: Oldenbourg Verlag-Akademie Verlag, 1999, pp. 32-61 (Mormann however ignores James' influence on Neurath). For a brief mention of the connection Jerusalem-James see Nancy Cartwright, Jordi Cat, Lola Fleck, Thomas Uebel, *Otto Neurath: Philosophy between Science and Politics*, Cambridge: Cambridge University Press 1996, p. 94 n. 10.

26 Otto Neurath, "Protokollsätze", in: *Erkenntnis*, III, 1932, p. 206. Regarding Neurath's "anti-fondationalistic Pragmatism" see Thomas Uebel, *Vernunftkritik und Wissenschaft*, op. cit., pp. 88, 101 as well as Thomas Uebel, "Otto Neurath, the Vienna Circle and the Austrian Tradition", in: A. O'Hear (Ed.), *German Philosophy since Kant*, Cambridge: Cambridge University Press 1999, pp. 257, 267.

physical and chemical hypotheses, so many classifications, each one of them good for so much and yet not good for everything, that the notion that even the truest formula may be a human device and not a literal transcript has dawned upon us. We hear scientific laws now treated as so much 'conceptual shorthand', true so far as they are useful but not farther. Our mind has become tolerant of symbol instead of reproduction, of approximation instead of exactness, of plasticity instead of rigor[27].

We may well ask if this and similar statements can be read as providing another reason for looking at 19th century philosophy of science in the nostalgic, but also fruitful way proposed by Michael Heidelberger. I would like to suggest that James and some of its supporters such as Vailati or Jerusalem provide the occasion for a stimulating case study that offers to us a good opportunity to achieve new insights into the past and, starting from a reconsideration of this neglected interaction, into the future of the history of philosophy of science.

Università degli Studi di Torino
Dipartimento di filosofia
Via S. Ottavio, 20
10124 Torino
Italy
massimo.ferrari@unito.it

27 William James, "Humanism and Truth", in: *The Meaning of Truth. A Sequel to Pragmatism*, Cambridge-Massachusetts: Harvard University Press 1975, p. 206

Part II
Formal Methods

Vincenzo Crupi and Stephan Hartmann

Formal and Empirical Methods in Philosophy of Science

Abstract

This essay addresses the methodology of philosophy of science and illustrates how formal and empirical methods can be fruitfully combined. Special emphasis is given to the application of experimental methods to confirmation theory and to recent work on the conjunction fallacy, a key topic in the rationality debate arising from research in cognitive psychology. Several other issues can be studied in this way. In the concluding section, a brief outline is provided of three further examples.

1. Introduction

Philosophers of science use a plurality of apparently divergent methods. This claim can easily be substantiated by looking into one of the relevant journals: one realizes that some authors use the traditional method of conceptual analysis, other engage in formal modelling, conduct case studies and – more recently – experiments, or consult the history of science in considerable detail. But how do these methods relate to each other? Is one of them the right one?

Pluralistic cautions would suggest that multiple methodological approaches are legitimate. In fact, we would like to stress that a combination of two or more methods may be particularly fruitful in some cases. Carnap, for example, combined formal methods (i.e., based on logic and probability theory) with conceptual analysis to arrive at an explication of the notion of confirmation. And authors in the tradition of Kuhn and Feyerabend use case studies from the history of science to challenge philosophical models of scientific reasoning such as Popper's falsificationism. In this essay we would like to explore how formal methods and experiments can be combined.

Experiments are all the rage in contemporary philosophy (Knobe and Nichols 2008, Stotz 2009). In epistemology, people's intuitions about Gettier cases have famously been tested. In ethics, aspects of the freedom of will debate are studied experimentally. Philosophers of language also test our intuitions about the reference of proper names. This list could easily be continued. Interestingly, the results of these studies are often surprising when compared with the corresponding intuitions of professional philosophers.

While many of these experiments are used to test philosophers' intuitions (or hypotheses), it is worth noting that experiments have other functions besides test-

F. Stadler (ed.), *The Present Situation in the Philosophy of Science*,
The Philosophy of Science in a European Perspective, DOI 10.1007/978-90-481-9115-4_7,

ing, as Hacking (1983) reminded us. Experiments may, for example, inspire new hypotheses, and this holds for experimental research in traditional domains as well as in philosophy. Usually, these hypotheses are not put forward in a theoretical vacuum: they may relate to an existing theoretical framework, and so some tinkering may have to be done to fit the new hypothesis (or a modified version of it) into the theoretical framework (or a modified version of it). In short, experimental data may provide guidance and insight in theory-construction in a number of ways.

This essay is meant to illustrate the claims above. It focuses on experiments and experimental phenomena which are directly related to work done by formal epistemologists. More specifically, we will look at two case studies. Section 2 focuses on confirmation theory and the recent empirical work in this field. Section 3 discusses the conjunction fallacy, which is of considerable importance as the rationality debate lurks in the background. Finally, we will outline a list of open problems suggesting promising lines of research to be pursued further in the future.

2. Case-study I: Confirmation

Hypothesis testing and confirmation have been central issues in the philosophy of science for decades. Early accounts based on logic and essentially qualitative notions have struggled to deal with a number of puzzles, including the "tacking" problem, Hempel's paradoxes, Goodman's new riddle, the variety of evidence, and the Duhem-Quine thesis. Importantly, such issues have been shown to receive a more effective treatment in *quantitative* terms within a Bayesian approach to confirmation and scientific reasoning (see Earman 1992, pp. 63-86, for a now classical discussion in this vein). A quantitative approach also seems to be up to a general real-world challenge: judgments concerning the *amount* (or degree) of support that a piece of information brings to a hypothesis are commonly required in scientific research as well as in other domains (medicine, law). Thus, a central aim of philosophy of science and epistemology is to provide a proper foundation to such judgments.

Bayesianism arguably is a major theoretical perspective in contemporary discussions of reasoning in science as well as in other domains (e.g., Bovens and Hartmann 2003, Howson and Urbach 2006, Oaksford and Chater 2007). Bayesian theorists postulate a probabilistic analysis of many sorts of ordinary and scientific reasoning by endorsing a subjective reading of probability, i.e., by using probabilities to model degrees of subjective belief. Within this framework, contemporary Bayesians commonly identify confirmation with an increase in the probability of a hypothesis h provided by a piece of evidence e as compared to the initial probability of h (i.e., with evidence e not being given). A natural way to measure confirmational strength then amounts to a function mapping relevant probability values of h

and e onto a number which is either positive, null or negative depending on $p(h|e)$ being higher, equal or lower as compared to $p(h)$. Among traditional proposals meeting this basic constraint are the following:[1]

- the difference measure: $D(h,e) = p(h|e) - p(h)$
- the (log) ratio measure: $R(h,e) = \log[p(h|e)/p(h)]$
- the (log) likelihood ratio measure: $L(h,e) = \log[p(e|h)/p(e|\neg h)]$

More recent variants include the following:[2]

$$S(h,e) = p(h|e) - p(h|\neg e)$$

$$Z(h, e) = \begin{cases} \dfrac{p(h \mid e) - p(h)}{1 - p(h)} & \text{if } p(h \mid e) \geq p(h) \\ \dfrac{p(h \mid e) - p(h)}{p(h)} & \text{if } p(h \mid e) < p(h) \end{cases}$$

Quantitative Bayesian accounts of confirmation can usefully merge with (and profit from) various technical and theoretical refinements and extensions of the Bayesian framework, such as the use of Bayesian networks and probability updating upon uncertain evidence (e.g., Crupi, Festa and Mastropasqua, 2008). Importantly, quantitative measures such as those listed above allow *ordinal* judgments concerning confirmational strength, such as: "hypothesis h receives more empirical support by e_1 than by e_2" or "e confirms h_1 to a greater extent than h_2". One open problem here is that – as both Fitelson (1999) and Festa (1999) emphasized – alternative confirmation measures are *not* generally ordinally equivalent (for a proof concerning a whole set of measures including the five above, see Crupi, Tentori and Gonzalez 2007, p. 231). Indeed, their implied rankings crucially diverge in various interesting classes of cases. This is known as the "problem of measure sensitivity" (Fitelson 1999).

In philosophical quarters, two opposite kinds of reactions can be identified concerning the plurality of confirmation measures (Steel 2007). On the one hand, one can bite the bullet and take a largely pluralistic stance: different measures reflect different aspects of the confirmation relation (e.g., Joyce 2004; Huber 2008; also see Crupi, Festa and Buttasi forthcoming). So far, however, it is not quite clear how this form of pluralism relates to actual scientific practice and real-world

1 These classical measures trace back to Carnap (1950/1962, p. 361), Keynes (1921, pp. 150-155) and Alan Turing (as reported by Good, 1950, pp. 62-63), respectively.

2 Measure $S(h,e)$ has been independently introduced by Christensen (1999) and Joyce (1999). As pointed out in Crupi, Festa and Buttasi (forthcoming), $Z(h,e)$ can be seen as a measure of the relative reduction of uncertainty. It has been explicitly advocated by Crupi, Tentori and Gonzalez (2007). Other occurrences include Rescher (1958, p. 87), Shortliffe and Buchanan (1984, pp. 248 ff.), Cooke (1991, p. 57) and Mura (2006, 2008).

judgments of confirmational strength. On the other hand, one may want to argue in favour of the specific properties of one particular measure on independent, often intuitive, grounds. As a matter of fact, though, intuitions diverge among scholars. So much so that various conflicting measures have been defended in this way (see, for instance, Milne 1996, and Fitelson 2001).

Interestingly, the plurality of confirmation measures has been addressed empirically in recent times, fostering a novel line of experimental investigation in the psychology of reasoning. The basic idea has been to see whether and how empirical data of naïve reasoners' judgments sort out alternative proposals from the literature in philosophy of science and formal epistemology. Two recent papers (Tentori et al. 2007; Crupi, Tentori and Gonzalez 2007) report results from the first attempts to test the descriptive adequacy of alternative Bayesian measures of confirmation with an urn setting experiment and naïve participants (university students). The results seem highly interesting. To begin with, this study provides the first neat demonstration that probabilistic confirmation (as contrasted to probability *tout court*) does belong to the repertoire of the human mind. Second, it shows that the theoretical divergence among confirmation measures is of psychological significance, as competing accounts do yield different degrees of predictive accuracy. In particular, measure Z scored as the most accurate predictor of elicited confirmation judgments, with a slight advantage over the theoretically appealing competitors of likelihood ratio based measures. Crupi, Tentori and Gonzalez (2007) also presented normative reasons in favour of Z, suggesting that it might eventually be singled out on descriptive and normative grounds alike. (It should be noticed that a rather similar scenario is emerging with regards to another sophisticated concept from contemporary Bayesian epistemology, which is also strictly connected to confirmation, i.e., *coherence*. We'll come back to the latter issue in Section 4. below)

The interaction with formal philosophy of science also extends to more traditional branches of experimental research in cognitive psychology. A case in point is represented by the "selection task" (Wason 1966, 1968). Since the very beginning, this widely known experimental paradigm was directly inspired by earlier accounts of hypothesis testing in the philosophy of science. And indeed, cognitive psychologists have recently stressed and explored tight connections with Hempel's celebrated raven paradox (McKenzie and Mikkelsen 2000). For long thought to elicit a basic form of "confirmation bias" and irrational behaviour (see, e.g., Manktelow and Over 1993, Stich 1990, Stein 1996), the selection task has then been reanalyzed through a sophisticated Bayesian account of information search, by which participants' responses have been said to be not only vindicated, but also actually explained as arising from cognitive processes reflecting rational data selection (Oaksford and Chater 1994; see Oaksford and Chater 2003, and the references therein for major contributions to the lively debate on this issue). Indeed, models of the value of information (see Nelson 2005) yield important theoretical connections with probabilistic confirmation. So much so that other similar accounts of the same experimental phenomenon explicitly resort to standard Bayesian measures of confirmation (Nickerson 1996, and Fitelson forthcoming). Finally, Bayesian confirmation measures have also occurred in debates on normative and behavioural aspects of "probative value" in legal contexts (see Davis and Follette 2002, 2003, and Kaye and Koehler 2003) as well as in the psychological literature on causal induction (see Perales and Shanks 2003).

3. Case-study II: The Conjunction Fallacy

The final remarks in the previous section document a growing trend to rely on tools from formal epistemology and philosophy of science for a better insight into long standing phenomena and puzzles in the empirical study of human cognition. A further rich source of relevant considerations arises from the recent literature on yet another largely known phenomenon: the conjunction fallacy. In an often quoted illustration, Tversky and Kahneman (1983) had participants faced with the description of a character, Linda (31 years old, single, outspoken and very bright, with a major in philosophy and concerns about discrimination, social justice and pacifism), ranking the conjunction "Linda is a bank teller and is active in the feminist movement" as more probable than "Linda is a bank teller". From then on, a number of studies have reported that, under certain conditions, people may judge a conjunction of hypotheses as more probable than one of its conjuncts, contrary to elementary principles of probability theory.

The conjunction fallacy has become a key topic in debates on the rationality of human reasoning and its limitations. The phenomenon prompted an enormous amount of research work in psychology and beyond, like many others achievements from Tversky and Kahneman's research programme, which readily and steadily attracted interest from philosophers (Levi 1985; Stich 1990; Samuels, Stich and Bishop 2002) and also established the new interdisciplinary field of behavioural economics (Camerer, Loewenstein and Rabin 2003).

For more than two decades, psychologists have discussed and empirically explored the subtleties of the conjunction fallacy effect, ultimately showing that its robustness and recurrence deserve explanation in a satisfactory account of human reasoning and judgment (see Sides et al. 2002; Tentori, Bonini and Osherson 2004; Wedell and Moro 2008).

A formal analysis accounting for results obtained in the Linda problem has been presented by Bovens and Hartmann (2003, pp. 85-88). Briefly put, the proposal is the following. Suppose "Linda is a bank teller" and "Linda is a feminist bank teller" are reports of two distinct sources of information s_1 and s_2 which are not perfectly reliable. Linda's description may well suggest that source s_1 is less reliable than s_2. But then, probability theory is consistent with the statement that the probability of "bank teller" conditional on the relatively low reliability of s_1 is lower than the probability of "feminist bank teller" conditional on the relatively high reliability of s_2. Bovens and Hartmann (2003) submit that this is what participants' responses express. (See Hintikka 2004, for an independent argument along similar lines). More recently, Hartmann and Meijs (forthcoming) provided a more sophisticated variant of this account, and plan to put it to empirical test.

A different approach has been taken by Crupi, Fitelson and Tentori (2008), following and extending some earlier suggestions from both the psychological and philosophical literature (e.g., Tenenbaum and Griffiths 2001; Levi 2004). While recognizing that conjunction fallacy results document a genuine error in probabilistic judgment, these authors have outlined an explanatory framework based on the notion of confirmation, meant in terms of Bayesian confirmation theory (see Section 2. above). By a close analysis of previous empirical results (Osherson et al. 1990, and Lagnado and Shanks 2002), they argued that the participants' fallacious probability judgments might reflect the assessment of confirmation relations among the evidence provided and the hypotheses at issue in the experimental scenarios. Moreover, extending an earlier result by Sides et al. (2002), they showed that, in a whole class of cases including the Linda example along with others, Bayesian quantitative models of inductive confirmation imply that the evidence provided does support the conjunctive statement more than the single conjunct. Roughly, this class of cases is identified by the evidence provided (e.g., Linda's description) confirming the added conjunct ("feminist") but not the isolated one ("bank teller") (see Tentori and Crupi 2009, for original data in support of this account; Tentori and Crupi forthcoming, and Schupbach forthcoming for a debate; and Atkinson, Peijnenburg, and Kuipers 2009 for some further relevant results).

The latter confirmation-theoretic reading of the Linda problem is one way to flesh out the otherwise esoteric statement by Tversky and Kahneman themselves that "*feminist bank teller* is a better hypothesis about Linda than *bank teller*" (1983, p. 311). A different strategy to fill in the blanks of this noteworthy remark has been provided Cevolani, Crupi and Festa (forthcoming), suggesting that assessments of *expected verisimilitude* may also crucially contribute to conjunction fallacy results. Indeed, these authors proved that, under very weak and plausible assumptions, "feminist bank teller", while less likely to be true than "bank teller", may well be more likely to be close to the whole truth about Linda in a well-defined formal sense. As it can be seen, proposed explanations of the conjunction fallacy based on core notions from the philosophy of science have literally flourished in recent times. For a further example, Shogenji (forthcoming) should be mentioned, who employs a probabilistic and quantitative theory of epistemic justification to account for the phenomenon.

4. Open Problems and concluding remarks

In this section, we sketch three open problems from the philosophy of science that might well gain from a combination of formal and empirical methods.

a. Justification and Coherence

According to the coherence theory of justification, a set of propositions (e.g., a scientific theory) is justified if the respective propositions cohere with each other. But what does it mean that propositions cohere with each other? And how can one measure how much they cohere? To address these questions, various measures of coherence have been proposed (Bovens and Hartmann 2003, Douven and Meijs 2007). However, there is no consensus in the literature as to what the right measure is: different authors appeal to different intuitions or stress different formal requirements, and none of the measures on the market satisfies all of them. This, in turn, lead to a deadlock of the debate, which needs to be resolved. Inspired by the successful work on confirmation measures (see Section 2), empirical investigation seems to provide a promising perspective. It will help us to understand better which role coherence plays in people's actual judgments, and which (if any) of the proposed measures is psychologically realistic. Empirical studies may also foster the construction of alternative measures, or the refinement of existing ones. Notably, original experimental procedures to compare different quantitative accounts of coherence have been recently devised by Harris and Hahn (forthcoming).

b. Scientific explanation

The debate about scientific explanation is in a similar situation. Here we find a spectrum of different theories, supported by altogether different philosophical background beliefs. In this situation, empirical studies may stir the debate in a new direction. More specifically, the following philosophically relevant questions seem to be worth addressing experimentally: Which role do simplicity, probability and coherence play in explanations? And how do people assess the strength of an explanation? The resulting findings will help evaluating existing theories of explanation and may inspire new ones. For some preliminary work in this direction, see Lombrozo (2006) and Schupbach and Sprenger (2009).

c. Social epistemology and philosophy of science

Social epistemology studies the social aspects of science from an epistemological point of view. To do so, case studies have been conducted and formal models have been constructed (see, e.g., Lehrer and Wagner 1981, and Hartmann, Martini and Sprenger 2009). These studies should be accompanied by empirical investigations, as some "empirical input" is needed to answer questions such as the following: What is the best way to proceed when different scientists disagree?

Philosophers and decision theorists developed a host of models that reflect certain ideals of rationality. These models are typically a priori, i.e., they do not include any empirical information about the deliberation process. Conducting experiments will help us to better understand how deliberation works and, eventually, how deliberation should work if the goal of the committee in question is to make the right decision.

Most of the examples above illustrate ongoing trends of research, fostering (and requiring) much further work to provide fully established results. For our present purposes, however, two connected remarks can be firmly put forward. First, the theoretical toolbox of researchers empirically investigating human cognition and behaviour is being expanded from basic probability theory to more advanced formal notions with distinct philosophical origins. Second, a number of recent and current empirical investigations have a potential to provide a fresh look on traditional concerns addressed by formal epistemology and philosophy of science.

Acknowledgements
Work supported by a grant from the Spanish Department of Science and Innovation (FFI2008-01169/FISO).

References

Atkinson, D., Peijnenburg, J., and Kuipers, T. (2009), "How to confirm the conjunction of two disconfirmed hypotheses", *Philosophy of Science*, 76, pp. 1-21.

Bovens, L. and Hartmann, S. (2003), *Bayesian epistemology*, Oxford University Press, Oxford (UK).

Camerer, C., Loewenstein, G. and Rabin, M. (2003), *Advances in Behavioral Economics*, Princeton University Press, Princeton (NJ).

Carnap, R. (1950/1962), *Logical Foundations of Probability*, University of Chicago Press, Chicago.

Cevolani, G., Crupi, V., and Festa, R. (forthcoming), "The whole truth about Linda: Probability, verisimilitude, and a paradox of conjunction", in C. Sinigaglia and M. D'Agostino (eds.), *Selected papers from the 2007 Conference of the Italian Association for Logic and Philosophy of Science*, College Publications, London.

Christensen, D. (1999), "Measuring Confirmation", *Journal of Philosophy*, 96, pp. 437– 461.

Cooke, R.M. (1991), *Experts in Uncertainty. Opinion and Subjective Probability in Science*, Oxford University Press, Oxford (UK).

Crupi, V., Festa, R. and Buttasi, C. (forthcoming), "Towards a grammar of confirmation", in M. Dorato, M. Rèdei and M. Suárez (eds.), *Selected papers from the First Conference of the European Association for the Philosophy of Science,* Springer, Berlin.

Crupi, V., Festa, R. and Mastropasqua, T. (2008), "Bayesian confirmation by uncertain evidence: A reply to Huber (2005)", *The British Journal for the Philosophy of Science*, 59, pp. 201-211.

Crupi, V., Fitelson, B. and Tentori, K. (2008), "Probability, confirmation and the conjunction fallacy", *Thinking and Reasoning*, 14, pp. 182-199.

Crupi, V., Tentori, K. and Gonzalez, M. (2007), "On Bayesian measures of evidential support: Theoretical and empirical issues", *Philosophy of Science*, 74, pp. 229-252.

Davis, D. and Follette, W.C. (2002), "Rethinking probative value of evidence: Base rates, intuitive profiling and the postdiction of behavior", *Law and Human Behavior*, 26, pp. 133–158.

Davis, D. and Follette, W.C. (2003), "Towards an empirical approach to evidentiary ruling", *Law and Human Behavior*, 27, pp. 661–684.

Douven, I. and Meijs, W. (2007), "Measuring coherence", *Synthese*, 156, pp. 405-425.

Earman, J. (1992), *Bayes or Bust?*, MIT Press, Cambridge (MA). Joyce, J. (1999), *The Foundations of Causal Decision Theory*, Cambridge University Press, Cambridge (UK).

Festa, R. (1999). "Bayesian Confirmation", in M. Galavotti & A. Pagnini (eds.), *Experience, Reality, and Scientific Explanation*, Dordrecht, Kluwer, pp. 55–87.

Fitelson, B. (1999), "The plurality of Bayesian measures of confirmation and the problem of measure sensitivity", *Philosophy of Science*, 66, pp. S362–S378.

Fitelson, B. (2001), "A Bayesian account of independent evidence with applications", *Philosophy of Science*, 68, pp. S123-40.

Fitelson, B. (forthcoming), "Bayesian confirmation theory and the Wason selection task", *Synthese*.

Good, I.J. (1950), *Probability and the Weighing of Evidence*, Griffin, London.

Hacking, I. (1983), *Representing and Intervening*, Cambridge University Press, Cambridge (MA).

Harris, A.J.L. and Hahn, U. (forthcoming), "Bayesian rationality in evaluating multiple testimonies: Incorporating the role of coherence", *Journal of Experimental Psychology: Learning, Memory, and Cognition.*

Hartmann, S. and Meijs, W. (forthcoming) "Walter the banker: The conjunction fallacy reconsidered", *Synthese*.

Hartmann, S., Martini, C. and Sprenger, J. (2009), "Consensual decision-making among epistemic peers", *Episteme*, 6, pp. 110-129.

Hintikka, J. (2004), "A fallacious fallacy?", *Synthese*, 140, pp. 25-35.

Howson, C. and Urbach, P. (2006), *Scientific Reasoning: The Bayesian Approach*, Open Court, La Salle.

Huber, F. (2008), "Milne's argument for the log-ratio measure", *Philosophy of Science*, 75, pp. 413-420.

Joyce, J. (1999), *The Foundations of Causal Decision Theory*, Cambridge University Press, Cambridge (UK).

Joyce, J. (2004), "Bayes's theorem", in E. N. Zalta (ed.), *The Stanford Encyclopedia of Philosophy* (Summer 2004 Edition), URL = http://plato.stanford.edu/archives/sum2004/
entries/bayes-theorem/.

Kaye, D.H. and Koehler, J.J. (2003), "The misquantification of probative value", *Law and Human Behavior*, 27, pp. 645–659.

Keynes, J. (1921), *A Treatise on Probability*, Macmillan, London.

Knobe, J. and Nichols, S. (2008), *Experimental Philosophy*, Oxford University Press, New York.

Lagnado, D.A. and Shanks, D.R. (2002), "Probability judgment in hierarchical learning: A conflict between predictiveness and coherence", *Cognition*, 83, pp. 81-112.

Lehrer, K. and Wagner, C. (1981), *Rational consensus in science and society: A philosophical and mathematical study*, Reidel, Dordrecht.

Levi, I. (1985), "Illusions about uncertainty", *British Journal for the Philosophy of Science*, 36, pp. 331-340.

Levi, I. (2004), "Jaakko Hintikka", *Synthese*, 140, pp. 37-41.

Lombrozo, T. (2006), "The structure and function of explanations", *Trends in Cognitive Sciences*, 10, pp. 464-470.

Manktelow, K.I., and Over, D.E. (1993), *Rationality: Psychological and philosophical perspectives*, Routledge, London.

McKenzie, C.R.M. and Mikkelsen, L.A. (2000), "The psychological side of Hempel's paradox of confirmation", *Psychonomic Bulletin & Review*, 7, pp. 360-366.

Milne, P. (1996), "Log[$p(h/eb)/p(h/b)$] is the one true measure of confirmation", *Philosophy of Science*, 63, pp. 21–6.

Mura, A. (2006), "Deductive probability, physical probability and partial entailment", in M. Alai and G. Tarozzi (eds.), *Karl Popper Philosopher of Science*, Rubbettino, Soveria Mannelli, pp. 181-202.

Mura, A. (2008), "Can logical probability be viewed as a measure of degrees of partial entailment?", *Logic & Philosophy of Science*, 6, pp. 25-33.

Nelson, J.D. (2005), "Finding useful questions: On Bayesian diagnosticity, probability, impact and information gain", *Psychological Review*, 112, pp. 979-999.

Nickerson, R.S. (1996), "Hempel's paradox and Wason's selection task: Logical and psychological puzzles of confirmation", *Thinking and Reasoning*, 2, pp. 1-31.

Oaksford, M. and Chater, N. (1994), "A rational analysis of the selection task as optimal data selection", *Psychological Review*, 101, pp. 608-631.

Oaksford, M. and Chater, N. (2003), "Optimal data selection: Revision, review, and reevaluation", *Psychonomic Bulletin & Review*, 10, pp. 289-318.

Oaksford, M. and Chater, N. (2007), *Bayesian Rationality: The Probabilistic Approach to Human Reasoning*, Oxford University Press, Oxford (UK).

Osherson, D.N., Smith, E.E., Wilkie, O., Lopez, A. and Shafir, E. (1990), "Category-based induction", *Psychological Review*, 97, pp. 185-200.

Perales, J.C. and Shanks, D.R. (2003), "Normative and descriptive accounts of the influence of power and contingency on causal judgement", *Quarterly Journal of Experimental Psychology*, 56A, pp. 977-1007.

Rescher, N. (1956), "A theory of evidence", *Philosophy of Science*, 25, pp. 83-94.

Samuels, R., Stich, S. and Bishop, M. (2002), "Ending the rationality wars: How to make disputed about human rationality disappear", in E.Renee (ed.), *Common sense, reasoning and rationality*, Oxfor University Press, New York, pp. 236-268.

Schupbach J. (forthcoming), "Is the conjunction fallacy tied to probabilistic confirmation?", *Synthese*.

Schupbach, J. and Sprenger, J. (2009), "The logic of explanatory power", Working Paper, Tilburg Center for Logic and Philosophy of Science.

Shogenji, T. (forthcoming), "The degree of epistemic justification and the conjunction fallacy, *Synthese*.

Shortliffe, E.H. and Buchanan, B.G. (1984), "A model of inexact reasoning in medicine", in B.G. Buchanan and E.H. Shortliffe (eds.), *Rule-Based Expert Systems*, Wesley, Reading (MA), pp. 233-262.

Sides, A., Osherson, D., Bonini, N., and Viale, R. (2002), "On the reality of the conjunction fallacy", *Memory & Cognition*, 30, pp. 191-198.

Steel, D. (2007), "Bayesian confirmation theory and the likelihood principle", *Synthese*, 156, pp. 55-77.

Stein, E. (1996), *Without good reason*, Oxford University Press, Oxford (UK).

Stich, S. (1990), *The fragmentation of reason*, MIT Press, Cambridge (MA).

Stotz, K. (ed.) (2009), *Experimental Philosophy and Philosophy of Science*, in *Studies in History and Philosophy of Science A*, 40 (Special Issue).

Tenenbaum, J.B. and Griffiths, T.L. (2001), "The rational basis of representativeness", in J.D. Moore and K. Stenning (eds.), *Proceedings of 23rd Annual Conference of the Cognitive Science Society*, pp. 1036-1041.

Tentori, K. and Crupi, V. (2009), "Explaining the conjunction fallacy: Probability *vs.* confirmation", Annual Conference of the Cognitive Science Society 2009.

Tentori, K. and Crupi, V. (forthcoming), "How the conjunction fallacy is tied to probabilistic confirmation: Some remarks on Schupbach (2009)", *Synthese*.

Tentori, K., Bonini, N. and Osherson, D. (2004), "The conjunction fallacy: A misunderstanding about conjunction?", *Cognitive Science*, 28, 467-477.

Tentori, K., Crupi, V., Bonini, N. and Osherson, D. (2007), "Comparison of confirmation measures", *Cognition*, 103, pp. 107-119.

Tversky, A. and Kahneman, D. (1983), "Extensional *vs.* intuitive reasoning: The conjunction fallacy in probability judgment", in T. Gilovich, D. Griffin, and D. Kahneman (eds.), *Heuristics and Biases: The Psychology of Intuitive Judgment*, Cambridge University Press, New York, pp. 19-48.

Wason, P. (1966), "Reasoning", in B. Foss (ed.), *New Horizons in Psychology*, Penguin, Harmonsworth (UK), pp. 135-151.

Wason, P. (1968), "Reasoning about a rule", *Quarterly Journal of Experimental Psychology*, 20, pp. 273-281.

Wedell, D.H. and Moro, R. (2008), "Testing boundary conditions for the conjunction fallacy: Effects of the response mode, conceptual focus, and problem type", *Cognition*, 107, pp. 105-136.

Vincenzo Crupi
Department of Philosophy
University of Turin, via Sant'Ottavio 20
10124 Turin
Italy
vincenzo.crupi@unito.it

Stephan Hartmann
Tilburg Center for Logic and Philosophy of Science
Tilburg University
5000 LE Tilburg
The Netherlands
S.Hartmann@uvt.nl

VINCENT F. HENDRICKS

THE BANE OF TWO TRUTHS

ABSTRACT

A common view among methodologists is that truth and convergence are related in such a way that scientific theories in their historical order of appearance contribute to the convergence to an ultimate ideal theory. It is not a fact that science develops accordingly but rather a hypothetical thought experiment to explain why science develops at all. Here, a simple formal model is presented for scrutinizing the relations between two truths and convergence.

TYPES OF CONVERGENCE

Typically convergence arguments are viewed as supportive frames for some version of scientific realism. Scientific realism was orginally associated with a Platonic idea. The things that are really here in the world are the unworldly forms; reality is beyond the sensory experience and that is what episteme is to grasp. Convergence, if anything, is Eureka-convergence to the truth about the realm of forms. Contemporary scientific realism relaxes the Platonic metaphysics. Realism insists on a theory independent reality and for the reality to be independent of theory is not the same as transcending all possible experience. "Truth" is the epistemological correlate to the ontological "being". Since reality is independent of the scientific theories about it no guanrantees can be provided to the effect that one will ever find the truth let alone know the truth. But there are indications that science is on the truth-track as scientific inquiry is not to be identified with any form of arbitrary inquiry. It is a self-correcting, error-eliminating and technologically sophisticated endeavour which, as time goes by, obtains better and better epistemological accuracy. The better the accuracy, the closer to the truth and the existence of the entities claimed by different sciences. For instance what has been labelled as the *convergence argument of experimental results points* to the independent oucomes of experiments all supporting the same theoretical state of affairs. If the entities and posits postulated in the hypotheses don't refer to anything at all in the real world, but rather are made-up constructions then one should be tempted to think that the observable outcomes should rather diverge than converge. But the fact that they converge exactly witness verisimilitude or truth rather than a cosmic coincidence and insofar, according to Putnam [Putnam78b], [Putnam 78a], renders realism the only tenable metaphysical position not making the general success of science a miracle.

F. Stadler (ed.), *The Present Situation in the Philosophy of Science*,
The Philosophy of Science in a European Perspective, DOI 10.1007/978-90-481-9115-4_8,

The realist is willing to wait, even willing to wait all the way to limit. This may very well turn out to a substantial wait and is still consistent with the possibility of dissapointment. As time does not guarantee that the theory independent reality conforms the the scientific theories about it in an isomorphic way. The general anti-realistic attitude on the other hand is that such a guarantee is to be provided because anti-realism declares bogus the realistic idea of a world whose intrinsic structural nature is strictly independent of the scientific theories about it. Peirce's version of pragmatism is an instance of anti-realism because the ideal theory converged to licences the definition of truth which again is exhausted in limiting consensus. Limiting consensus is detectable. Another pragmatist W. James holds a similar view

> The absolutists in this matter say that we not only can attain to knowing the truth, but we can also know when we have attained to knowing it; while the empiricists think that although we may attain it, we cannot infallibly know when. [James 60], pp. 95.

Both scientific realism and anti-realism may subscribe to a notion of convergence; it's just a matter of what one claims to converge to. Consider positivism. Even given its anti-realistic garments it may still be affiliated with a view of convergence since the cumulative nature of science should eventually result in the true picture of the world. But the world of what? Positivistic rage over metaphysics and the verificationistic criterion of meaning would conclude that convergence to the truth only includes convergence to the *world of experience* and its possible course since that is the only world existing to science. There is Kantian support for this view because of the world in itself hypotheses are neither true nor false hence convergence to true (or false) hypotheses only applies to the world of experience. One of the most prominent contemporary neo-positivists and anti-realists Bas van Fraassen has launched a very influential argument as to why attention should be restricted to the world of experience and observables. By extending positivistic semantics he accuses realism of committing itself to something that it cannot produce namely "literally true" descriptions and theories of the world. Most theories are metaphorical accounts, but metaphorically true accounts do not sum up to literary truth and what is really there. Instead science should only be asked to deliver what it can deliver: empirically adequate theories [van Frassen 80]. The empirical adequacy only requires that there be at least one model under which all the observational sentences are true; this suffices to "save the phenomena". Now, what van Fraassen essentially does is to replace the cognitive goal of truth with the more lenient one of empirical adequacy and this substitution does not obscure sound scientific inquiry. It does restrict convergence to the world of experience however.

So it seems that depending on whether one is a realist or a anti-realist may choose to converge in two different ways. The two ways are dubbed:

- *Epistemical convergence*
- *Ontological convergence*

Primitively speaking convergence solely means for the method to stabilize to some hypothesis and not perform any subsequent mind-changes concerning the hypothesis never minding what the future brings regarding the subject matter. One may *converge epistemically* if one converges in all possible worlds in the background knowledge extending the empirical values of the world observed so far. It is called epistemic convergence because what constitutes the background knowledge in principle is free to fluctuate in accordance with one assumes to be viable alternatives to the actual world. Suppose on the other hand that the cognitive goal requires for its satisfaction the method to identify the actual world. This is a stronger notion of convergence since it forces the method to make a conjecture that encapsulates the way the world really is from here to eternity. This is essentially what *ontological convergence* amounts to; the prospect of which many epistemologists and philosophers of science have remained rather skeptical:

> One often hears that successive theories grow ever closer to, or approximate more and more closely to, the truth. Apparent generalizations like that refer ... to its ontology, to the match, that is, between the entities with which the theory populates nature and what is 'really there'.
>
> There is, I think, no theory independent way to reconstruct phrases like 'really there': the notion of a match between the ontology of a theory and its 'real' counterpart in nature now seems to me to be illusive in principle. ... if the position is relativism, I cannot see that the relativist loses anything needed to account for the nature and development of the sciences. [Kuhn 70], pp. 206.

Recall that when a scientist is working within a particular accepted constellation of ideas, methods, tools, exemplars, instrumentation and instrumental techniques but also general theoretical and *metaphysical* assumptions the scientist is said, according to Kuhn, to be working within a specific paradigm. In the period of normal science the scientists will articulate the paradigm in their ongoing attempt to account and accommodate the behavior of some relevant aspects of the real world as revealed through the results of experimentation. Parts of the articulation should go to show the adequacy of the paradigm as a representation of the world and insofar prove its relative ontological adequacy. Hence one may speak of ontological convergence to some hypothesis. But the ontological convergence is only *relative* to a given paradigm. A new paradigm may include other incommensurable ontological assumptions so there is no unique notion of ontological convergence *trans-paradigmatically*. Hence, according to Kuhn, in comparing trans-paradigmatically one may only speak of *epistemic* convergence in some particular paradigm. This lack of a unique ontological convergence to a hypothesis across paradigms has often motivated critics of Kuhn to label his position as relativistic.

Relativism may mean many things including the subjectivist's view that whatever some method believes is true for that method and truth is exhausted in what that method chooses to believe. Subjectivism is however more an extreme type of relativism than a general characterization of the position. Relativism is not equivalent to arbitrariness. A sanitary version of relativism advocates the systematic dependency of truth or correctness on some parameter that the method is able to tune or joggle. But:

> The adoption of one parameter as opposed to another is arbitrary, but the truth relative to a given parameter is not. [Kelly 96], pp. 380.

If convergence to the truth was understood subjectively then true convergence would be a fairly trivial and uninteresting matter. The method would just have to believe something and fix this belief as truth. This would go for both epistemical and ontological convergence. But even a sober relativist commited to the view that there is no such thing as ontological convergence across paradigms may still place his money on ontological convergence relative to some asserted paradigm and its metaphysical presuppositions. However standing outside some paradigm looking in, only amounts to possible epistemical convergence to some hypothesis.

Anti-realists cannot comply with the dictum of the extreme realist. When a scientist converges to some hypothesis or theory he does not assert the truth of it, only displays it and claims a variety of virtues for it which for their part may fall very short of truth. The lack of truth does not obscure the aim of science (across possible paradigms) which is to tell a reasonable or empirical adequate story about the world. Relativism does not imply subjectivism and if truth both depends upon things over which the scientist has control and upon things over which he has no control, then the scientist may not know exactly how truth depends upon what he chooses to do. It may still may be hard for the scientist to converge to a true (relative to the prevalent paradigm) hypothesis even if he is of a relativistic mind – just as hard as it may be for the (moderate) realist.

All but the extreme realist and the extreme relativist (*i. e.*, subjectivist) may legitimately operate with a notion of epistemical convergence relative to the background knowledge of empirical possibilities without in any way committing themselves to ontological convergence. Essentially whether one favors possible epistemical or ontological convergence depends upon the adopted definition of truth of a hypothesis in a world.

2 Knowledge and Truth

The above discussion on convergence and truth did not make much of an attempt to clarify what truth essentially is. And for good reason: Truth is, and has always been, philosophically an odd beast. A great number of definitions and theories pertaining to truth have been advanced over the years and we are not about to launch yet a new definition or theory. The cognitive goal or correctness relation of science may be theoretical truth or something like it but in principle could be substituted for a host of others as van Fraassen has done with empirical adequacy above. Thus, let's merge and distill two the most dominant views regarding correctness in the aim of distinguishing betweeen two distinct cognitive goals reflecting insights that philosophers and epistemologists have had regardless of whether they are believers in truth as such or not.

2.1 The Trouble with Truths

Paul Benacerraf provides an instructive sketch of the general problem. In discussing the concept of mathematical truth Benacerraf quoted at some length notes that even though the discussion revolves around mathematical truth, also broader epistemological issues concerning knowledge of mahematics and knowledge in general, are at stake:

> It is my contention that two quite distinct kinds of concerns have separately motivated accounts of the nature of mathematical truths: (1) the concern for having a homogeneous semantical theory in semantics for the propositions of mathematics parallel the semantics for the rest of the language, and (2) the concern that the account that the mathematical truth mesh with a reasonable epistemology. [Benacerraf 96], p. 14.

Benacerraf continues to argue that attempts to clarify the former often enough involve neglecting or even violating the latter and vice versa. The two conflict:

> Since I believe further that both concerns must be met by any adequate account, I find myself deeply satisfied with any package of semantics and epistemology that purports to account for truth and knowledge both within and outside of mathematics. For as I will suggest, accounts of truth that treat mathematical and non-mathematical discourse in relevant similar ways do so at the cost of leaving it unintelligble how we can have any mathematical knowledge whatsoever: whereas those which attribute to mathematical propositions the kinds of truth conditions we can clearly know to obtain, do so at the expense of failing to connect these conditions with any analysis of the sentences which shows how the assigned conditions are conditions of their truth. [Benacerraf 96], p. 14.

Now, truth and semantics should hook up with epistemology in such a way that an explanation is furnished of how knowledge of the truth is possible to obtain. On the semantical side, Benacerraf suggests a Tarskian approach:

> I take it that we have only one such account: Tarski's, and that its essential feature is to define truth in terms of reference (or satisfaction) on the basis of a particular kind of syntactico-semantical analysis of the language, and thus that any putative analysis of mathematical truth must be an analysis of a concept which is a truth concept at least in Tarski's sense. [Benacerraf 96], p. 19.

On the epistemological half it is argued that:

> To put it more strongly, the concept of mathematical truth, as explicated, must fit into an over-all account of knowledge in a way that makes intelligble how we have the mathematical knowledge that we have. An acceptable semantics for mathematics must fit an acceptable epistemology. [Benacerraf 96], p. 19.

On these two counts, Benacerraf and van Fraassen's express congruent views from their respecive positions of mathematics and philosophy of science. van Fraassen's formal formulation of empirical adequacy relies heavily on model theoretic properties and he notes with respect to the language of science in general that:

> What we should try to do here is to characterize (fragments of) scientific language by means of the concepts of formal semantics but in such a way that the model structures derive in an obvious way from the models of scientific theories. [van Fraassen 80], p. 199.

The reason why the model structures should be derived in an obvious way from the models of scientific theories is exactly that the scientific theories should be empirically adequate, hence the models making the theories so adequate should be within reach of the *acceptable epistemology* (in reference to Benacerraf above) called (constructive) empiricism:[1]

> One relation a theory may have to the world is that of being true, of giving a true account of the facts. It may at first seem trivial to assert that science aims to find true theories. But coupled with the preceding view of what theories are like, the triviality disappears. Together they imply that science aims to find a true description of unobservable processes that explain the observable ones, and also of what are possible states of affairs, not just of what is actual. Empiricism has always been a main philosophical guide in the study of nature. But empiricism requires theories only to give a true account of what is observable, counting further postulated structures as a means to that end. ... So from an empiricist point of view, to serve as the aims of science, the postulates need not be true, except in what they say about what is actual and empirically attestable. [van Fraassen 80], p. 199.

Now if we on top of these two requirements add in the Kuhnian variable to the equation which should yield knowable truth, truth must be such that it conforms to the paradigmatic structure and the relativistic dimension:

> The world that the student then enters is not, however, fixed once and for all by the nature of the environment, on the one hand, and of science on the other. Rather it is determined jointly by the environment and the particular normal-scientific tradition that the student has been trained to pursue. [Kuhn 70], p. 111-112.

In conclusion, the trouble with truth is then twofold:

1. *The truth should be knowable.*

2. *The truth should respect the paradigmatic structure of science.*

Below, a formal concept of truth called *epistemic truth* is presented which essentially is a mixture of van Fraassen's empirical adequacy, Kuhnian paradigmatics

1 Observe that it is not claimed that Benacerraf holds the same (constructive) empirical view on mathematics that van Fraassen holds towards science. It is only claimed that they express the same sort of conditions to be met by any theory of truth and knowledge.

and which also restricts the scope of truth and falsity to the world of experience or phenomena as Kant requires:

> Our critical deduction by no means excludes the things of that sort (nouemena), but rather limits the principles of the Aesthetic[2] in such a way that they shall not extend to all things (as everything would then be turned into mere appearance) but that they shall hold good only of objects of possible experience. Hereby, then, beings of the understanding are admitted, but with the inculcation of this rule which admits of no exception: that we neither know, nor can know anything determinate whatever about these pure beings of the understanding, because our pure concepts of the understanding as well as our pure intuitions extend to nothing but the objects of possible experience, consequently, to mere things of sense; and as soon as we leave this sphere, these concepts retain no meaning whatsoever. [Kant 64], p. 57-58.

Benacerraf speaks of what he calls the "standard view" of mathematical truth and knowledge which he finds unacceptable since:

> As I have suggested above, the principal defect of the standard view is that it appears to violate the requirement that our account of mathematical truth be susceptible to integration into our over-all account of knowledge. [Benacerraf 96], p. 22.

The standard view is an extreme realism of a Platonic nature which attributes a standard model consisting of "independent" objects to classical theories expressed in a first order language. For instance, while Peano axiomatized number theory, Frege ontologically reduced the natural numbers to sets that are all extensions or purely logical concepts. Frege's Platonistic conception of numbers as unituitable objects is of course very anti-Kantian. Kant was not of the opinion that mathematics is Platonistic. For Kant, mathematics construct its objects in the "pure intuitions" of space and time. Then the mathemtical objects are the *a priori* forms of transcendentally ideal empirical objects. Now the Kantian combination of epistemic empiricism with ontological idealism explains the physical applicability of mathematics and licences scientific legitimacy to mathematical procedures.

To do justice to a whole tradition we choose to introduce yet another notion of truth which blends Platonic forms with Kantian noumenon.

3 Types of Truth

The formal model of inquiry is rendered from [Hendricks 01], [Hendricks 03], and [Hendricks 07]. An evidence stream ε is an ω-sequence of natural numbers, *i. e.*, $\varepsilon \in \omega^\omega$. Hence, an evidence stream $\varepsilon = (a_0, a_1, a_3..., a_n, ...)$ consists of code numbers of evidence, *i. e.*, at each stage i in inquiry a_i is the code number of all evidence acquired at this stage. Continue to define a possible world. A possible world is a pair consisting of a data stream ε and a time n, (ε, n), such that $\varepsilon \in \omega^\omega$

2 I. e. the principles of sensibility, particularly, space and time.

and $n \in \omega$. The set of all possible worlds $\mathcal{W} = \{(\varepsilon, n) \mid \varepsilon \in \omega^\omega, n \in \omega\}$. Let $(\varepsilon \mid n)$ denote the finite initial segment of a world (ε, n). Furthermore $\omega^{<\omega}$ denotes the set of all finite initial segments of elements in ω. Let $[\varepsilon \mid n]$ denote the set of all infinite evidence streams that extends $(\varepsilon \mid n)$. Refer to the finite initial segment $(\varepsilon \mid n)$ as the *handle* with *fan* $[\varepsilon \mid n]$. The world-fan is defined as $\widetilde{[\varepsilon \mid n]} = [\varepsilon \mid n] \times \omega$. The background knowledge of accessible possible worlds is defined as the set of all worlds that extends $(\varepsilon \mid n)$, *i. e.* background knowledge $[\varepsilon \mid n]_{\mathcal{K}} = \{(\widetilde{[\varepsilon \mid n]} \mid \mathcal{K}\}, \mathcal{K} \subseteq \mathcal{W}$. Note that as time goes by, the background knowledge concentrates ever tighther around the actual world course.

$$[\varepsilon \mid n+k]_{\mathcal{K}} \subset \ldots \subset [\varepsilon \mid n+3]_{\mathcal{K}} \subset [\varepsilon \mid n+2]_{\mathcal{K}} \subset [\varepsilon \mid n+1]_{\mathcal{K}} \subset [\varepsilon \mid n]_{\mathcal{K}}.$$

This is by all means an interesting characteristic because it points to how background knowledge may be understood from a broader epistemological and scientific point of view. The handle is the raw evidence that the world presents to science which scientists then identify as X-ray radiation, Zeeman effects, electromagnetic phenomena, etc. These phenomena simply exist but the interpretation of them may fluctuate as Kuhn pointed out. The phenomena exist independently of whether they are actually identified or not – they don't step into existence just because they are discovered, even though they step into epistemic existence for the scientist. In consequence, to denounce the handle is to denounce the world. Kuhnian paradigmatics are not to be identified with irrationalism such that some new paradigm simply denounces the existence of certain phenomena but just points to the fact that these phenomena may be interpreted in different ways. The fan represents the ways in which the phenomena may be interpreted according to the paradigm. Now any later paradigm must take the indisputable existence of these phenomena into account. Science evolves because it as times goes discovers explanations to a great variety of indisputable phenomena. From this point of view it makes sense to say that one background knowledge is included in another later background knowledge exactly because the background knowledge respects the phenomena that the world has shown science up until "now", yet the interpretation of the phenomena is allowed to diverge.

The method that the scientist applies conjectures hypotheses in response to the evidence seen so far. In accordance with standard practice identify hypotheses with sets of possible worlds, *i. e.* the set of all empirical hypotheses $\mathcal{H} = P(\omega^\omega \times \omega)$ such that an empirical hypothesis h is a member of $\mathcal{H}$. Finally a scientific discovery method is a function from finite evidence sequences to hypotheses: $\delta : \omega^{<\omega} \longrightarrow \mathcal{H}$.

3.1 Epistemic Truth

Recall that the background knowledge consists of the empirical possible values the world may take from a given time onward. When truth has to be knowable based on Tarski-like semantics. Hence one way to understand truth is that truth is a relation between obtaining between some held hypothesis and the set of possible empirical world courses relative to the background knowledge. A concept of truth for *phenomena* called *epistemic truth.*

Epistemic truth

$$\textit{Hypothesis } h \textit{ is } \text{true}^E \textit{ in world } (\varepsilon, n)$$
$$(i.e., (\varepsilon, n) \text{ validates } h) \Leftrightarrow [\varepsilon \mid n]_{\mathcal{K}} \cap h \neq \emptyset.$$

The epistemic notion maintains that h is true in (ε, n) just in case there is a non-empty intersection between the background knowledge $[\varepsilon \mid n]_{\mathcal{K}}$ and the hypothesis h. In other words (ε, n) agrees with h up until and including n and from there on, the fan of (ε, n), *i. e.*, the background knowledge $[\varepsilon \mid n]_{\mathcal{K}}$ must only have a non-empty intersection with h, though no single world is be picked out. Some may object that this notion is consistency with the evidence rather than a concept of truth. Realize though that h is considered to be epistemically true in a world (ε, n) if and only if:

- *the hypothesis (or proposition, i. e. set of possible worlds) corresponding to* h *is verified by evidence up to* n*,*
- *possibly the hypothesis will be verified by future evidence in the sense that there exists possible worlds in the proposition corresponding to* h *which are consistent with existing evidence.*

Epistemic truth is hence more than consistency with the evidence but still a rather weak conception of correctness as it only guarantees possible truth in the future. It is evident in the epistemic notion of truth that

$$\text{if } (\varepsilon, n) \; validates \; h \text{ then } (\varepsilon, n-1) \; validates \; h.$$

But not the other way around, *i. e.* (ε, n) $validates$ h does not imply $(\varepsilon, n + 1)$ $validates$ h. What does hold however is the following which goes to show how a notion of possibility is inherent in the notion of epistemic truth:

$$\text{if } (\varepsilon, n) \; validates \; h \text{ then } \exists(\tau, n+1) \in [\varepsilon \mid n]_{\mathcal{K}} : (\tau, n+1) \; validates \; h,$$

and even

$$\text{if } (\varepsilon, n) \; validates \; h \text{ then } \exists \tau \in [\varepsilon \mid n] \forall k \in \omega : (\tau, k) \; validates \; h. \qquad (\blacktriangle)$$

If anything is, this seems to be what van Fraassen means by empirical adequacy:

To believe a theory is to believe that one of its models correctly represents the world. Think of the models as representing the possible worlds allowed by the theory; one of these possible worlds is meant to be the real one. To believe the theory is to believe that exactly one of its models correctly represents the world (not just to some extent, but in all respects). Therefore, if we believe of a family of theories that are all empirically adequate, but each goes beyond the phenomena, then we are still free to believe that each is false, and hence their common part is false. For that common part is phrasable as: one of the models of one of the theories correctly represents the world. [van Fraassen 80], p. 47.

It is also immediate that epistemic truth is more than consistency with the evidence. Consistency with the evidence simply requires that

$$(\varepsilon \mid n) \cap h \neq \emptyset$$

while epistemic truth requires two conditions to be met:

1. *h is consistent with the evidence up to n, i. e.,* $(\varepsilon \mid n) \cap h \neq \emptyset$,
2. *it is not possible to provide a counterexample to h given*
   (a) *all the information acquired up until and including n and*
   (b) *the background knowledge $[\varepsilon \mid n]_{\mathcal{K}}$.*

Epistemic truth is also a verificationistic and conservative truth-concept while a hypothesis is considered true if it is not falsified by evidence up till now. A hypothesis is epistemically false if it is actually falsified by existing evidence, *i. e.*

$$[\varepsilon \mid n]_{\mathcal{K}} \cap h = \emptyset \; (i.e.\ [\varepsilon \mid n]_{\mathcal{K}} \subseteq \overline{h}).$$

3.2 Metaphysical Truth

A concept of truth for *nouemena* dictates that truth isn't held relative to the background knowledge of possible empirical values but rather demands identification of the only world (an sich) in which the hypothesis in question holds:
Metaphysical truth

Hypothesis h is true^M *in world* (ε, n) $(i.e., (\varepsilon, n)\ validates\ h) \Leftrightarrow (\varepsilon, n) \in h.$

For the metaphysical truth of a simple empirical hypothesis h it is also clear that

$$(\varepsilon, n)\ validates\ h\ iff\ (\varepsilon, k)\ validates\ h \text{ for all } k.$$

4 Truthful Simultaneity

Finally, it is worthwhile to examine the interrelations between epistemic and metaphysical truth of hypotheses since these interrelations have implications with respect to both the intrinsic characterization of the notion of truth and furthermore carry implications as to the inquiry method's epistemic performance.

First of all, it is impossible to have a hypothesis which is metaphysically true but epistemically false, since

$$(\varepsilon, n) \in h \; and \; [\varepsilon \mid n]_{\mathcal{K}} \cap h = \emptyset. \tag{0}$$

is a contradiction, which of course is as it should be.

However, the following three relations are consistent possibilities of alethic simultaneity:

$$(\varepsilon, n) \in h \; and \; [\varepsilon \mid n]_{\mathcal{K}} \cap h \neq \emptyset. \tag{1}$$

(1) is clearly a possible scenario since a hypothesis may well be metaphysically and epistemically true simultaneously.

Now consider the case where a hypothesis is simultaneously metaphysically and epistemically false, *i. e.*:

$$(\varepsilon, n) \notin h \; and \; [\varepsilon \mid n]_{\mathcal{K}} \cap h = \emptyset. \tag{2}$$

In this case, there exists falsifying evidence at a time m earlier than n such that:

$$\exists m < n : [\varepsilon \mid m]_{\mathcal{K}} \cap h \neq \emptyset \; and \; [\varepsilon \mid m+1]_{\mathcal{K}} \cap h = \emptyset. \tag{2*}$$

Finally consider the case in which a hypothesis is metaphysically false but at the same time epistemically true:

$$(\varepsilon, n) \notin h \; and \; [\varepsilon \mid n]_{\mathcal{K}} \cap h \neq \emptyset. \tag{3}$$

In this case there exists falsifying evidence at some point in the *future*.That is,

$$\exists m \geq n : [\varepsilon \mid m]_{\mathcal{K}} \cap h \neq \emptyset \; and \; [\varepsilon \mid m+1]_{\mathcal{K}} \cap h = \emptyset. \tag{3*}$$

However, it is not possible for the method to find falsifying evidence prior to $m+1$. Furthermore, a method δ may know h at time n since it will conjecture h as true in all extensions of $(\varepsilon \mid n)$. This is essentially a *closed world assumption*: Method δ assumes h to be epistemically true unless the accumulated evidence at a given time in the world history shows otherwise. This is also in agreement with usual praxis in statistical methodology. If existing evidence does not refute your hypothesis then accept it.

You may have epistemic knowledge, but you cannot be sure that it extends to metaphysical knowledge. But then again, who can when you are just an epistemic agent equipped only with background knowledge? That's the bane of two truths.

References

[Benacerraf 96] Benacerraf, P. (1996). "Mathematical Truth", *Journal of Philosophical Logic*, vol. 70, **19**: 661–679.

[Hendricks 01] Hendricks, V. F. (2001). *The Convergence of Scientific Knowledge*. Dordrecht: Springer.

[Hendricks 03] Hendricks, V. F. (2003)."Active Agents", *Journal of Logic, Language and Information*, **12**: 469–495.

[Hendricks 07] Hendricks, V. F. (2006). *Mainstream and Formal Epistemology*. New York: Cambridge University Press..

[Hintikka 62] Hintikka, J. (1962). *Knowledge and Belief*. Cornell: Cornell University Press.

[James 60] James, W. (1960). "The Will to Believe", *Essays in Pragmatism*. Hafner Publishing Company.

[Jammer 66] Jammer, M. (1966). *The Conceptual Development of Quantum Mechanics*. McGraw-Hill Inc.

[Kant 64] Kant, I. (1977). *Kritik der reinen Vernunft*. Translated by N. Kemp Smith, *Critique of Pure Reason*. London: Macmillan.,

[Kelly 96] Kelly, K. (1996). *The Logic of Reliable Inquiry*. New York: Oxford University Press.

[Kuhn 70] Kuhn, T. (1970). *The Structure of Scientific Revolutions*. Chicago: University of Chicago Press.

[Pauli 46] Pauli, W. (1946). "Exclusion Principle and Quantum Mechanics", *Editions du Griffon*. Neuchatel.Macmillian.

[Putnam 78a] Putnam, H. (1978). "Reference and Understanding", in [Putnam78c], p. 97-119.

[Putnam 78b] Putnam, H. (1978). "Realism and Reason". In [Putnam78c], p. 123-140.

[Putnam 78c] Putnam, H. (1978). *Meaning and Moral Sciences*. London: Routledge and Kegan Paul.

[van Fraassen 80] van Fraassen, B. (1980). *The Scientific Image*. Oxford: Clarendon Press.

Department of Philosophy
University of Copenhagen
Njalsgade 80
DK-2300 Copenhagen S
Denmark
vincent@hum.ku.dk

THOMAS MÜLLER

FORMAL METHODS IN THE PHILOSOPHY OF NATURAL SCIENCE

What is the proper place of formal methods in philosophy of natural science, or in philosophy more broadly speaking? The idea that philosophy should proceed formally ("more geometrico", as in the title of Spinoza's *Ethica*) has been around for some time, but both the attitude towards formal methods and the understanding of formal methods itself has changed. Mathematical logic has succeeded geometrical demonstration as the paradigm of formal precision, and in technical areas such as foundations of mathematics and logic, Frege's and Russell's logicist programmes indicate early peaks of the application of these methods. The idea of employing such formal-logical methods in philosophy more generally was championed by the logical empiricism of the 1920s and 1930s. Wrestling with the methodological foundations of their discipline in an attempt to exclude what they perceived to be nonsense, some at the time even sought recourse in a purely formal-logical foundation for philosophy. Frege's student Carnap in his programmatic paper on "the old and the new logic" (Carnap, 1930, 26) put the matter thus: "To pursue philosophy means nothing but: clarifying the concepts and sentences of science by logical analysis."[1]

As the philosophical sub-discipline of philosophy of science is to a large extent historically continuous with logical empiricism, it is no wonder that the newly emerging field of philosophy of science – which mostly meant: philosophy of natural science – in the 1950s centered around an array of formal-logical methods.

This attitude towards formal methods has not remained unchallenged: the 1960s saw a historicist turn in philosophy of science that has led to a fairly critical attitude towards formal methods. As Kuipers (2007, viii) remarks, "many philosophers do not like to be associated with the logical empiricists". In this paper I will argue that the availability of new formal methods and an increased sensitivity for the uses and limitations of formal approaches makes possible a fresh case for the usefulness of formal methods in philosophy of science and particularly in the philosophy of natural science. Formal methods also play an integral part in the methodology of conceptual modeling that lies behind a number of recent success stories in that and in related areas of philosophy. Individual contributions of the ESF Network's Team A, which centers on formal methods, all testify to the usefulness of that methodological outlook.

Before arguing for these claims starting in section 2, I will set the stage by expanding a bit on the historical background of the question of the place of formal methods in philosophy of science.

1 German original: "Philosophie betreiben bedeutet nichts Anderes als: die Begriffe und Sätze der Wissenschaft durch logische Analyse klären."

F. Stadler (ed.), *The Present Situation in the Philosophy of Science*,
The Philosophy of Science in a European Perspective, DOI 10.1007/978-90-481-9115-4_9,

1 Philosophy of natural science in the 20th century

Depending on one's outlook, philosophy of natural science can be viewed as an old subject, or as a rather new one. Certainly philosophical reflection accompanied the development of the New Science in the early modern period, and there are good reasons for viewing philosophy of science as a historically unified enterprise with roots in the 13th century, or even in Aristotle. This historical lineage is the subject of the flourishing field of history of philosophy of science. On the other hand, the current academic sub-discipline of philosophy of science is a development of the late 19th and the early 20th centuries – Ernst Mach in 1895 was the first person to hold a chair in philosophy of science at Vienna, and the *Verein Ernst Mach*, subsequently the *Vienna Circle*, together with the *Berlin Circle* in the 1920s and 1930s were the birthplace of logical empiricism, which played a key role in forming and establishing the discipline of philosophy of science. As already mentioned, this more recent historical lineage is crucially important when it comes to the role of formal methods.

Logical empiricism was, broadly speaking, an attempt at turning philosophy into a respectable scientific discipline. In the eyes of the propounders of this doctrine this meant to abolish metaphysics, where no clear scientific standards were discernible, and instead to embrace strict standards of reasoning, the strictest of which, apparently sufficient even for strengthening the foundations of mathematics, were made possible by the development of modern formal logic. As the quote from Carnap given above indicates, logical analysis of science would be all that was left of serious philosophy.

On the other hand, the idea of a formal study of science can also be linked to the widespread formal self-understanding of science. The idea that proper science needs to be mathematical has been strong since the 17th century – witness Galileo's image of the book of nature being written in the language of mathematics, or Kant's later pronouncement that a purported science was a science only insofar as it was mathematical.[2] It appears only natural that such a subject should be approached by tools equally mathematical or formal. The use of formal methods in 20th century philosophy of (natural) science thus appears as a confluence of two mutually supporting ideas: the logical empiricists' idea of logical analysis as *the* tool of philosophy, and the commonsensical idea of studying that which of itself is formal by formal means. Based on the demand for the unity of science characteristic of logical empiricism, the deployment of formal methods was then assumed to spread to other areas as well. The messy details of actual science notwithstanding, unified science was to be rationally reconstructed using the formal methods of logic – and that of course meant: of the logic of the time.[3]

2 Cf. Galilei (1623) and Kant (1786).

3 The history is of course more tangled than this sketch suggests. It should not be forgotten that the "left wing" Vienna Circle besides Carnap also included philosophers like Neurath, who proposed a pragmatic approach to the philosophy of science including psychological and sociological studies, cf. Uebel (2001) on Neurath (1932). This idea

Philosophy of science developed as a subject proper mainly in the U.S., following the emigration of many of the leading logical empiricists due to the rise of Nazism.[4] In the 1950s, the field consolidated around a positivist orthodoxy, leading to compendia such as Nagel's *The Structure of Science* (1961). Formal accounts of explanation, confirmation, theory reduction, laws of nature, and other key concepts had been worked out by then. The cracks were however already beginning to show: the adequacy of those formal accounts appeared doubtful.[5]

Initially, logical empiricism could respond to criticisms about the descriptive adequacy of proposed accounts by pointing to their status as first steps in a research program. When the account of scientific concepts remained questionable vis-à-vis actual practice over decades, however, it appeared that the research program had failed to deliver. Historical and sociological studies of actual science such as Kuhn's (1962) *Structure of Scientific Revolutions* (published in the logical empiricists' own book series) were seen as more important than logical constructions that increasingly seemed to be built of thin air.

This sketch of the historical background may help to explain the generally critical attitude towards formal methods that is, or at least was, prevalent among many philosophers of science.[6] Determining the proper place of formal methods in philosophy of natural science nowadays means to be aware of this historical baggage, and to take up the challenge of showing how the criticism leveled against logical empiricism's deployment of formal methods can be met.

2 The Uses and Success of Formal Methods in Recent Philosophy of Natural Science

Despite the mentioned criticism, formal methods never vanished from philosophy of science. Many of the early formal-logical accounts – e.g., the deductive-nomological account of explanation – have always remained important for the field, not at least in teaching the subject, and not just because of their historical significance, but also because they remain systematically significant due to their clarity and exactness.

There are however also many *new* success stories of the deployment of formal methods in the philosophy of natural science. I will argue that nowadays, formal methods have their proper place right in the center of philosophy of science, and that we can identify two factors that explain their successful return: the develop-

however had little impact on the development of the subject of philosophy of science in the years after the Second World War.

4 The historical context of logical empiricism is described in detail in the essays of Stadler, Hoffmann and Reisch in Richardson and Uebel (2007).

5 Cf., however, Feigl (1970) for a dissenting view on the relevanceof actual scientific practice.

6 For a more detailed overview, cf., e.g., Richardson (2007).

ment of new formal methods on the one hand, and the adoption of the methodology of conceptual modeling on the other.

2.1 Conceptual modeling

The philosophy of science of the 1950s focused on a mostly static view of the metamethodological embedding of formal methods. Explication of key concepts was considered to be a matter of logical analysis of what was there. More recent applications of formal methods however mostly occur in a dynamic setting. This move is usefully described in Kuipers (2007), who tells a story of refined ways of concept application. In a similar vein, but from a broader perspective, I would like to describe the respective metamethodological change as a move towards *conceptual modeling*.

In science and engineering, *mathematical modeling* has long been seen as one of the most fundamental methodologies, and one of growing importance. Mathematical modeling presupposes quantitative and computational methods. However, a slight generalization of the same methodology that may be called *conceptual modeling* is ubiquitous also in non-quantitative research areas. This methodology and its uses are described in more detail in Löwe and Müller (2009). Briefly, conceptual modeling is an iterative process through which a stable reflexive equilibrium is reached between a concept or a collection as concepts, X, as explanandum and a (somewhat) formal representation of it. Each iteration towards the equilibrium involves three steps:

1. *Formal representation.* Guided by either a pretheoretic understanding of X or the earlier steps in the iteration, one develops a (more or less) formal representation of the explanandum.

2. *Phenomenology.* With a view towards step 3, one collects evidence in the range of the explanandum that is ideally able either to corroborate or to question the current formal representation.

3. *Assessment.* In the light of the results from step 2, one assesses the adequacy of the representation. If this assessment is positive, the modeling cycle is left – no further iteration is necessary since an equilibrium has been reached. Otherwise, the representation has to be changed, and a new iteration is started at step 1.

This method obviously covers mathematical modeling as employed in the sciences and in engineering, where the formal representation typically comes with a numerical mathematical model that allows for quantitative predictions. In the case of philosophy, the scheme usefully generalizes the methods of "conceptual analysis" or of "logical analysis" as invoked by Carnap: it leaves room for a dynamical, iterative approach, and it is not confined to a fixed set of formal means of representation. The examples from philosophy of natural science given below testify to the usefulness of that method.

2.2 Example success stories: new formal methods

Formal methods are nowadays not limited to the traditional field of formal logic – which by itself has expanded vastly, providing for modal, temporal and other logics and giving much formal insight into the important notion of a model, or a structure. The methods also include a significant amount of probability theory and aspects of game theory, graph theory, computer simulations and other techniques of formal modeling. It should also be emphasized that in this development, philosophy of science does not play the merely passive role of employing off-the-shelf techniques developed in other disciplines, but has also led to the development of new techniques.[7]

In the following short descriptions of formal success stories, the contrast is always between the way matters were seen within the original paradigm of logical empiricism focusing on inferential relations among sentences and logical analysis, and new approaches based on an extended array of formal methods and pursued in a modeling framework.

No originality is claimed for the accounts of the employment of formal methods given here. These accounts are rather meant to illustrate my main point, which is that we are witnessing a return of the fruitful employment of formal methods in philosophy of science. Consequently the following sketches will be rather brief. Other examples connected with the work done in the ESF Network's Team A could easily be added, e.g., work on Bayesian methods in confirmation (Fitelson and Hawthorne, 2005; Huber, 2005), or on social aspects of science (Hartmann and Bovens, 2008; Dietrich, 2006; Pigozzi, 2006).

Reduction vs. intertheoretic relations What is the relation between a scientific theory and the theory that historically takes its place – like, e.g., the Newtonian theory of universal gravitation superseding Galileo's law of falling bodies? The new theory should at least account for the same empirical facts as the old one. Thus, within the logical empiricist paradigm of theories as collections of general statements, it seemed that some relation of logical derivability or reduction would be appropriate: the new theory should allow one to derive all empirical statements of the old one, plus some more. It is easy to see that this idea breaks down even in the case of the example of Galileo vs. Newton (ironically used as an illustration by Nagel (1961)): In the earth's non-uniform gravitational field, the Galilean law is only an approximation to what Newton's theory predicts.

The move to present-day probabilistic methods has proved to be promising. Rather than focus on the "reduction" of one theory by another, a wider picture of intertheoretic relations emerges. That picture also includes the data the theories account for and thus remains much closer to actual scientific practice (Batterman, 2008; Hartmann, 2008). Methodologically, the move from theory reduction to a

7 Cf., e.g., Leitgeb (2009), who also echoes the earlier programmatic paper of van Benthem (1982). Cf. also Horsten and Douven (2008) for a state-of-the-art survey.

Bayesian account of intertheoretic relations exemplifies concrete work in conceptual modeling.

Quantum logic: old and new The quantum logic of Birkhoff and von Neumann (1936) was an attempt at reading off a "new logic" from the mathematical structure of quantum mechanics. Initially the idea was to find an interpretation of propositional connectives like conjunction and negation that would be a formal counterpart to operations on the set of subspaces of a Hilbert space that constitutes the state space of a quantum system. A fascinating possibility was that the "true" logic could turn out to be different from classical propositional logic – and for empirical reasons.

Present-day logic paints a different picture, and again, the conceptual modeling paradigm captures this development. Quantum logic never came to replace classical logic (signaling inadequacy in the assessment step) – but the logic community has also become much more open towards the idea that there could be different logics, each suited to a specific domain.[8] Furthermore, there are new tools within logic that can be fruitfully employed in a study of quantum mechanics (there are more options for a fresh start of the modeling cycle). In fact dynamic logics seem to be very well suited for a description of quantum operations studied in quantum information theory (Baltag and Smets, 2008). Thus, advanced formal methods allow one to leave old normative questions (about "the" logic) behind and work towards a better understanding of science as actually practised.

Determinism and indeterminism of theories The question of whether a given scientific theory is deterministic or not, was approached mostly informally before Montague (1962) introduced a model theoretic approach. In this field many advanced methods of mathematical physics have been employed, and the formal technical level of discussion is very high (witness Earman, 2007). In fact here the deployment of formal methods has significantly advanced other discussions, too, in that the importance of precise definitions of, e.g., the notion of state has been recognised. Questions of theory determinism or indeterminism are furthermore relevant not just for philosophy of science, but also for science itself.

2.3 A proper place for modality in the philosophy of natural science?

In the sketches just given I have stressed the involvement of new formal methods that go beyond the traditional toolbox of logical empiricism, and the importance of a broadened understanding of what one is doing in employing formal methods via the method of conceptual modeling. I will now take a closer look at my last example, viz., determinism and indeterminism or, more broadly, the involvement of modality in the philosophy of natural science.

8 Carnap's Principle of Tolerance (Carnap, 1937, 51f.) already points in that direction.

Determinism is a modal notion: it signifies the absence of open possibilities. Modality arguably plays a role in many other concepts of science, too: laws of nature, essences and natural kinds, causation and intervention, and probability. My suggestion is that the time is ripe for taking modality seriously in philosophy of natural science.

Even though modality is studied formally nowadays, this was not so in the early days of logical empiricism. From that doctrine's point of view, there were two problems about modality in science. Firstly, modality was interpreted as *logical modality*, where logical possibility just means the absence of formal contradiction – but this is not the notion of modality that is needed to analyse the mentioned scientific concepts. The notion of logical possibility is too broad: many things that are physically impossible are still logically possible (think, e.g., of going faster than the speed of light). Secondly, modality apparently has poor empiricist credentials. This continues to stand in the way of a fruitful employment of modal notions in philosophy of science. After all, mere possibilities – possibilities that are not actualised – are empirically inaccessible because they are unreal, so how could they be important for empirical science?

The first important step towards an employment of modality in philosophy of science is to take a lead from the discussion about different modalities. This discussion developed out of formal research into the semantics of modal logic since the 1950s. Initially one may view this semantic enterprise as a quest for a formal representation of *the* meaning of "possibly" and "necessarily". The semantics that was established, the so-called Kripke semantics that spells out the modalities in terms of relations among possible worlds, showed however that there is much leeway in specifying different modal logics with different semantics. The initial assessment of this fact was rather critical: among all those options, it seemed that one still had to find the right one to specify what "possibly" and "necessarily" *really* meant. This assessment has changed in the meantime, and the many options for a semantics of modality are now seen as a good thing. It has become common to acknowledge a number of different kinds of modality: there isn't just logical modality, but there are various other kinds of modality that may have different formal properties and a different metaphysical status. In terms of the modeling paradigm, this means that a larger range of formal ways of spelling out aspects of modality has become available. It will be best to explain some of these options in terms of possibility; the consequences for the dual modality of necessity follow immediately.[9]

As mentioned, there is logical possibility: the absence of formal contradiction. This notion is rather broad. Famously Ramsey pointed out to Wittgenstein that his *Tractatus* theory, which relied on logical possibility in postulating the independence of elementary propositions, was flawed because it could not, e.g., account for the rather straightforward impossibility of the same patch's being both red and green – no formal contradiction is involved here, since "red" and "green" just fig-

9 Possibility and necessity are dual in the following sense: It is necessary that *p* if and only if it is not possible that non-*p*.

ure as two different predicates, and it is logically possible for one and the same thing to fall under any number of different predicates. The colour overlap in question is however clearly impossible in another sense. It has become common to speak of *metaphysical* (im)possibility here, and to base philosophical arguments on metaphysical rather than logical modality. For philosophy of science, however, a notion of *physical* possibility seems to play an even more important role. Physical possibility is often taken to be what laws of nature express, and insofar as science is a quest for the laws of nature, science is really about physical possibility. Determining the place of modality in philosophy of science thus comes down to modeling physical modality.

2.3.1 Modeling physical modality

Questions about the interrelation of various kinds of modality are notoriously difficult to resolve. There are arguments in favour of modal monism (the claim that there is one single fundamental modality, to which all other modal notions can be reduced), but also in favour of modal pluralism (the claim that there are different irreducible modalities). Thus, the question of whether physical possibility is just a restricted version of logical or metaphysical possibility has been debated: e.g., Fine (2005) argues convincingly that physical and metaphysical modality are independent and indeed believes that they are both fundamental, thus providing an argument in favour of modal pluralism.

My conviction is that physical possibility is not fundamental, and that a fruitful explanation of the use of possibility in philosophy of science needs to refer to a different notion of possibility: *real possibility*, also known as historical possibility because of its link with temporality.[10] The peculiarities of that notion of possibility are best explained via some of its specific formal properties.

2.3.2 The formalities of real possibility

The formalities of real possibility have been worked out since the 1950s. Prior's *Time and Modality* (1957) set the agenda for research into the interrelation between modality and tense, whose formal similarities as sentence-modifying operators had by then just been recognized. Prior (1967) and subsequently Thomason (1970) developed models for so-called "branching time" in which the tempo-modal notion of an open future serves as the basis for a semantics of both the tenses and the modalities of real possibility and real necessity. In a model of branching time, possible courses of events, also called *histories*, are maximal linear subsets of a

10 Fine, in the mentioned work, explicitly excludes real ("historical") modality from his discussion, but gives no reason for this (cf. Fine, 2005, 237n4). This strikes me as odd, since he himself has contributed to the development of the formalities of real possibility; cf. Prior and Fine (1977).

branching tree of open possibilities.[11] A modern description of the branching time framework is given by Belnap et al. (2001, Chap. 6-8).

In terms of formal properties, real possibility is special because of its interaction with the tense operators. We will employ the standard formalisations of "F" for the future operator "it will be the case that" (the past tense "it was the case that" is accordingly symbolized as "P"), and "◊" and "□" for the modal operators "possibly" and "necessarily", respectively. A specific aspect of real possibility is the satisfiability of the formulae

$$\Diamond p \,\&\, \mathrm{F}\neg\Diamond p \tag{F1}$$

and

$$\Diamond p \,\&\, \neg\mathrm{F}\Diamond p, \tag{F2}$$

which express the temporality of real possibility. (F1) says that some p that is now possible, will at some future point in time not be possible any more – a fact that we know all too well, as witnessed by the fact that we sometimes complain about missed opportunities. (F2) is even stronger, saying that p, which is now possible, will cease to be possible immediately in the future – it's now or never, so to speak. Instances of this are also well known.

These formulae are *not* satisfiable if "◊" is read as logical or as metaphysical possibility; those modal notions are abstract, without any link with the passage of time. What is logically possible now will remain so forever, and has in fact always been logically possible – if those temporal determinations make any sense at all.[12] For further formal properties of real possibilities based on branching time, cf. again Belnap et al. (2001).

The mentioned formal framework of branching time has been extended in order to overcome one of its major shortcomings: While real possibility is possibility in a concrete and thus concretely localised situation, branching time does not capture that *spatial* aspect. In the extended formal framework of *branching space-times* (BST; Belnap 1992) this aspect is explicitly recognised, as histories (possible courses of events) in that framework do not have the form of a single temporal chain of events, but of a single space-time. In BST it is therefore possible to express the fact that something that is possible here now, is not possible

11 Technically, a history in a model of branching time is a maximal linear subset of the tree, i.e., a subset in which any two elements are comparable and which is maximal with respect to that property. Such a set corresponds to a complete path through the tree.

12 This question is mirrored in the case of mathematics, where there are different opinions as to whether "It is now the case that $2+2=4$" makes any sense at all. – Do not be misled by the fact that, e.g., a logical possibility may be *instantiated* as a real possibility, which then *is* temporal. E.g., it is logically possible that crows fly, and it may be really possible that a certain concrete crow that is now before you should fly within the next five minutes. This, however, is not the same as the mentioned abstract logical possibility, but also depends on many local and temporal factors, e.g., the state of the crow's feathers and the air pressure.

now somewhere else.[13] Belnap's BST is the most advanced formal framework for studying real possibility available to date, and it has been used in a number of applications to problems of metaphysics, philosophy of language, and philosophy of physics.[14]

2.3.3 A model for physical possiblity based on real possibility

Physical possibility, the modal notion that determines the laws of nature, belongs to the same group of abstract, a-temporal modalities as logical and metaphysical possibility: what is physically possible now, has always been physically possible and will remain so forever.[15] Real possibility, on the other hand, is possibility in a concrete, indexically specifiable situation: it is right there before us. The main question about the interrelation of real vs. physical possibility is how scientific practice, which is based on real, concrete experiments and observations, can help us gain access to abstract physical possibility. This question is similar to the question about the interrelation of theory and observation in the sciences, but phrasing it in terms of possibilities gives it an importantly different twist.

Real possibilities rule in the lab and in scientific work generally: Every concrete run of an experiment reveals one of the outcomes that are *really* possible in the given, concrete situation – including, in almost all cases, the real possibility that the experiment may fail due to some sort of interference. Even though experiments thus primarily reveal something about real possibilities, they can sensibly be seen as probes of physical possibility, too. At least that is what experiments are designed for: Generally speaking, in an experiment one wants to find out not about the really, but about the physically possible outcomes, together with their probabilities, of an experimental set-up with given, experimenter-controlled initial conditions. One will therefore disregard certain runs as not pertinent to the question about physical possibility (e.g., because somebody kicked the apparatus), even though the pertinence of these runs for the issue of real possibility cannot be questioned. One will also smooth out the observed distribution of results in various ways. Details vary by case – here a connection with Bogen and Woodward's (1988) data/phenomena distinction suggests itself: physical possibilities appear as phenomena distilled from real possibility figuring as data, with all the well-known

13 In view of BST's compatibility with relativity theory, the "now" of course has to be taken with a grain of salt. Technically, possibilities are linked to space-time locations in BST, in a manner that is fully compatible with the absence of a notion of absolute simultaneity in special relativity theory.

14 Cf., e.g., Belnap (2005) for causation, Weiner and Belnap (2006) and Müller (2005) for objective single-case probabilities, Müller et al. (2008) for modal correlations, Placek and Müller (2007) for counterfactuals, and Müller and Placek (2001) as well as Placek (2009) for Bell-type correlations.

15 At least this is so if one disregards scenarios in which the laws of nature change over time. I will ignore such scenarios in what follows. The point about abstractness would remain in any case.

idiosyncracies of that step. It is generally acknowledged that there is no formal way of inferring phenomena from data.

Physical possibilities as summed up in laws of nature and physical theories are thus determined via the notion of real possibility that has primacy in scientific practice. In concrete runs of experiments, real possibilities are actualized. Both the concrete initial situation of the respective runs and the concrete outcomes are then described via a number of variables, giving rise to stable, repeatable phenomena. The aim of the experimenter in such a description is to record all salient variables, not everything at all. Physical possibilities (which in a given case may be physical necessities) are then arrived at from real possibilities: so-called laws of nature are established as generalisations covering many experiments, and considerations of saliency again play a crucial role here, as in any case in which phenomena are inferred from data. Statements about laws of nature on this account have an unquestionable modal content: they simply report what is physically possible or necessary, and they are based in real possibilities.

Acknowledgements
I would like to thank the audience and my co-symposiasts at the ESF Conference *The Present Situation in the Philosophy of Science*, Vienna, 18 December 2008, for helpful discussions. This text also draws on material from a related paper given at the *Workshop on Formal Methods in Philosophy*, Kraków, 24 August 2008. Support by the *Deutsche Forschungsgemeinschaft* is gratefully acknowledged.

References

Baltag, A. and Smets, S. (2008), A dynamic-logical perspective on quantum behaviour. *Studia Logica*, 89 (2): 187-211.

Batterman, R. (2008), Intertheory relations in physics, in: Zalta, E. N. ed., *The Stanford Encyclopedia of Philosophy*. http://plato.stanford.edu/archives/fall 2008/entries/physics-interrelat/.

Belnap, N. (1992), Branching space-time. *Synthese*, 92: 385-434.

Belnap, N. (2005), A theory of causation: *Causae causantes* (originating causes) as inus conditions in branching space-times. *British Journal for the Philosophy of Science*, 56: 221-253.

Belnap, N., Perloff, M. and Xu, M. (2001), *Facing the Future. Agents and Choices in Our Indeterminist World*. Oxford: Oxford University Press.

Birkhoff, G. and von Neumann, J. (1936), The logic of quantum mechanics. *Annals of Mathematics*, 37: 823-843.

Bogen, J. and Woodward, J. (1988), Saving the phenomena. *Philosophical Review*, 97 (3): 303-352.

Carnap, R. (1930), Die alte und die neue Logik. *Erkenntnis*, 1: 12-26.

Dietrich, F. (2006), Judgment aggregation: (im)possibility theorems. *Journal of Economic Theory*, 126 (1): 286-298.

Earman, J. (2007), Aspects of determinism in modern physics, in: Butterfield, J. and Earman, J., eds., *Handbook of the Philosophy of Physics*, pp. 1369-1434. Amsterdam: Elsevier.

Fine, K. (2005), *Modality and Tense*. Oxford: Oxford University Press.

Fitelson, B. and Hawthorne, J. (2005), How Bayesian confirmation theory handles the paradox of the ravens, in: Eells, E. and Fetzer, J., eds., *The Place of Probability in Science*. Chicago: Open Court.

Galilei, G. (1623), *Il Saggiatore*. Rome: Giacomo Mascardi. English translation as "The Assayer", in: Drake, S. and O'Malley, C. D., eds., *The Controversy on the Comets of 1618*. Philadelphia: University of Pennsylvania Press, 1960.

Hartmann, S. (2008), Between unity and disunity: A Bayesian account of intertheoretic relations. Forthcoming.

Hartmann, S. and Bovens, L. (2008), Welfare, voting and the constitution of a federal assembly, in: Galavotti, M., Scazzieri, R. and Suppes, P., eds., *Reasoning, Rationality and Probability*, pp. 61-76. Stanford: CSLI.

Horsten, L. and Douven, I. (2008), Formal methods in the philosophy of science. *Studia Logica* 89 (2): 151-162.

Huber, F. (2005), What is the point of confirmation? *Philosophy of Science*, 72: 1146-1159.

Kant, I. (1786), *Metaphysische Anfangsgründe der Naturwissenschaft*. English translation by M. Friedman: *Metaphysical Foundations of Natural Science*, Cambridge: Cambridge University Press, 2004.

Kuhn, T. S. (1962), *The Structure of Scientific Revolutions*. International Encyclopedia of Unified Science. Chicago: University of Chicago Press.

Kuipers, T. A. F. (2007), Explication in philosophy of science, in: Kuipers, T. A. F., ed., *Handbook of the Philosophy of Science. General Philosophy of Science – Focal Issues*, pp. vii-xxiii. Amsterdam: Elsevier.

Leitgeb, H. (2009), Logic in general philosophy of science: Old things and new things, in: Hendricks, V., ed., *PHIBOOK, Yearbook of Philosophical Logic*. Copenhagen: Automatic Press.

Löwe, B. and Müller, T. (2009), Data and phenomena in conceptual modelling. *Synthese*, forthcoming.

Montague, R. (1962), Deterministic theories, in: Willner, D., ed., *Decisions, Values and Groups*, pp. 325-370. Oxford: Pergamon Press. reprinted in *Formal Philosophy*, ed. R. H. Thomason, New Haven, CT: Yale University Press, 1974, pp. 303-359.

Müller, T. (2005), Probability theory and causation: A branching space-time analysis. *British Journal for the Philosophy of Science*, 56: 487-520.

Müller, T., Belnap, N. and Kishida, K. (2008), Funny business in branching space-times: infinite modal correlations. *Synthese*, 164, 141-159.

Müller, T. and Placek, T. (2001), Against a minimalist reading of Bell's theorem: lessons from Fine. *Synthese*, 128: 343-379.

Nagel, E. (1961), *The Structure of Science*. New York: Hartcourt, Brace and World.

Neurath, O. (1932), Soziologie im Physikalismus. *Erkenntnis*, 2: 393-431.

Pigozzi, G. (2006), Belief merging and the discursive dilemma: an argument-based account to paradoxes of judgment aggregation. *Synthese*, 152 (2): 285-298.

Placek, T. (2009), On propensity-frequentist models for stochastic phenomena with applications to Bell's theorem, in: Czarnecki, T., Kijania-Placek, K., Kukushkina, V., and Woleński, J. (eds.), *The Analytical Way. Proceedings of the 6th European Congress of Analytic Philosophy*. London: College Publications 2009.

Placek, T. and Müller, T. (2007), Counterfactuals and historical possibility. *Synthese*, 154: 173-197.

Prior, A. N. (1957), *Time and Modality*. Oxford: Oxford University Press.

Prior, A. N. (1967), *Past, Present and Future*. Oxford: Oxford University Press.

Prior, A. N. and Fine, K. (1977), *Worlds, Times and Selves*. London: Duckworth.

Richardson, A. (2007), Thomas Kuhn and the decline of Logical Empiricism, in: Richardson and Uebel (2007), pp. 346-369.

Richardson, A. and Uebel, T. E., eds. (2007), *The Cambridge Companion to Logical Empiricism*. Cambridge: Cambridge University Press.

Thomason, R. H. (1970), Indeterminist time and truth-value gaps. *Theoria (Lund)*, 36: 264-281.

Uebel, T. (2001), Carnap and Neurath in exile: Can their disputes be resolved? *International Studies in the Philosophy of Science*, 15: 211-220.

van Benthem, J. (1982), The logical study of science. *Synthese*, 51: 431-472.

Weiner, M. and Belnap, N. (2006), How causal possibilities might fit into our objectively indeterministic world. *Synthese*, 149: 1-36.

Department of Philosophy
Utrecht University
Heidelberglaan 6
3584 CS Utrecht
The Netherlands
Thomas.Mueller@phil.uu.nl

FRANZ DIETRICH AND CHRISTIAN LIST

THE PROBLEM OF CONSTRAINED JUDGMENT AGGREGATION

Group decisions must often obey exogenous constraints. While in a preference aggregation problem constraints are modelled by restricting the set of feasible alternatives, this paper discusses the modelling of constraints when aggregating individual yes/no judgments on interconnected propositions. For example, court judgments in breach-of-contract cases should respect the constraint that action and obligation are necessary and sufficient for liability, and judgments on budget items should respect budgetary constraints. In this paper, we make constraints in judgment aggregation explicit by relativizing the rationality conditions of consistency and deductive closure to a constraint set, whose variation yields more or less strong notions of rationality. This approach of modelling constraints explicitly contrasts with that of building constraints as axioms into the logic, which turns compliance with constraints into a matter of logical consistency and thereby conflates requirements of ordinary logical consistency (such as not to affirm both a proposition and its negation) and requirements dictated by the environment (such as budgetary constraints). We present some general impossibility results on constrained judgment aggregation; they are immediate corollaries of known results on (unconstrained) judgment aggregation.

1 INTRODUCTION

The theory of judgment aggregation asks by which aggregation procedure a group of individuals can or should arrive at collective acceptance/rejection judgments on a given set of interconnected propositions (e.g., List and Pettit 2002, Pauly and van Hees 2006, Dietrich 2006, Nehring and Puppe 2008). A classic illustration is given by the 'doctrinal paradox' (Kornhauser and Sager 1986). Suppose a three-member court has to make collective judgments on three connected propositions:

a: The defendant did action X.

b: The defendant had a contractual obligation not to do action X.

c: The defendant is liable for breach of contract.

Suppose further that legal doctrine imposes the constraint that action and obligation (the two *premises*) are necessary and sufficient for liability (the *conclusion*), in short $c \leftrightarrow (a \wedge b)$. It can then happen that the majority judgments on the two premises (a and b) conflict with the majority judgment on the conclusion (c), relative to that constraint. Suppose, for example, the first judge holds both a and b to be true; the second holds a but not b to be true; and the third holds b but not a to

F. Stadler (ed.), *The Present Situation in the Philosophy of Science*,
The Philosophy of Science in a European Perspective, DOI 10.1007/978-90-481-9115-4_10,

be true. If each judge individually respects the constraint that $c \leftrightarrow (a \wedge b)$, then the majority judgments – in support of a and b and against c – violate the given constraint, as shown in Table 1.

	a	b	c
Individual 1	True	True	True
Individual 2	True	False	False
Individual 3	False	True	False
Majority	True	True	False

Table 1: The doctrinal paradox

The conflict may disappear if we modify the constraint. For example, the majority judgments $\{a, b, \neg c\}$ pose no problem if a and b are considered necessary but not sufficient for liability (so that the constraint is $c \rightarrow (a \wedge b)$ instead of $c \leftrightarrow (a \wedge b)$), or if we introduce a third premise d (so that the constraint is $c \leftrightarrow (a \wedge b \wedge d)$), or if we drop the constraint altogether.

Our aim in this paper is to investigate judgment aggregation on general agendas of propositions with general sets of constraints. This framework is suitable for modelling not only the court example but also many other judgment aggregation problems. Judgments on budget items, for example, are required to respect budgetary constraints. If propositions a, b and c state, respectively, that spending on education, healthcare and defense should be increased, then a budgetary constraint could stipulate that not all three can be accepted together, formally $\neg(a \wedge b \wedge c)$. Judgments on binary ranking propositions such as 'x is preferable to y', 'y is preferable to z' and 'x is preferable to z' are connected by constraints such as transitivity or acyclicity. Judgments of biologists on whether two organisms fall into the same species are constrained by the assumption that belonging to the same species is an equivalence relation.

We explain how constraints between propositions can be naturally incorporated into the judgment aggregation model. Constraints have of course played a role in earlier work, particularly in the computer science literature under the label 'integrity constraints' (e.g., Konieczny and Pino-Perez 2002). See also the notion of 'context' in Nehring and Puppe (2008) and that of the 'axioms' in Dietrich (2007).

We present two general impossibility theorems that depend on the nature of those constraints. The results are corollaries of results in Dietrich and List (2007a), but have a somewhat different interpretational flavour. They are also closely related to results by Dokow and Holzman (forthcoming) and prior results by Nehring and Puppe (2002).

To illustrate our approach, we apply our two theorems to the aggregation of judgments on binary relations (which can represent various forms of comparisons), distinguishing between different constraint sets on such binary relations. In particular, we consider strict orderings, acyclic binary relations and equivalence relations. This application generalizes earlier results by List and Pettit (2001/2004),

Dietrich (2007), Dietrich and List (2007a) and Nehring and Puppe (2008) on the representation of preference aggregation in the judgment aggregation model (a related result drawing on the 'property space' framework is Nehring 2003). A comprehensive bibliography on judgment aggregation can be found online (List 2004-7). This paper draws extensively on our prior work in Dietrich and List (2008a).

2 The Model

We consider a group of individuals $N = \{1, 2, \ldots, n\}$ ($n \geq 2$). The propositions on which judgments are made are represented in logic (following List and Pettit 2002, 2004; we use Dietrich's 2007 generalized model).

2.1 Logic

A *logic* is an ordered pair $(\mathbf{L}, \vdash)$, where (i) $\mathbf{L}$ is a non-empty set of sentences, called *propositions*, closed under negation (i.e., if $p \in \mathbf{L}$ then $\neg p \in \mathbf{L}$, where $\neg$ denotes 'not'), and (ii) $\vdash$ is an *entailment relation*, where, for each set $S \subseteq \mathbf{L}$ and each proposition $p \in \mathbf{L}$, $S \vdash p$ is read as 'S entails p' (we write $p \vdash q$ to abbreviate $\{p\} \vdash q$).[1] A set $S \subseteq \mathbf{L}$ is *inconsistent* if $S \vdash p$ and $S \vdash \neg p$ for some $p \in \mathbf{L}$, and *consistent* otherwise. We require the logic to satisfy the following minimal conditions:[2]

(L1) For all $p \in \mathbf{L}$, $p \vdash p$ (self-entailment).

(L2) For all $p \in \mathbf{L}$ and $S \subseteq T \subseteq \mathbf{L}$, if $S \vdash p$ then $T \vdash p$ (monotonicity).

(L3) $\emptyset$ is consistent, and each consistent set $S \subseteq \mathbf{L}$ has a consistent superset $T \subseteq \mathbf{L}$ containing a member of each pair $p, \neg p \in \mathbf{L}$ (completability).

In standard propositional logic, $\mathbf{L}$ contains propositions such as a, b, $a \wedge b$, $a \vee b$, $\neg(a \rightarrow b)$ (where $\wedge$, $\vee$, $\rightarrow$ denote 'and', 'or', 'if-then', respectively). The set $\{a, a \rightarrow b\}$ entails proposition b, for example, whereas the set $\{a \vee b\}$ does not entail a. Examples of consistent sets are $\{a, a \rightarrow b, b\}$ and $\{a \wedge b\}$, examples of inconsistent ones $\{a, \neg a\}$ and $\{a, a \rightarrow b, \neg b\}$.

1 Formally, $\vdash \subseteq \mathcal{P}(\mathbf{L}) \times \mathbf{L}$, where $\mathcal{P}(\mathbf{L})$ is the power set of $\mathbf{L}$.

2 Alternatively we may assume three conditions on the consistency notion (jointly equivalent to L1-L3): (C1) All sets $\{p, \neg p\} \subseteq \mathbf{L}$ are inconsistent; (C2) subsets of consistent sets $S \subseteq \mathbf{L}$ are consistent; (C3) L3 holds. In many (non-paraconsistent) logics, the notion of entailment is uniquely determined by that of consistency (via $A \vdash p \Leftrightarrow [A \cup \{\neg p\}$ is inconsistent$]$), so that the two notions are interdefinable. If we restrict attention to logics with interdefinability, or if we are ultimately interested only in whether judgments are consistent (not in whether they are deductively closed), we can use the system of consistent sets rather than the relation $\vdash$ as the primitive logical notion (and assume C1-C3). For details see Dietrich (2007).

2.2 Agenda

The *agenda* is the set of propositions on which judgments are made, defined as a non-empty subset $X \subseteq \mathbf{L}$ expressible as $X = \{p, \neg p : p \in X_+\}$ for a set $X_+ \subseteq \mathbf{L}$ of unnegated propositions. Notationally, we assume that double negations cancel each other out (i.e., $\neg\neg p$ stands for p).[3] In the three-member court example, $X = \{a, \neg a, b, \neg b, c, \neg c\}$.

2.3 Constraints

A *constraint set* is a consistent subset $C \subseteq \mathbf{L}$. It is meant to represent logical interconnections that are stipulated to hold between propositions. In the three-member court example, $C = \{c \leftrightarrow (a \wedge b)\}$. We say that a set $S \subseteq \mathbf{L}$ *entails* a proposition $p \in \mathbf{L}$ *relative to* C, formally $S \vdash_C p$, if $S \cup C \vdash p$. We say that a set $S \subseteq \mathbf{L}$ is *consistent relative to* C if $S \cup C$ is consistent, and *inconsistent relative to* C otherwise. Hereafter we refer to C*-entailment* and C*-(in)consistency*. The relationship between C-(in)consistency and C-entailment is analogous to that between (in)consistency and entailment *simpliciter*, which can be seen as the special cases of C-(in)consistency and C-entailment for $C = \emptyset$. A set $S \subseteq \mathbf{L}$ is *minimally* C*-inconsistent* if S is C-inconsistent but every proper subset of S is C-consistent. A proposition $p \in \mathbf{L}$ is C*-contingent* if $\{p\}$ and $\{\neg p\}$ are C-consistent. Informally, a C-contingent proposition is one whose truth or falsity is not settled by the constraints in C alone.

2.4 Individual judgment sets

Each individual i's *judgment set* is the set $A_i \subseteq X$ of propositions that he or she accepts. On a belief interpretation, A_i is the set of propositions believed by individual i to be true; on a desire interpretation, the set of propositions desired by individual i to be true. A judgment set A_i is

- C*-consistent* if, as just defined, $A_i \cup C$ is consistent;
- C*-deductively closed* if it contains all propositions $p \in X$ such that $A_i \cup C \vdash p$ (i.e., $A_i \vdash_C p$);
- *complete* if it contains a member of each proposition-negation pair $p, \neg p \in X$.

A *profile* is an n-tuple $(A_1, \ldots, A_n)$ of individual judgment sets.

3 Strictly speaking, when we use the symbol $\neg$ hereafter, we mean a modified negation symbol $\sim$, where $\sim p := \neg p$ if p is unnegated and $\sim p := q$ if $p = \neg q$ for some q. This convention is to ensure that $p \in X$ implies $\neg p \in X$.

2.5 *Aggregation functions*

An *aggregation function* is a function F that maps each profile $(A_1, \ldots, A_n)$ from some domain of admissible ones to a collective judgment set $F(A_1, \ldots, A_n) = A \subseteq X$, the set of propositions that the group as a whole accepts. The judgment set A can be interpreted as the set of propositions collectively believed to be true or as the set collectively desired to be true. Below we impose minimal conditions on aggregation functions (including on the domain of admissible profiles and the co-domain of admissible collective judgment sets). Standard examples of aggregation functions are

- *majority voting*, where $F(A_1, ..., A_n)$ is the set of propositions $p \in X$ for which the number of individuals with $p \in A_i$ exceeds that with $p \notin A_i$;
- *dictatorships*, where $F(A_1, ..., A_n) = A_i$ for some antecedently fixed individual $i \in N$; and
- *inverse dictatorships*, where $F(A_1, ..., A_n) = \{\neg p : p \in A_i\}$ for some antecedently fixed individual $i \in N$.

3 Why Explicit Constraints?

We could avoid explicit reference to constraints by building them into the logic. Indeed, whenever the logic $(\mathbf{L}, \vdash)$ satisfies L1, L2 and L3, then so does the logic $(\mathbf{L}, \vdash_C)$ induced by the constraint set C. C-consistency in $(\mathbf{L}, \vdash)$ translates into standard consistency in $(\mathbf{L}, \vdash_C)$, and C-deductive closure in $(\mathbf{L}, \vdash)$ translates into standard deductive closure in $(\mathbf{L}, \vdash_C)$. This is in fact the only insight needed to translate existing theorems into theorems with explicit constraints. Why, then, should we use explicit constraints at all?

3.1 *A first argument*

First of all, constraints introduce a different perspective on the notion of consistency. For a judgment set to be logically inconsistent is somewhat different and perhaps more dramatically 'irrational' than to be merely C-inconsistent, i.e., incompatible with the given constraints. If constraints are built into the logic, the distinction between these two kinds of inconsistency disappears: all inconsistencies are by definition logical ones.

3.2 *A second argument*

The nature of the appropriate set of constraints is often unclear or controversial. For example, what are the correct budgetary constraints or legal constraints when a government cabinet makes decisions? It may thus be interesting to vary the constraint set C, so that we can express the fact that a judgment set is C-consistent yet C'-inconsistent (for distinct $C, C' \subseteq \mathbf{L}$). If reaching C-consistent collective judgments turns out to be unrealistic, the group might look for C'-consistent collective judgments for a 'less ambitious' constraint set C', say a proper subset $C' \subsetneq C$. There is a long tradition in social choice theory of considering differently strong rationality constraints on preferences: one may or may not require completeness, one may or may not require full transitivity etc. As discussed later, each set of rationality conditions on preferences corresponds to a particular constraint set.

3.3 A third argument

If it is unclear for some proposition $p \in \mathbf{L}$ whether or not it should constrain the group decision, a natural move is to put it into the agenda X (rather than into the constraint set C): i.e., to let the group decide whether or not p should constrain the judgments on the (other) propositions in the agenda. For instance, the 'legal doctrine' in the introductory court example or the condition of a balanced budget might be made part of the agenda X rather than of the constraint set C.

When a constraint becomes a proposition under decision, its correct logical representation becomes crucial. Let us illustrate this point using the two examples just mentioned. First, consider the court example, and suppose the 'legal doctrine' (that action and obligation are necessary and sufficient for liability) is not imposed on the judges but put up for decision. One might be tempted to represent the legal doctrine as a material biconditional $c \leftrightarrow (a \wedge b)$. This, however, is a problematic representation. Consider the resulting agenda $X = \{a, \neg a, b, \neg b, c, \neg c, c \leftrightarrow (a \wedge b), \neg(c \leftrightarrow (a \wedge b))\}$. When a judge rejects the legal doctrine, what he or she rejects is in fact not the material biimplication $c \leftrightarrow (a \wedge b)$; indeed, he or she may well believe that a, b and c are all true or all false (so that $c \leftrightarrow (a \wedge b)$ holds). Rather the judge rejects the binding nature of a and b for c. One might say, the judge rejects a *subjunctive* biconditional between c and $a \wedge b$, or perhaps that he or she rejects the proposition $\blacksquare(c \leftrightarrow (a \wedge b))$, where $\blacksquare$ is a modal necessity operator ('necessarily, i.e., in all possible worlds, it is the case that...'). If the legal doctrine is represented using a subjunctive biconditional or modal necessity operator, negating the resulting proposition becomes logically consistent with assigning arbitrary truth values to a, b and c, so that the previous problem is avoided.

Similarly, suppose a government faces a decision problem, and suppose a balanced budget is not imposed as a constraint but represented by a proposition p in the government's agenda X. One might be tempted to specify p as the disjunction $\vee_{q \in S} q$, where each proposition $q \in S$ describes a way in which the budget can be balanced (such as 'low spending on education and average spending on social security and...'). The problem here is similar to that just identified in the court example. An individual who rejects the requirement that the budget *must* be balanced may still hold other beliefs that entail a balanced budget (i.e., that entail $\vee_{q \in S} q$): he or she may see no *necessity* of a balanced budget yet favour low total spending for other reasons. A more appropriate representation of the balanced budget requirement might be to let p be the proposition $O(\vee_{q \in S} q)$, where O is a deontic 'ought' operator ('it is required that...'). Since $O(\vee_{q \in S} q)$ states that the budget *ought* to be balanced, it becomes consistent (in standard deontic logic) to negate $O(\vee_{q \in S} q)$ while asserting $\vee_{q \in S} q$, which removes the problem that arises when p is defined as $\vee_{q \in S} q$.

However, if a constraint is not part of the agenda but part of the constraint set C, its misrepresentation is less problematic. The reason is that propositions in C cannot be negated, and often the logical interconnections induced by the (non-negated) constraints in the form of C-consistency and C-deductive closure do not change if these constraints are misspecified in the sense just illustrated. In the court example, for instance, the material biimplication $c \leftrightarrow (a \wedge b)$ imposes exactly the same constraints on a, b and c as a subjunctive one, and also as the proposition $\blacksquare(c \leftrightarrow (a \wedge b))$, namely that a, b, c can only have truth values (T, T, T) or (F, F, F) or (T, F, F) or (F, T, F). For this reason, when giving concrete examples of constraint sets in this paper we usually omit modal or deontic necessity operators and do not address the nature of (bi)conditionals. For instance, when we later consider the transitivity constraint on preferences, we model it as the statement that 'preferences *are* transitive', not the statement that 'preferences *are necessarily* transitive' (and similarly for other constraints on preferences).[4]

4 Impossibilities of Aggregation under Constraints

Can we find attractive aggregation functions? The answer to this question depends on two things. First, it depends on what conditions we impose on the aggregation function. If, for example, we do not seek to achieve C-consistency at the collective level (for an appropriate C), majority voting may be a perfectly fine solution. Likewise, in the absence of any democratic requirements, a dictatorship of one individual arises as a possibility, which generates C-consistent and complete judgment sets. If the only democratic requirement is non-dictatorship and we allow collective judgments to be incomplete but retain their C-consistency and C-deductive closure, then oligarchies arise as a solution; here, any proposition is accepted if and only if all members of a fixed set $M \subseteq N$ of 'oligarchs' accept it.[5]

Second, the question of whether we can find attractive aggregation functions depends on how the propositions in the agenda are logically connected, which in

4 On subjunctive implications in judgment aggregation, see Dietrich (forthcoming); on modal operators for representing legal prescriptions, see List (2006) and Dietrich (2007).

5 More precisely, an oligarchy F is defined by $F(A_1, .., A_n) = \cap_{i \in M} A_i$ for all profiles $(A_1, ..., A_n)$ in the universal C-domain (as defined below), where $M \subseteq N$ is a fixed non-empty set of 'oligarchs'. Oligarchies generate C-consistent and C-deductively closed (but usually incomplete) collective judgment sets, as the intersection of C-consistent and C-deductively closed sets is C-consistent and C-deductively closed. To avoid dictatorship, there must be at least two oligarchs; if all individuals are oligarchs, F is unanimity rule, an anonymous rule with considerable collective incompleteness.

turn depends on the constraint set C. More constraints can often make aggregation problems harder to solve. If the court in the original example did not have to respect the constraint that action and obligation are necessary and sufficient for liability ($c \leftrightarrow (a \wedge b)$), then the majority judgments resulting from the individual judgments in Table 1 would not be considered inconsistent.

Let us address these questions in general terms. Consider some given agenda X and constraint set C. The theorems to be presented here are C-relativized versions of existing theorems from Dietrich and List (2007a). We choose to focus on theorems that require complete and C-consistent (hence also C-deductively closed) collective judgment sets. But one could equally obtain theorems that require merely C-consistent and C-deductively closed (possibly incomplete) collective judgment sets (by adapting results by Dietrich and List 2008b and Dokow and Holzman 2006), or theorems that require just C-consistent collective judgment sets (by adapting recent results by Dietrich and List 2007b).

4.1 An impossibility of systematic aggregation

In this subsection, we require the aggregation function to satisfy the following conditions:

Universal C-domain. The domain of F is the set of all possible profiles of C-consistent and complete individual judgment sets on the agenda X.

Collective C-rationality. F generates C-consistent and complete collective judgment sets on the agenda X.

Systematicity. For any propositions $p, q \in X$ and profiles $(A_1, \ldots, A_n), (A_1^*, \ldots, A_n^*) \in Domain(F)$, if [for all individuals i, $p \in A_i$ if and only if $q \in A_i^*$] then [$p \in F(A_1, \ldots, A_n)$ if and only if $q \in F(A_1^*, \ldots, A_n^*)$].

Universal C-domain requires that the aggregation function accept as admissible any possible profile of fully rational individual judgment sets respecting the constraints in the set C. Collective C-rationality requires that the aggregation function produce as output a fully rational collective judgment set respecting the same constraints. Systematicity requires, first, that the collective judgment on each proposition depend only on individual judgments on that proposition, and, second, that the pattern of dependence be the same for all propositions. The first part of the condition requires the aggregation to be propositionwise (as captured by the *independence* condition defined later), and the second part adds a *neutrality* requirement.

Call agenda X *minimally C-connected* if it satisfies the following conditions:

(i) X has a minimal C-inconsistent subset Y with $|Y| \geq 3$, and

(ii) X has a minimal C-inconsistent subset Y such that $(Y\backslash Z) \cup \{\neg z : z \in Z\}$ is C-consistent for some subset $Z \subseteq Y$ of even size.[6]

It is easy to see that the agenda $X = \{a, \neg a, b, \neg b, c, \neg c\}$ in the three-member court example with constraint set $C = \{c \leftrightarrow (a \wedge b)\}$ satisfies minimal C-connectedness. On the other hand, if C were the empty set, the agenda $X = \{a, \neg a, b, \neg b, c, \neg c\}$ would not be minimally C-connected: it would violate both (i) and (ii). Thus the question of whether or not an agenda is minimally C-connected depends crucially on the strength of the constraint set C.

The following is a corollary of Dietrich and List's (2007) Theorem 1 (which in turn generalizes earlier results on systematicity by List and Pettit 2002 and Pauly and van Hees 2006):

Theorem 1. For a minimally C-connected agenda X, every aggregation function F satisfying universal C-domain, collective C-rationality and systematicity is a (possibly inverse) dictatorship.

The agenda condition of Theorem 1 (minimal C-connectedness) is tight if the agenda is finite or the logic is compact (and $n \geq 3$ and X contains at least one C-contingent proposition), i.e., minimal C-connectedness is also necessary, and not merely sufficient, for characterizing (possibly inverse) dictatorships by the conditions of Theorem 1.[7] The same holds for the agenda conditions of the other theorems stated below.

There are two ways in which Theorem 1 can be turned into a characterization of dictatorships as opposed to possibly inverse ones. One way is to impose an additional unanimity condition on the aggregation function:

Unanimity. For any unanimous profile $(A, \ldots, A) \in \mathit{Domain}(F)$, $F(A, \ldots, A) = A$.

Theorem 1a. For a minimally C-connected agenda X, every aggregation function F satisfying universal C-domain, collective C-rationality, systematicity and unanimity is a dictatorship.

6 This clause is for finite X equivalent to a C-relativized version of Dokow and Holzman's (forthcoming) *non-affineness* condition: the set of admissible yes/no views on the propositions in X (corresponding to C-consistent and complete judgment sets on X) is a non-affine subset of $\{0, 1\}^X$.

7 If X is not minimally C-connected, there exists an aggregation function that satisfies universal C-domain, collective C-rationality and systematicity and is not a (possibly inverse) dictatorship. Let M be a subset of $\{1, ..., n\}$ of odd size at least 3. If part (i) of minimal C-connectedness is violated, then majority voting among the individuals in M satisfies all requirements. If part (ii) is violated, the aggregation rule F with universal C-domain defined by $F(A_1, ..., A_n) := \{p \in X :$ the number of individuals $i \in M$ with $p \in A_i$ is odd$\}$ satisfies all requirements. The second example is based on Dokow and Holzman (2005).

The other way to obtain a characterization of dictatorships from Theorem 1 is to impose an additional asymmetry condition on the agenda. Call agenda X *C-asymmetric* if there exists a C-inconsistent subset $Y \subseteq X$ such that $\{\neg y : y \in Y\}$ is C-consistent.

Theorem 1b. For a minimally C-connected and C-asymmetric agenda X, every aggregation function F satisfying universal C-domain, collective C-rationality and systematicity is a dictatorship.

4.2 An impossibility of general propositionwise aggregation

The above condition of systematicity is a strong condition on an aggregation function, which goes well beyond the requirement of propositionwise aggregation by adding the (neutrality-type) requirement of equal treatment of all propositions. We now ask whether we can obtain a characterization of dictatorships using the weaker condition of propositionwise aggregation, the so-called independence condition, which drops the neutrality part of the systematicity condition.

Independence. For any proposition $p \in X$ and profiles $(A_1, \ldots, A_n)$, $(A_1^*, \ldots, A_n^*) \in Domain(F)$, if [for all individuals i, $p \in A_i$ if and only if $p \in A_i^*$] then [$p \in F(A_1, \ldots, A_n)$ if and only if $p \in F(A_1^*, \ldots, A_n^*)$].

Let us define the agenda condition of *C-path-connectedness*, building upon Nehring and Puppe's (2002) condition of *total blockedness.*[8] For any $p, q \in X$, we write $p \vdash_C^* q$ if $\{p, \neg q\} \cup Y$ is C-inconsistent for some $Y \subseteq X$ that is C-consistent with p and with $\neg q$.[9] Now an agenda X is *C-path-connected* if

(iii) for every C-contingent $p, q \in X$, there exist $p_1, p_2, ..., p_k \in X$ (with $p = p_1$ and $q = p_k$) such that $p_1 \vdash_C^* p_2$, $p_2 \vdash_C^* p_3$, ..., $p_{k-1} \vdash_C^* p_k$.

The agenda in the three-member court example above is minimally C-connected but not C-path-connected, but as shown below, preference aggregation problems can be represented by agendas that are both minimally C-connected and C-path-connected. Call an agenda *strongly C-connected* if it is C-path-connected and satisfies (ii). It then follows (for finite X or a compact logic) that X also satisfies (i) and hence that it is minimally C-connected as well.

Theorem 2. For a strongly C-connected agenda X, every aggregation function F satisfying universal C-domain, collective C-rationality, independence and unanimity is a dictatorship.

8 The relationship between C-path-connectedness and total blockedness arises when $C = \varnothing$. For a compact logic, $\varnothing$-path-connectedness is equivalent to total blockedness; generally, $\varnothing$-path-connectedness is weaker than total blockedness.

9 For non-paraconsistent logics (in the sense of L4 in Dietrich 2007), $\{p, \neg q\} \cup Y$ is C-inconsistent if and only if $\{p\} \cup Y \vdash_C q$.

This result is the C-relativized version of a result proved independently by Dietrich and List (2007) and Dokow and Holzman (forthcoming).[10] Both of these results extend a prior result by Nehring and Puppe (2002) with an additional monotonicity condition on F.

Finally, all results in this section continue to hold under generalized definitions of minimal and strong C-connectedness.[11]

5 AN APPLICATION: BINARY RELATIONS

To illustrate the results above, we apply them to the aggregation of binary comparisons, such as betterness judgments or judgments of (a given type of) equivalence. Such judgments are given by a binary relation over a set of objects to be compared, e.g., policy alternatives, job candidates or organisms to be classified into species. How can binary relations be represented in the judgment aggregation model? We use the following construction, drawing on List and Pettit (2001/2004), Dietrich (2007) and Dietrich and List (2007).

5.1 A simple predicate logic

We consider a predicate logic with constants $x, y, z, ... \in K$ (representing objects), variables v, w, v_1, v_2 , ... (ranging over objects), identity symbol $=$, a binary relation symbol P (representing the comparative relation in question), logical connectives $\neg$ (not), $\wedge$ (and), $\vee$ (or), $\rightarrow$ (if-then), and universal quantifier $\forall$. Formally, $\mathbf{L}$ is the smallest set such that

- $\mathbf{L}$ contains all propositions of the forms $\alpha P \beta$ and $\alpha = \beta$, where α and β are constants or variables, and
- whenever $\mathbf{L}$ contains two propositions p and q, then $\mathbf{L}$ also contains $\neg p$, $(p \wedge q)$, $(p \vee q)$, $(p \rightarrow q)$ and $(\forall v)p$, where v is any variable.

We drop brackets when there is no ambiguity.

5.2 Constraint sets

We consider some alternative constraint sets. We begin with the constraint set on fully rational strict preferences, the paradigmatic binary relation in social choice

10 Dokow and Holzman restrict the agenda to be finite (with only contingent propositions) and for this case show the tightness of the agenda assumptions (if $n \geq 3$).

11 In the definitions of minimal and strong C-connectedness, (i) and (ii) can be weakened, namely to the C-relativised versions of the conditions (i*) and (ii*) given in Dietrich (2007). All theorems presented survive the weakening, and the agenda assumptions of Theorems 1, 1a and 1b become tight even for infinite X in a non-compact logic (again provided that X contains a contingent proposition and $n \geq 3$). The weakened conditions become equivalent to the original ones for finite X or a compact logic.

theory:

$$C_{\text{fully rational}} = \left\{ \begin{array}{l} (\forall v_1)(\forall v_2)(v_1 P v_2 \rightarrow \neg v_2 P v_1) \\ (\forall v_1)(\forall v_2)(\forall v_3)((v_1 P v_2 \wedge v_2 P v_3) \rightarrow v_1 P v_3) \\ (\forall v_1)(\forall v_2)(\neg v_1{=}v_2 \rightarrow (v_1 P v_2 \vee v_2 P v_1)) \end{array} \right\}^{12}.$$

The three displayed propositions in $C_{\text{fully rational}}$ are the constraints of asymmetry, transitivity and connectedness. To represent weak preferences rather than strict ones, $C_{\text{fully rational}}$ needs to be redefined as the set of rationality conditions on weak preferences (i.e., reflexivity, transitivity and connectedness); see also Dietrich (2007).[13]

Contrast this with the constraint set on merely acyclic (but not necessarily fully rational) strict preferences, representing a weaker notion of rationality:

$$C_{\text{acyclic}} = \left\{ \begin{array}{l} \neg(a_1 P a_2 \wedge \ldots \wedge a_{m-1} P a_m \wedge a_m P a_1) \\ \qquad : a_1, \ldots a_m \in K \text{ pairwise distinct}, m \geq 1 \end{array} \right\}^{14}.$$

The propositions in C_{acyclic} rule out any cycle of any length $m \geq 1$. In particular, irreflexivity is enforced (take $m = 1$). Transitivity, however, is not required. Thus the set $\{xPy, yPz, \neg xPz\}$, while inconsistent relative to $C_{\text{fully rational}}$, is consistent relative to C_{acyclic}.

Next we consider the constraint set on equivalence relations, suitable for classifying objects:

$$C_{\text{equivalence}} = \left\{ \begin{array}{l} (\forall v)(v P v) \\ (\forall v_1)(\forall v_2)(\forall v_3)((v_1 P v_2 \wedge v_2 P v_3) \rightarrow v_1 P v_3) \\ (\forall v_1)(\forall v_2)(v_1 P v_2 \rightarrow v_2 P v_1)) \end{array} \right\}^{15}.$$

The three displayed propositions in $C_{\text{equivalence}}$ are the constraints of reflexivity, transitivity and symmetry. While this constraint set would obviously not be imposed when P represents a *preference* relation (since 'better than' is neither reflexive nor symmetric), it may be imposed on a relation of *equal suitability* between job candidates (since 'is as suitable as' is plausibly an equivalence relation) or on the relation of *belonging to the same species* among organisms.

---

12 For technical reasons, the constraint set also contains, for each pair of distinct constants x, y, the condition $\neg x{=}y$.

13 Transitivity and connectedness are as defined above. Reflexivity can be stated by the proposition $(\forall v)(vPv)$. For aesthetic reasons, one might also replace the predicate symbol P by R in the logic.

14 Again, the constraint set also contains, for each pair of distinct constants x, y, the condition $\neg x{=}y$.

15 Again, the constraint set also contains, for each pair of distinct constants x, y, the condition $\neg x{=}y$.

Each of these constraint sets C induces its own notions of C-consistency and C-deductive closure.

5.3 The agenda

The *binary-relation agenda* is the set X of all propositions of the form xPy, $\neg xPy \in \mathbf{L}$, where x and y are constants. The question of which agenda condition is met by the binary-relation agenda depends crucially on the given constraint set. The following lemma holds:

Lemma 1. The binary-relation agenda X (with $|K| \geq 3$) is
(a) strongly C-connected when $C = C_{\text{fully rational}}$;
(b) minimally, but not strongly, C-connected when $C = C_{\text{acyclic}}$;
(c) minimally, but not strongly, C-connected when $C = C_{\text{equivalence}}$.

In part (a), the C-path-connectedness part is a variant of a lemma by Nehring (2003); for instance, $xPy \vdash_C^* xPz$ because $\{xPy, yPz\} \vdash_C xPz$ (where $x, y, z \in K$ are pairwise distinct). In parts (a) and (b), minimal C-connectedness holds since any cycle $Y = \{xPy, yPz, zPx\} \subseteq X$ defines a minimal C-inconsistent set, which becomes C-consistent by negating two elements. In part (c), minimal C-connectedness holds because any set of type $Y = \{xPy, yPz, \neg xPz\} \subseteq X$ (with x, y, z pairwise distinct) is minimally C-inconsistent and becomes C-consistent by negating any two members.

By this lemma, Theorems 1, 1a and 1b apply to the binary relation agenda for any of the three constraint sets C. This allows the conclusion that it is impossible to aggregate preference relations – whether fully rational or just acyclic – or equivalence relations in a systematic and non-degenerate way, unless we restrict the domain of individual inputs or allow some kind of collective irrationality (such as incomplete collective judgment sets).

By part (a), the stronger impossibility of Theorem 2 applies when the constraint set is $C_{\text{fully rational}}$. It is impossible to aggregate fully rational preference relations in an independent, unanimity preserving and non-dictatorial manner, again unless we restrict the domain of individual inputs or allow collective irrationality. The latter is precisely Arrow's famous theorem on the aggregation of preferences (in the case where indifference between distinct options is excluded).

In conclusion, the present approach allows us to derive a large number of general results on aggregation problems with various constraints in a simple unified framework. An interesting question for future research is how the results are affected when different constraints are imposed at individual and collective levels, for example, when the constraints on collective judgments are weaker than those on individual ones or vice-versa.

References

Dietrich F. (2006), Judgment Aggregation: (Im)Possibility Theorems. *Journal of Economic Theory* 126(1): 286-298

Dietrich F. (2007), A generalised model of judgment aggregation. *Social Choice and Welfare* 28(4): 529-565

Dietrich (forthcoming), The possibility of judgment aggregation on agendas with subjunctive implications, *Journal of Economic Theory*

Dietrich F., List, C. (2007a), Arrow's theorem in judgment aggregation. *Social Choice and Welfare* 29(1): 19-33

Dietrich F., List C. (2007b), Judgment aggregation with consistency alone, Working paper, London School of Economics

Dietrich F., List, C. (2008a), Judgment aggregation under constraints, in: *Economics, Rational Choice and Normative Philosophy*, T. Boylan and R. Gekker (eds.), London (Routledge)

Dietrich F., List, C. (2008b), Judgment aggregation without full rationality, *Social Choice and Welfare* 31(1): 15-39

Dokow E., Holzman R (forthcoming) Aggregation of binary evaluations, Working paper, Technion Israel Institute of Technology

Dokow E., Holzman R. (2006), Aggregation of binary evaluations with abstentions. Working paper, Technion Israel Institute of Technology

Konieczny S., Pino-Perez R. (2002), Merging information under constraints: a logical framework. *Journal of Logic and Computation* 12: 773-808

List C. (2004-7), Judgment aggregation: a bibliography on the discursive dilemma, doctrinal paradox and decisions on multiple propositions. Available at http://personal.lse.ac.uk/list/

List C. (2006), Republican Freedom and the Rule of Law. *Politics, Philosophy and Economics* 5(2): 201-220

List C., Pettit P. (2002), Aggregating Sets of Judgments: An Impossibility Result. *Economics and Philosophy* 18: 89-110

List C., Pettit P. (2001/2004), Aggregating Sets of Judgments: Two Impossibility Results Compared. Social and Political Theory Paper W20 (technical report ID 931), Australian National University; *Synthese* 140(1-2): 207-235

Nehring K. (2003), Arrow's theorem as a corollary. *Economics Letters* 80: 379-382

Nehring K., Puppe C. (2002), Strategyproof Social Choice on Single-Peaked Domains: Possibility, Impossibility and the Space Between. Working paper, University of California at Davis

Nehring K., Puppe C. (2008), Consistent Judgment Aggregation: The Truth-Functional Case. *Social Choice and Welfare* 31: 41-57

Pauly M., van Hees M. (2006), Logical Constraints on Judgment Aggregation. *Journal of Philosophical Logic* 35: 569-585

Franz Dietrich
London School of Economics
Department of Philosophy, Logic and Scientific Method
Houghton Street
London WC2A 2AE
United Kingdom
F.Dietrich@lse.ac.uk

Christian List
Departments of Government and Philosophy
London School of Economics
London WC2A 2AE
U.K.
C.List@lse.ac.uk

GABRIELLA PIGOZZI

AGGREGATION PROBLEMS AND MODELS: WHAT COMES FIRST?

ABSTRACT

The aggregation of consistent individual judgments on logically interconnected propositions into a collective judgment on the same propositions has recently drawn attention in law, philosophy, economics and computer science. Despite the apparent simplicity of the problem, reasonable aggregation procedures, such as propositionwise majority voting, cannot ensure a consistent collective outcome. The literature on judgment aggregation has been influenced by earlier work in social choice theory. As preference aggregation investigated in social choice theory, judgment aggregation studies aggregation functions under specific conditions. These are derived from properties of the preference aggregation realm. In this paper we argue that judgment aggregation problems are intrinsically different from preference aggregation ones. Thus, imposing exogenous models and properties is detrimental to a deep understanding of the specificity of judgment aggregation.

1 INTRODUCTION

A judgment is a yes/no position on a proposition. Judgment aggregation is a novel discipline that studies how consistent individual judgments on propositions displaying a logical form can be aggregated into a consistent collective judgment on the same propositions. The propositions are of two kinds: *premise* and *conclusion*. The premises serve as supporting reasons to derive a certain judgment on the conclusion. Suppose that a three judges court has to make a decision on whether a person is liable of breaching a contract (proposition R, or *conclusion*). The legal doctrine states that a person is liable if and only if there was a contract (*premise* P), and there was a conduct constituting breach of such a contract (*premise* Q). Hence, the legal rule can be formally expressed as follows $(P \wedge Q) \leftrightarrow R$. Suppose now that the three judges express their judgments according to Table 1 [12, 13].

The aim of judgment aggregation is to examine how the individual judgments on the conclusion and on the reasons supporting that conclusion should be combined to obtain a collective view on the conclusion while providing reasons in support of that decision. Majority voting appears to be a natural candidate. However, propositionwise majority voting (consisting in the separate aggregation of the votes for each proposition P, Q and R via majority rule) results in an inconsistent

F. Stadler (ed.), *The Present Situation in the Philosophy of Science*,
The Philosophy of Science in a European Perspective, DOI 10.1007/978-90-481-9115-4_11,

	P	Q	$R = (P \wedge Q)$
Judge 1	yes	no	no
Judge 2	no	yes	no
Judge 3	yes	yes	yes
Majority	yes	yes	no

Table 1: Doctrinal paradox. Premises: P = There was a contract, Q = There was conduct constituting breach of such a contract. Conclusion: $R = (P \wedge Q)$ = The defendant is liable.

collective outcome: a majority supports P and Q and yet a majority supports $\neg R$. This is a violation of the rule $(P \wedge Q) \leftrightarrow R$ and it is an instance of the so-called *doctrinal paradox*.[1] This means that the court is paralyzed in its decision. A majority deems the defendant not liable, but the judges cannot provide reasons for this conclusion as a majority of them agrees that there was a contract and (another) majority deems that there was a conduct constituting breach of such a contract. The essence of the paradox is that, despite the fact that the individuals are logically consistent and submit judgments that obey the decision rule, the group supports an inconsistent judgment set.

The relevance of the judgment aggregation riddles goes beyond the specific court example, because it applies to all situations in which individual binary evaluations on premises and conclusion need to be combined into a group decision. To illustrate that judgment aggregation is a more general problem than those arising in courts, Pettit introduced the term of *discursive dilemma* [20]. Judgment aggregation problems can be found outside of the legal domain, paradoxes can arise using other decision rules than the conjunctive one of the court example, and other aggregation rules than majority. For instance, an example given in the literature with the same logical structure as the original contract paradox is due to Bovens and Rabinowicz [2] and illustrates the case of a committee that agrees that someone is worthy of tenure if he is worthy of tenure on teaching and worthy of tenure on research.

The literature on judgment aggregation has been influenced by earlier work in social choice theory [1, 25]. Preference aggregation is a sub-field of social choice theory and investigates how individual preferences can be aggregated into a collectively preferred alternative. List and Pettit [15, 16] gave the first formal model of judgment aggregation based on propositional logic[2] combined with an axiomatic approach in the social choice tradition. Later Dietrich and List explored the relations between the famous Arrow's impossibility theorem of preference ag-

1 It is important to mention that the problem of aggregating individual judgments is not restricted to majority voting, but it applies to *all* aggregation procedures satisfying some desirable conditions. For a survey on this discipline, the reader is referred to [17].

2 Dietrich [6] shows that judgment aggregation problems can be expressed in more expressive logics than the propositional one.

gregation and an impossibility result in judgment aggregation. They show that preference aggregation can be embedded into judgment aggregation, thus concluding that judgment aggregation is a more general model than Arrow's model of preference aggregation [7].

When we chose a model, we inevitably privilege some features over others. For this reason, models are not only research tools, but they should themselves be the legitimate object of inquiry. In this paper we claim that, though the similarities between preference and judgment aggregation are undeniable, framing judgment aggregation into a classical social choice theory approach has not served to fully understand the specificity of the novel aggregation dilemma. Moreover, the comparison of the doctrinal paradox with well-known voting paradoxes shows that the possibility of group inconsistent outcomes should not be the only concern for judgment aggregation.

The paper proceeds as follows. In Section 2 we recall the benchmark paradox of preference aggregation, i.e. the Condorcet paradox, and argue that the differences between preference and judgment aggregation are such to call for different formal models. Section 3 discusses two well-known voting paradoxes, the Ostrogorski paradox and the paradox of multiple elections, in relation with the discursive dilemma. On the one hand, we maintain that the logical connections between propositions do not exhaust the specificity of judgment aggregation instances. On the other hand, we argue that also other conundrums than the discursive dilemma should occupy judgment aggregation. Finally, Section 4 contains conclusive remarks.

2 JUDGMENT AGGREGATION AND PREFERENCE AGGREGATION

One of the questions that social choice theory addresses is how individual preferences on a given set of alternatives should be combined into a collective preference. Probably the most famous method for the aggregation of preferences is the one proposed in the 18th century by the Marquis de Condorcet. Given a set of individual preferences, we compare each of the alternatives in pairs. For each pair we determine the winner by majority voting, and the final social ordering is obtained by a combination of all partial results. Unfortunately, this method led to the first aggregation problem, known as the *Condorcet paradox*: the pairwise majority rule can lead to cycles in the collective ordering. In other words, this ordering cannot be used to select an overall preferred candidate. Consider three voters who express their preferences in the following way, where $x > y$ means that x is (strictly) preferred to y:

Voter 1: $x > y > z$
Voter 2: $y > z > x$
Voter 3: $z > x > y$

Since there are two people who prefer x to y, two voters who prefer y to z and two people who prefer z to x, we obtain the following social ordering: $x > y > z > x$. This outcome cannot be accepted as it is cyclic.

The lesson is that, when we combine individual choices into a collective one, we may lose something that held at the individual level, like transitivity or logical consistency. Kornhauser [11] notices that the doctrinal paradox resembles the Condorcet paradox, but clarifies that the two paradoxes are not equivalent. Indeed, as stated by List and Pettit:

> [W]hen transcribed into the framework of preferences instances of the discursive dilemma do not always constitute instances of the Condorcet paradox; and equally instances of the Condorcet paradox do not always constitute instances of the discursive dilemma. ([16], pp. 216-217)

Besides the impossibility of a one-to-one mapping between instances of the discursive dilemma and those of the Condorcet paradox, for Kornhauser and Sager [12] the distinctive difference between preference and judgment aggregation is that, when an individual expresses a preference, she "speaks only to her own values and advantage" ([12], p. 85). So two people may disagree in their preferences without either of them being wrong. On the other hand, when a person makes a judgment, she is stating her opinion about the truth. If two people disagree in their judgments, they acknowledge that they may be wrong. If preferences are neither true or false, judgments are. Thus, expressing preferences and judgments are diverse activities. Indeed, judgment aggregation procedures can be evaluated with respect to their truth-tracking properties, as in [14, 2, 9]

It is worth mentioning that, a decade before judgment aggregation made its appearance, Sen criticized the classical framework of social choice for being too general and abstract, and thus unable to capture the specificities of different types of aggregation:

> It can be argued that some of the difficulties in the general theory of social choice arise from a desire to fit essentially different classes of group aggregation problems into one uniform framework and from seeking excessive generality. An alternative is to classify these problems into a number of categories and to investigate the appropriate structure for each category. ([24], p. 53)

In particular, Sen distinguishes between the aggregation of individual interests and of individual judgments. A committee decision is concerned with the aggregation of the views of its members on some alternative proposals, whereas the exercise of interest aggregation is concerned with the aggregation of the personal welfare levels of the different people involved. Sen's notion of individual judgments is thus that of judgment rankings of alternative policies, rather than of logically connected propositions. Sen claims that essentially diverse approaches are needed for different classes of aggregation problems. The conclusion is a re-examination of the impossibility results in social choice theory, which would go beyond the purpose of this article. What interests us here is the warning to consider the kinds of information to be combined in several contexts besides the formal similarities that such situations may display. Preference and judgment aggregation are fundamentally diverse exercises in the nature of the information they combine and also because their inputs have different structures, as we show in the next section.

2.1 The independence condition

Arrow's famous impossibility theorem [1] shows that the culprit of the Condorcet paradox is not the pairwise majority voting. Indeed the problem is more general as there exists no function that assigns a collective preference ordering to a set of individual preference orderings, and that meets some minimal conditions. Similarly, List and Pettit [15, 16] show that the doctrinal paradox does not depend on the specific choice of the aggregation procedure. Rather, by rephrasing Arrow's properties in the logic based model of judgment aggregation, they prove a general impossibility theorem stating that there exists no aggregation function that satisfies a minimal set of desirable conditions.[3]

The relations between the two aggregation frameworks have been explored by List and Pettit [16] and, later and more thoroughly, by Dietrich and List [7],who prove that Arrow's theorem (for strict preferences) is a corollary of one of the impossibility results in judgment aggregation. So Dietrich and List conclude that the Arrowian preference aggregation is a special case of judgment aggregation. A proposal for a unification of the two frameworks has been lately put forward by Grossi [8], who shows a correspondence between preference aggregation and a subclass of judgment aggregation problems in a many-valued logic.

Undeniably the two frameworks share some features. Nevertheless, formal translations of one model into the other should not overshadow the differences of the aggregation problems the two disciplines aim to capture. As recalled in Section 2, not only the Condorcet paradox and the discursive dilemma are not equivalent, but they also (as maintained by Kornhauser and Sager) combine different types of information. In addition, we now want to illustrate that the inputs of preference and judgment aggregation have different structures. Hence, conditions that can be reasonably imposed in one framework result to be odd in the other.

One of the properties in Arrow's theorem is the *independence of irrelevant alternatives*, that warrants that the group ranking over any pair of alternatives depends solely on the personal rankings over the same pair of alternatives. The intuition is that the social ranking over, for example, x and y should be determined exclusively on how the individuals rank x compared to y and not on other (irrelevant) alternatives, like z. This requirement has been introduced in judgment aggregation as the *independence* condition. This ensures that the collective judg-

3 For other impossibility theorems that strengthened and expanded the original formulation, see [17].

ment on each proposition depends exclusively on the individual judgments on that proposition (and not on other – assumed to be independent – propositions).[4]

Yet, we believe that (beside the different types of information to be aggregated) the feature that most distinguishes judgment aggregation from other aggregation exercises amounts to making a decision on the conclusion while providing reasons in support of that decision. No equivalent distinction between premises and conclusion can be traced in preference aggregation. The two aggregation problems are structurally different, besides the fact that they combine different types of information. Importing the independence property into the models for judgment aggregation contributes to neglect this feature. Indeed, the distinction between premises and conclusion does not play a role in the logical models of judgment aggregation [22]. Such a division is crucial and undermines the independence condition since premises are often independent from each other, but they are never independent from the conclusion (and vice versa).[5] In order to aggregate premises and conclusions, we need to better understand their relations. This amounts to investigate the justification for the independence condition and eventually to propose weakened (or alternative) conditions, which may provide escape routes from the impossibility results that plague the discipline.[6]

3 The doctrinal paradox and other aggregation paradoxes

The discursive dilemma does not resemble only the Condorcet paradox. List remarks the analogies between the doctrinal paradox and other well-known voting paradoxes, and bounds the essence of the discursive dilemma to the logical relations among the propositions:

> The doctrinal paradox is related to Anscombe's paradox, or Ostrogorski's paradox [...]. Like the doctrinal paradox, these paradoxes are concerned with aggregation over multiple propositions. Unlike the doctrinal paradox, they do not incorporate explicit logical connections between the propositions. ([14], p. 4)

In this section the discursive dilemma is compared with the Ostrogorski's paradox and the paradox of multiple elections. The aim of the first comparison is to

---

4 Here we briefly recall also the other two conditions imposed on any judgment aggregation function F: *universal domain* and *anonymity*. The first ensures that F accepts as inputs all consistent and complete individual judgment sets, while anonymity guarantees that all individuals' judgments are given equal weight in the aggregation.

5 The urge for a theory of judgment aggregation on normatively defensible conditions has been claimed by Mongin [18]. Mongin recognizes that propositional formulas are not independent when they share propositional variables. This leads Mongin to weaken the independence condition. Nevertheless, his new independence condition is not weak enough to ensure possibility results.

6 How the distinction between premises and conclusion can be introduced in a weakened independence condition has been studied in [22].

show that, by focusing exclusively on the logical connections between propositions, other relevant aspects of the judgment aggregation problem may be missed. As we have seen, the doctrinal paradox is the benchmark example of judgment aggregation. However, as illustrated by the multiple election paradox, collective irrationality should not be the only worry of judgment aggregation.

Let us start with the *Ostrogorski's paradox*.[7] Ostrogorski was concerned with the democratic governance. He argued that individuals should vote directly for policies and not for political parties. The Ostrogorski's paradox hints at the distortions that political parties produce when individuals are not allowed to express their opinions directly on the policies [23, 10].

Consider a two-party contest (government and opposition) and three issues (economic, environmental, international). The two parties have opposite views on the issues, and each individual casts a vote (yes or no) depending on whether she wishes a policy change on that issue (so she agrees with the opposition on that issue), or the government represents her opinion on that matter (no policy change). Each voter votes for the government (resp. opposition) if she agrees with the government (resp. opposition) on a majority of the issues. Suppose there are five voters and that they vote as in Table 2.

	Econ	Env	Int	Party
Voter 1	no	yes	no	no
Voter 2	no	yes	no	no
Voter 3	yes	no	no	no
Voter 4	yes	yes	yes	yes
Voter 5	yes	yes	yes	yes
Majority	yes	yes	no	no

Table 2: Ostrogorski's paradox

Like the doctrinal paradox, the Ostrogorski's case is puzzling because, despite the individuals being rational, the collective outcome is inconsistent. If each voter votes for the party with which she agrees on a majority of issues, the government wins. However, the opposition represents the views of the majority of the voters on the issues, specifically on the economic and the environmental policies.

The interdependence of the propositions in the two paradoxes are not of the same kind. In the Ostrogorski's paradox it is a compound majority decision that binds parties and issues, whereas in the discursive dilemma the propositions are logically connected. Nevertheless, the policies-party division in the Ostrogorski's paradox is similar to the premises-conclusion one in the doctrinal paradox. The supported party/conclusion depends on the opinions expressed on the policies/

7 The analysis is based on [21], though the scope there was broader: an aggregation operator was recommended to deal with the Ostrogorski's and the doctrinal paradox.

premises, whereas the policies/premises are independent from each other (any combination of judgments is admissible).

The difficulty with the aggregation problems is that the set of propositions on which most group members agree is not guaranteed to be a candidate for the collective decision. The set can fail to satisfy the dependence relations among the items even though each member consistently expressed her judgments or votes. The comparison with the Ostrogorski's paradox shows that the kind of dependence between the items of the agenda to be voted on does not need to be of a logic kind to lead to a paradoxical outcome.

How compelling are the dependence relations among the issues for a paradox to arise? To answer this question it is instructive to consider a last voting paradox traditionally studied in social choice theory: the *paradox of multiple elections*. In a multiple election voters are requested to vote on several issues, and they may vote on an issue without knowing the result of the election on the previous one. Brams, Kilgour and Zwicker [4] introduce the paradox as follows:

> Consider a referendum in which voters can vote either Y or N on each proposition on the ballot. The paradox of multiple elections occurs when the combination of propositions that wins receives the fewest votes, or is tied for fewest. ([4], p. 213)

An instance of this phenomenon is in Table 3. If votes are aggregated separately on each issue, the winning combination is yes-yes-yes. Although yes-yes-yes is socially acceptable, no single individual voted that combination. Two people voted yes-yes-no, two voted yes-no-yes, four individuals voted no-yes-yes and, finally, two voted yes-no-no. Thus, the social outcome is a sequence of alternatives that does not correspond to the vote of any individual. To put it more dramatically, the outcome does not reflect the will of any member of the electorate.

Voter 1	yes	yes	no
Voter 2	yes	yes	no
Voter 3	yes	no	yes
Voter 4	yes	no	yes
Voter 5	no	yes	yes
Voter 6	no	yes	yes
Voter 7	no	yes	yes
Voter 8	no	yes	yes
Voter 9	yes	no	no
Voter 10	yes	no	no
Majority	yes	yes	yes

Table 3: The paradox of multiple elections

Even more alarming is the "complete-reversal paradox" where, for example, the issue by issue aggregation selects yes-yes-yes-yes whereas the combination that receives the most votes is the 'opposite' no-no-no-no (and yes-yes-yes-yes receives no vote).[8]

Brams, Kilgour and Zwicker provide necessary and sufficient condition for the paradox to appear and we refer the reader to their paper for the full analysis. Here we want to draw the attention to the following question: should an outcome like the ones in the paradox of multiple elections be acceptable in judgment aggregation? Suppose that propositionwise majority voting selects a collective judgment set that satisfies the decision rule, but that does not correspond to any of the individuals' judgment sets. Should this be a source of concern for judgment aggregation? If we find the discrepancy between the winner under proposition aggregation and the one chosen by combination aggregation disconcerting in the paradox of multiple elections, we should even more find so when to be aggregated are judgment sets instead of (independent) issues.

We believe that collective irrationality should not be the only concern of judgment aggregation, and the discussion on how to aggregate individual judgment sets should not be limited to avoid the discursive dilemma. As we have seen, judgment aggregation would not reject a consistent combination of reasons and conclusion that no member voted for. Yet, like in the paradox of multiple elections, this may not be a desirable solution. Suppose that the three judges of Section 1 convict the defendant for reasons that none of them submitted. Would such a verdict be acceptable? In order to tackle this problem, a more holistic notion of responsiveness is required. This observation was one of the motivations for an investigation of judgment aggregation in abstract argumentation [5], where three aggregation operators that guarantee a social outcome 'compatible' with the individual judgments are introduced. A 'compatible' group decision is such that any group member is able to defend the group decision without having to argue against his own opinions. Not only a collective inconsistent outcome would not be accepted, but neither would be a group outcome that is not compatible with all the group members. The new models should aim at capturing the specificity of these riddles rather than flattening them to preference aggregation problems.

4 Conclusion

In this paper we have considered a recent aggregation problem and its paradox: the discursive dilemma. The doctrinal paradox appears when seemingly natural aggregation rules are applied to individual judgments on logically connected

8 Nurmi conceives that the paradox of multiple elections is particularly troublesome when the issues are related to each other (i.e. "nonseparable", in technical terms). Yet he claims that "If the issues to be voted upon are separable, then the paradox just amounts to pointing out that no individual is exactly like the electorate considered as a whole. It could even be argued that the occurrence of the paradox amounts to there being no Arrovian dictator. This certainly should not bother us very much." ([19], p. 85)

propositions to obtain a collective judgment on the same propositions. Judgment aggregation is a novel discipline whose relations with aggregation problems studied in social choice have been explored in the literature. Because of the similarities with preference aggregation and the resemblance of the doctrinal paradox with the Condorcet paradox, the formal models developed so far for judgment aggregation combine logic with an axiomatic approach in the Arrowian spirit.

Here we have maintained that the differences between preference and judgment aggregation deserve more attention in the hope to define models for judgment aggregation that capture the specificities of the problem. "All models are wrong, but some are useful", said the statistician George Box [3]. Models necessarily privilege some aspects over others, so a certain amount of idealization is unavoidable. They try to capture what the modeler believes to be the essence of a complex phenomenon, or at least its relevant aspects. However, even if all models were wrong, some are more adequate than others.

The fact that an aggregation procedure that is normatively defensible is not meaningful because may produce irrational outcomes is shared by the paradoxes in judgment and in preference aggregation. We argued that the similarities between the two areas do not justify the imposition of the same conditions on their aggregation procedures. First, judgment aggregation combines individual evaluations on propositions rather than individual preferences. As Kornhauser and Sager observed, unlike for the case of judgments, the latter do not convey any truth value. Second, judgment and preference aggregation differ in the type of inputs they accept. Judgment sets are vectors of yes/no items on premises and conclusion, where the premises are typically independent on each other while the conclusion depends on the premises. Such a distinction is missing in the preference realm and, by consequence (and unfortunately), in the formal models of judgment aggregation that have borrowed the independence condition from social choice theory. Third and finally, we raised the question whether the logical relations among propositions constitute the main feature of judgment aggregation. As the paradox of multiple elections shows, a group outcome that does not violate any logical constraint may be perceived as unacceptable if no member submitted it. Thus, focusing exclusively on the doctrinal paradox may overshadow other relevant issues for judgment aggregation.

Judgment aggregation and preference aggregation are different classes of problems and to expect them to fit into a uniform framework imposes excessive generality. We believe that the differences that we have highlighted here call for distinct models. Judgment aggregation problems display features that need to be properly modeled without imposing exogenous frameworks.

References

[1] K. Arrow. *Social Choice and Individual Values. Cowles Foundation Monograph Series*, 1963.

[2] L. Bovens and W. Rabinowicz. Democratic answers to complex questions. An epistemic perspective. *Synthese*, 150: 131-153, 2006.

[3] G. Box. Robustness in the strategy of scientific model building. In R. Launer and G. Wilkinson, editors, *Robustness in Statistics*, pages 201-236. Academic Press: New York, 1979.

[4] S. Brams, D. Kilgour, and W. Zwicker. The paradox of multiple elections. *Social Choice and Welfare*, 15(2): 211-236, 1998.

[5] M. Caminada and G. Pigozzi. On judgment aggregation in abstract argumentation. *Autonomous Agents and Multi-Agent Systems*, forthcoming.

[6] F. Dietrich. A generalised model of judgment aggregation. *Social Choice and Welfare*, 28(4): 529-565, 2007.

[7] F. Dietrich and C. List. Arrow's theorem in judgment aggregation. *Social Choice and Welfare*, 29(1): 19-33, 2007.

[8] D. Grossi. Unifying preference and judgment aggregation. In: *Proceedings of the Eighth International Conference on Autonomous Agents and Multi-Agent Systems (AAMAS'09)*, Budapest, Hungary, 2009.

[9] S. Hartmann, G. Pigozzi, and J. Sprenger. Reliable methods of judgment aggregation. *Working paper*, 2009.

[10] J. Kelly. The Ostrogorski paradox. *Social Choice and Welfare*, 6: 71-76, 1989.

[11] L. Kornhauser. Modeling collegial courts. II. Legal doctrine. *Journal of Law, Economics, and Organization*, 8: 441, 1992.

[12] L. Kornhauser and L. Sager. Unpacking the court. *Yale Law Journal*, 96: 82-117, 1986.

[13] L. Kornhauser and L. Sager. The one and the many: Adjudication in collegial courts. *California Law Review*, 81: 1-51, 1993.

[14] C. List. The probability of inconsistencies in complex collective decisions. *Social Choice and Welfare*, 24: 3-32, 2005.

[15] C. List and P. Pettit. Aggregating sets of judgments: An impossibility result. *Economics and Philosophy*, 18: 89-110, 2002.

[16] C. List and P. Pettit. Aggregating sets of judgments: Two impossibility results compared. *Synthese*, 140: 207-235, 2004.

[17] C. List and C. Puppe. Judgment aggregation: A survey. In: P. Anand, C. Puppe, and P. Pattanaik, editors, *Oxford Handbook of Rational and Social Choice*. Oxford University Press, 2009.

[18] P. Mongin. Factoring out the impossibility of logical aggregation. *Journal of Economic Theory*, 141(1): 100-113, July 2008.

[19] H. Nurmi. *Voting Paradoxes and How to Deal with Them*. Springer, 1999.

[20] P. Pettit. Deliberative democracy and the discursive dilemma. *Philosophical Issues*, 11: 268-299, 2001.

[21] G. Pigozzi. Two aggregation paradoxes in social decision making: The Ostrogorski paradox and the discursive dilemma. *Episteme*, 2(2): 119-128, 2005. 13

[22] G. Pigozzi, M. Slavkovik, and L. van der Torre. Conclusion-based procedure for judgment aggregation satisfying premise independence. In G. Bonanno, B. Lowe, and W. van der Hoek, editors, *Proceedings of the Eighth International Conference on Logic and the Foundations of Game and Decision Theory, LOFT 2008*, p. 35, Amsterdam, The Netherlands, 2008.

[23] D. Rae and H. Daudt. The Ostrogorski paradox: A peculiarity of compound majority decision. *European Journal of Political Research*, 4(4): 391-398, 1976.

[24] A. Sen. Social choice theory: A re-examination. *Econometrica: Journal of the Econometric Society*, pp. 53-89, 1977.

[25] A. K. Sen. *Collective Choice and Social Welfare*. Holden Day, 1970.

Individual and Collective Reasoning
University of Luxembourg
Computer Science and Communication
rue R. Coudenhove-Kalergi 6
L-1359 Luxembourg
Luxembourg
gabriella.pigozzi@uni.lu

Part III
Philosophy of the Natural and Life Sciences

Marcel Weber

Life in a Physical World: The Place of the Life Sciences

1. Physicalism, Biology, and Reductionism: State of the Art

Debate about the place of the life sciences within the empirical sciences has often centered around the issues of physicalism and reductionism.[1] Given that some form of physicalism is correct, why is biological science not physical science? Why do biological theories appear to be autonomous and irreducible to physical theories? And what is the nature of biological laws or regularities, assuming that the fundamental interactions that govern the physical world also are at work in living organisms? These are some of the oldest and most extensively discussed questions concerning the biological sciences. While philosophers of science of a Logical Empiricist bent first tried to defend the view that biological theories such as those of classical Mendelian genetics are in principle reducible to physical-chemical theories,[2] an anti-reductionist consensus emerged during the 1970s.[3] This consensus was mainly based on the argument that genetic concepts such as dominance or the gene concept itself cannot be redefined in an extensionally equivalent way in terms of molecular concepts. The reason for this is thought to lie in the functional character of biological concepts. This means that certain theoretically significant properties in biology are individuated by their causal role, not some intrinsic structural property. But the molecular realizers of these causal roles are highly heterogeneous at the molecular level; in others words, the realizers don't have a theoretically significant molecular property in common that could be used to eliminate the higher-level terms. Therefore, higher-level concepts in biology remain explanatorily indispensable; they have autonomous explanatory value that cannot be reproduced by molecular theories alone. Thus, on this by

1 See, e.g., Ernst Mayr, *This is Biology*. Cambridge, Mass.: Harvard University Press 1997.

2 Kenneth F. Schaffner, "Approaches to Reduction", in: *Philosophy of Science* 34, 1967, pp. 137-147; Kenneth F. Schaffner, "The Watson-Crick Model and Reductionism", in: *British Journal for the Philosophy of Science* 20, 4, 1969, pp. 325-48.

3 David L. Hull, *Philosophy of Biological Science*. Englewood Cliffs: Prentice Hall 1974; Philip Kitcher, "1953 and All That. A Tale of Two Sciences", in: *The Philosophical Review* 93, 3, 1984, pp. 335-373; Alexander Rosenberg, *The Structure of Biological Science*. Cambridge: Cambridge University Press 1985; cf. C. Kenneth Waters, "Why the Anti-Reductionist Consensus Won't Survive the Case of Classical Mendelian Genetics", in: *PSA 1990*, East Lansing, Mich.: Philosophy of Science Association 1990, pp. 125-139.

F. Stadler (ed.), *The Present Situation in the Philosophy of Science*,
The Philosophy of Science in a European Perspective, DOI 10.1007/978-90-481-9115-4_12,

now received view in philosophy of biology, biological theories are irreducible for basically the same reason that most philosophers accept as the definitive refutation of mind-brain reductions.

Thus, philosophers of biology have reached similar conclusions as philosophers of mind have in regard of the issue of mind/brain-reductionism.[4] However, Jaegwon Kim[5] has argued that, in the philosophy of mind, this consensus is based on an inadequate model of reduction, namely Ernest Nagel's.[6] He proposed an alternative scheme according to which reduction does not consist in first connecting the terms of the theory to be reduced to those of the reducing theory by way of biconditional bridge principles (as Nagel's model assumes or is widely taken to assume), followed by the derivation of the laws of the theory to be reduced from the laws of the reducing theory. Instead, Kim argues that successful reductions must first give a functional characterization of the referents of the terms of the theory to be reduced. Such a characterization specifies the set of things that come under a concept by stating the causes and/or the effects that these things have in their containing system. Next, scientists must identify the things that play these causal roles at the lower level. For example, Kim thinks that the case of genetics provides a paradigm for this kind of reduction. Genes were first identified by the causal roles they play in living organisms, namely causing heritable character differences, being segregated and assorted in accordance with Mendel's laws, etc. etc. Later, it was discovered that these causal roles are actually fulfilled by DNA sequences that code for protein and/or RNA molecules.[7] This *is* a reduction; nothing more is required. I take it that Kim does not require that scientist be able to state necessary and sufficient *physical* (molecular) conditions for some thing to instantiate a theoretically significant higher-level property, for if he did, his model would basically collapse into Nagel's (or what is usually taken to be Nagel's). All that he requires is that the realizers of the causal role that defines the higher-level property be somehow describable from the physical level (he admits that this may not always be possible, for example, he things it is not possible for qualia).

Kim's suggestion has not succeeded in displacing the anti-reductionist consensus in the philosophy of biology; in fact, it was hardly noticed by philosophers of biology. However, it is clear what their response would be: Even if Kim's new model of reduction is accepted, *that* the molecular realizers of some functionally

4 Donald Davidson, "Mental Events", in: L. Foster/J. Swanson (eds.), *Experience and Theory*. Amherst, Mass.: University of Massachusetts Press 1970; Jerry A. Fodor, "Special Sciences or the Disunity of Science as a Working Hypothesis", in: *Synthese* 28, 1974, pp. 97-115.

5 Jaegwon Kim, *Mind in a Physical World: An Essay on the Mind-Body Problem and Mental Causation*. Cambridge, Mass.: MIT Press 1998.

6 Ernest Nagel, *The Structure of Science. Problems in the Logic of Scientific Explanation*. London: Routledge and Kegan Paul 1961.

7 C. Kenneth Waters, "Genes Made Molecular", *Philosophy of Science* 61, 1994, pp. 163-185.

individuated biological concept can be described at the molecular level *alone* is exactly what is not possible according to the anti-reductionist consensus. On the standard argument from multiple realizability, such a description would involve an ungainly disjunctive predicate without any explanatory force. This is why the higher-level theories are explanatorily indispensable.

To this reply, a Kim-style reductionist could retort that the identification of classical genes with protein- and RNA-coding DNA sequences is not ungainly at all. Understanding what genes are at the molecular level is precisely what molecular biology has done for genetics, and if this does not account as a reduction, then nothing does. However, this reductionist response misses that reduction is supposed to at least conserve the explanatory achievements of the theory to be reduced, in addition to providing explanations that exceed those of the theory to be reduced. But this is not the case in the genetics/molecular biology case according to antireductionists.[8] Classical transmission genetics offers explanations of inheritance patters that basically cite the pairing and separation of chromosomes. These explanations abstract away from the "gory" molecular details that constitute these processes at bottom. Kitcher[9] draws an analogy here to Putnam's well known square peg-in a round hole-argument. According to this argument, there is a perfectly fine explanation of why square pegs don't fit in round holes that appeals only to these objects' geometrical shape. This explanation abstracts away from the composition of the objects and from any physical laws that these may obey. In fact, this explanation is more general than any explanation that appeals to the objects' composition. Kitcher suggests that this is analogous to the explanations that classical genetics give of inheritance patterns.

There are different replies that a reductionist can give to this argument. First, it can be argued that the theoretical content of classical genetics is not exhausted by patterns of gene transmission. Classical geneticists described genetic structures with the help of elaborate maps long before molecular techniques such as DNA-sequencing became available. For instance, it was possible to show that genes must be linear structures, a finding which was confirmed by the discovery of the way in which DNA encodes genetic information.[10] This fits nicely with Kim's model of reduction.

A second possible response is that Kitcher's argument – just like Putnam's – is a manifestation of a theoretically unfounded "explanatory Protagoreanism",[11] according to which "some human or other is the measure of all putative explanations, of those which do explain and of those which do not." While Kitcher's

8 Kitcher, *op. cit.*

9 Kitcher, *op. cit.*, p. 350

10 Marcel Weber, "Representing Genes: Classical Mapping Techniques and the Growth of Genetical Knowledge", in: *Studies in History and Philosophy of Biological and Biomedical Sciences* 29, 2, 1998, pp. 295-315.

11 Alexander Rosenberg, *Darwinian Reductionism. Or, How to Stop Worrying and Love Molecular Biology*. Chicago: The University of Chicago Press 2006, p. 35.

chromosomal mechanics explanations or Putnam's square peg-in-a-round-hole-explanation may seem perfectly satisfactory to some people, perhaps relative to certain pragmatic contexts, it is not an explanation that would satisfy a physicist or a molecular biologist. Science ought to do better than that, for example, by showing exactly what forces pull the chromosomes apart before a cell divides, or what forces repel the peg from the hole, taking into account their composition, of course. Here, the reductionism/antireductionism debate turns on divergent assumptions as to what constitutes a good explanation of a phenomenon – a matter on which, naturally, reductionists and antireductionists have different intuitions.[12]

These arguments and counter-arguments are well known and have been discussed in the literature *ad nauseam*. It is beyond the scope of this paper to present all the twists and turns of the reductionism/antireductionism debate, or even to lay out the various positions that have been defended, from strong forms of reductionism to non-reductive physicalism, emergentism, scientific pluralism, and so on.[13] Instead, what I would like to do here is to examine some novel arguments, which have received little attention. I think that both attempts, while perhaps not successful, contain some genuine insights with respect to the place of the life sciences in the conceptual landscape of the natural sciences.

The first view I want to critically review is an attempt to defend of a strong form of reductionism about biology that can be found in Alex Rosenberg's recent book.[14] I will show why Rosenberg's account fails, even though it contains a valuable insight concerning the role of the concept of function in biology, namely in the individuation of traits. Rosenberg thinks that this makes all of biology conceptually dependent on evolutionary theory, which is not generally thought to be reducible to more fundamental theories. As a result, an unbridgeable gap threatens between biology and physical theories. Rosenberg tries to close this gap by trying to show that evolutionary theory, at least natural selection theory, *is* fundamental. I shall criticize Rosenberg's position on two counts: First, I will show that the idea that natural selection theory is fundamental is problematic (Section 2). Second, I will argue that there was no problem for the reductionist in the first place, because there are ways of individuating organismic traits that do not depend on the concept of natural selection (Section 3).

The second view I will discuss here comes from outside the philosophy of biology, namely from general metaphysics and is it is not very recent, but it has been hardly noticed by philosophers of biology and of science: the view of biological

12 Paul Hoyningen-Huene, "Epistemological Reductionism in Biology: Intuitions, Explications and Objections", in: P. Hoyningen-Huene/F. M. Wuketits (eds.), *Reductionism and Systems Theory in the Life Sciences*. Dordrecht: Kluwer Academic 1989: pp. 29-44.

13 An important strand of this debate is critically reviewed in Thomas Reydon's contribution in this volume, namely the issue of natural kinds and its implications for reduction.

14 Rosenberg 2006, *op. cit.*

laws that has been developed by Michael Thompson.[15] He thinks that biological laws differ fundamentally from physical laws. While this claim is hardly new, the specific differences that Thompson sees between the two classes of laws have, to my knowledge, not been noticed in the philosophy of biology. Even though I disagree with some parts of Thompson's account, I believe that it merits serious discussion, which I shall attempt in Section 4.

2. Rosenberg's Defense of Reductionism and Why it Fails

In his recent book,[16] Rosenberg firmly adheres to the view that "nothing in biology makes sense except in the light of evolution." Evolutionary biologists such as Ernst Mayr[17] or Theodosius Dobzhansky,[18] who have defended this view, based their arguments on the assumption that a full understanding of organisms requires the identification of the *ultimate* causes of their characteristic properties. To use Mayr's favorite example, even if we fully understand the physiological mechanisms that induce migratory birds to flock together and embark on a long journey towards a warmer climate zone – i.e., the proximate cause – a full understanding of this behavior requires an account of what it was selected for in the birds' evolutionary past – i.e., the ultimate cause. On this received view, proximate and ultimate explanations are *complementary* and *conceptually independent*. This conceptual independence allows for the possibility of endorsing both reductionism about proximate biology and antireductionism about evolutionary biology. The latter kind of antireductionism is usually justified on grounds of the multiple realizability of fitness.[19]

However, according to Rosenberg, ultimate and proximate biology are *not* conceptually independent. How could this be? Why can't biologists pinpoint an organism's molecular, physiological, developmental etc. mechanisms independently of its evolutionary history? For Rosenberg, this has to do with the way in which biologists pick the *explananda*, in other words, that which they want to explain by discovering the underlying mechanisms. Let us say, for example, that biologists want to understand how chick embryos form wings. 'Wing' is a functional concept. In other words, the classification of some structure as a wing, including its exact delimitation from neighboring structures, involves an appeal to

15 Michael Thompson, "The Representation of Life", in: R. Hursthouse/G. Lawrence/W. Quinn (eds.), *Virtues and Reasons. Philippa Foot and Moral Theory*. Oxford: Clarendon 1995, pp. 247-296.

16 Rosenberg 2006, *op. cit.*

17 Ernst Mayr, "Cause and Effect in Biology", in: *Science* 134, 1961, pp. 1501-1506.

18 Theodosius Dobzhansky, "Biology, Molecular and Organismic", in: *American Zoologist* 4, 4, 1964, pp. 443-452.

19 Elliott Sober, *The Nature of Selection. Evolutionary Theory in Philosophical Focus*. Cambridge Mass.: MIT Press 1984.

function (flight in this case). Rosenberg argues that the salient concept of function here must be that of *proper* function,[20] that is, function as selected effect. A wing is a structure that was selected because it confers the ability to fly. It is a functional type, and "function" means proper function according to Rosenberg. The realizers of this functional type are *heterogeneous* because different structures with different evolutionary origins can confer the ability to fly. This is why there are also no natural kinds (essences) in the traditional sense in biology, Rosenberg argues. For selection is blind to essences (intrinsic structure).[21] The upshot is that the way in which an organism is divided into parts crucially depends on the theory of natural selection. Since proximate biology takes its requests for explanation from such divisions ("what mechanisms control the development of the chick wing?"), it is conceptually dependent on evolutionary biology.

This position with respect to functions and proximate biology seems to put Rosenberg in the difficult position that, in order to maintain his reductionism, he must show either that the theory of natural selection is reducible to more fundamental theories or that it is *itself* a fundamental theory. He chooses the second path: He argues that what he calls the "principle of natural selection" is itself a fundamental law. Here is one formulation of this alleged "principle":[22]

> $\forall x \forall y \forall E$ [If x and y are competing organisms in generation n, and x is fitter than y in E, then probably (there is some generation n', at which x has more descendants than y)]

There are alternative formulations, and Rosenberg is aware that this may not be the most general way of stating the principle. Rosenberg takes this to be an empirical law (in contrast to Sober,[23] who thinks that the principle of natural selection is *a priori*) and he understands fitness in terms of a probabilistic propensity.

Now for what is probably Rosenberg's boldest claim: He argues that the principle of natural selection is a *physical* law, or perhaps a *chemical* law (or both).

20 Ruth G. Millikan, "In Defense of Proper Functions", in: *Philosophy of Science* 56, 1989, pp. 288-302.

21 Thomas Reydon (in this volume) argues that selected effect functions are *not* multiply realizable, because they require that the function bearers stand in an appropriate historical (genealogical) relationship, which means that even something which plays the exact same causal role today would not count as an instance of the function if it evolved independently. To this, it could be replied that nothing prevents a certain organ to change its internal structure (its essence) in evolution while it continues to benefit from natural selection, so the set of things that has the same activity and stands in the appropriate genealogical relations would count as instances of the function. This would count as multiple realization. However, Reydon's point does seem to limit the multiple realizability of selected effect functions.

22 Rosenberg (2006), *op. cit.*, p. 160

23 Elliott Sober, "Two Outbreaks of Lawlessness in Recent Philosophy of Biology", in: *Philosophy of Science (Proceedings)* 64, 1998, pp. S458-S467.

In support of this claim, he argues that even things that are not considered to be alive obey this principle, for example, self-replicating molecules. He also offers a story why textbooks of physical chemistry do not normally cite this law, namely, because physical chemists normally ask different questions. But this doesn't prove that this isn't a fundamental law of nature according to Rosenberg.

Rosenberg needs this claim in order to make "natural selection safe for reductionism." The reason is, as I have already shown, is that Rosenberg thinks that natural selection via the concept of proper function provides the explananda for biological explanations, even outside of evolutionary biology.

I would like to address two critical points at Rosenberg's argument. The first concerns his claim that there exists a "principle of natural selection" which is a physical law. The second point challenges the claim that natural selection theory is needed for identifying the explananda for biological explanations.

First, let us consider Rosenberg's alleged "principle of natural selection". As stated, it is only applicable to populations with discrete generations. Evolutionary theorists use different fitness measures for populations with discrete generations and for age-structured populations with overlapping generations. In one of my own works, I argue that if there is a general principle of natural selection, then it is highly abstract and needs to be instantiated by specific models.[24] On this view, the theory of natural selection is a family of models ("semantic view" of theories) and its content is not appropriately expressed by a universally quantified claim. Universally quantified claims only come in when it comes to stating classes of natural systems to which the models *apply*. The general theory is merely some sort of a guideline for building specific models.

Rosenberg could reply that, perhaps, he has not correctly stated the fundamental principle of natural selection, but that his point that there exists such a principle and that it is a fundamental law of nature stands. However, I don't think that he can sustain this view. The reason is that there are no reasons to believe that there is a fundamental measure of evolutionary fitness. "Fitness" means different things, depending on the evolutionary problem that biologists are trying to solve. Sometimes, fitness is an absolute growth rate. Sometimes it is an absolute number for the surviving offspring. Sometimes it is a coefficient in a population genetic model that makes explicit assumption the genetic system (e.g., Mendelian inheritance). Fitness is predicated of genes, genotypes, individuals, and groups. So far, there is no unifying framework for evolutionary theory, and there are no reasons to think why there should be one. There are different evolutionary processes and different questions that one can ask about them. Any fitness measure can be useful for answering one kind of question, but not another.

24 Marcel Weber, *Die Architektur der Synthese. Entstehung und Philosophie der modernen Evolutionstheorie.* Berlin: Walter de Gruyter 1998, Ch. 6.

If there is no fundamental fitness measure, it follows that there is no general "principle of natural selection". And *a fortiori* there is also no fundamental law of nature about natural selection.[25]

I think that this failure of Rosenberg's attempt is exemplary for the whole of biology. Biology is not concerned with identifying laws of nature in the traditional sense. Its goal is rather to answer specific why-questions by using various conceptual tools, including in some cases mathematical models. The answers to such why-questions cannot generally be incorporated into some unified framework.[26]

As we have seen, the ultimate motivation for Rosenberg's account of biological laws was his goal of showing that biological traits could be both functional, in the proper role sense, and yet physical. In the following section, I shall examine if there are no other ways of how biological traits can be individuated.

3. An Alternative Account of Functions and Trait Individuation

As we have seen, Rosenberg based his defense of reductionism on the view that biological traits are individuated *functionally*, where "function" is understood in the sense of selected effect function or "proper" function. I think the first part of this claim is correct, however, there is a problem with the second.

On this view, some item X has a function F in organism S exactly if X does F and the fact that some earlier tokens of X have done X is a cause of X's presence in S. The way in which earlier tokens can cause the presence of some item in later generations, of course, is natural selection. Thus, Rosenberg's view is that natural selection is not only needed to explain why some organism S came to have a part X, but to speak of X as having some kind of *unity* in the first place. It is for this reason that Rosenberg thinks that the theory of natural selection is fundamental for the whole of biology. This, of course, includes behavioral biology. According to Rosenberg, the description of behavioral traits is laden and/or ought to be laden by theoretical hypotheses about selection history. A trait such as a wing is individuated by the fact that it was selected for flying, no matter what other capacities it may have (for instance, it's capacity of being flapped so as to distract or attract some other animal). On this view, descriptions of an organism's traits are laden by the theory of natural selection and assumptions about the evolutionary past.

25 Daniel Sirtes (personal communication) objects that this argument at best proves that there is no *single* fundamental principle; there still could be one for every type of evolutionary process. However, it seems to me that such a hodgepodge of principles – and there would have to quite a lot of them – would not deserve the status of "fundamental" principles, because they would all only be applicable to some restricted number of cases.

26 This claim is generally known as scientific pluralism, see Stephen H. Kellert/Helen E. Longino/C. Kenneth Waters (eds.), *Scientific Pluralism*. Minnesota Studies in Philosophy of Science, Vol. XIX. Minneapolis: University of Minnesota Press 2006.

Paul Griffiths[27] has argued that this view puts the cart before the horse. The parts of organisms and their causal capacities must be understandable *independently* of natural selection. Otherwise, the following regress threatens:

1. Selected effect functions are ascribed by causal analysis of the capacities of the parts of ancestral organisms and a determination of their fitness contribution.
2. Thus, we must already be able to individuate the parts. This cannot be done on the basis of the ancestors to the ancestral organisms, because this would generate a regress
3. But if we are able to individuate parts for ancestral organisms independently of their selection history, then this is possible for living organisms

So if natural selection is not fit for the individuation of organismic parts, what is? This turns out to be a very difficult question, and I can answer it only in outline.

In essence, I do not think that there is a *general* answer to this question. In other words, there is no unique principle of cutting up an organism into parts in the way that Plato suggested in the infamous passage of the *Phaedrus*, according to which a good scientist should carve nature at her joints. Clearly, Socrates's advice from the *Phaedrus* to proceed by trying not to splinter any parts, "as a bad butcher might do,"[28] is not helpful at all, for we have no theory-independent way of knowing when we have splintered something.

The explanandum is almost never neutral with respect to the explanans. Different theoretical models often come with different ways of classifying the phenomena. This has long been recognized for the physical sciences, for example, by Kuhn and Feyerabend, but few people (excepting Rosenberg) have noticed that the same holds for biology. Developmental biology, evolutionary biology, evo-devo, physiology, cell biology and so on have different ways of individuating phenomena.

However, I do want to argue that the concept of biological function is *often* involved when biologists cut up an organism into parts, including mechanisms. But the salient concept of function need not be that of selected effect functions. There are other concepts of function, and they can also fulfill the role that Rosenberg thinks only selected effect functions can play.

As an alternative, I suggest a modified version of causal role functions.[29] This account starts with Cummins's[30] analysis according to which functions are such

27 Paul Griffiths, "Function, Homology, and Character Individuation", in: *Philosophy of Science* 73, 1, 2006, pp. 1-25.

28 Plato, *Complete Works*, J. M. Cooper (ed.). Indianapolis: Hackett 1997, 265e (p. 542).

29 Marcel Weber, *Philosophy of Experimental Biology*. Cambridge: Cambridge University Press 2005; Marcel Weber, "Holism, Coherence, and the Dispositional Concept of Functions", in: *Annals in the History and Philosophy of Biology* 10, 2005, pp. 189-201.

30 Robert Cummins, "Functional Analysis", in: *Journal of Philosophy* 72, 1975, pp. 741-

capacities that are capable of explaining a capacity of some containing system. The paradigm is the heart's capacity to pump blood figuring in any adequate explanation of the circulatory system's capacity to transport nutrients, oxygen and blood cells through the body. According to Cummins, the pertinent capacity of the containing system is a matter of an interest-based choice to be made by the investigator. I have modified this account by suggesting that this systems capacity should be made dependent not on the investigator's interests, but on the role that the containing system itself plays in the self-reproduction of the *whole* organism. I argue that this is what turns Cummins-functions into *biological* functions. Cummins-functions can be applied to any kind of system. But only biological systems are capable of self-reproduction. In order for self-reproduction to occur, an organism's functions must work together. The specific contribution that some organ's causal capacities make to self-reproduction makes will depend on what other organs do. For example, if there were subsystems of an organism that would use the heart's heat production towards something that itself makes a contribution of self-reproduction, then the heart would (also) have the function of producing heat. It is the place that such a causal capacity plays in a whole network that gives it its function (perhaps much in the way in which a linguistic expression's meaning is given by the inferential role that the expression plays in a network of other expressions, as claimed by inferentialists and semantic holists).

I have argued that introducing such a global constraint on a system of functions might make the interest-dependence vanish, provided that there is exactly one way of laying a network of cooperating functions over an organism. Of course, this is hard to prove; but I suggest that it might be possible by using a notion of maximal explanatory coherence.[31]

Thus, contrary to what Rosenberg claims, dividing up an organism into different parts or traits can be done independently of its selection history. Whether there is one correct or natural way of doing this, however, is very difficult to say.[32] What seems clear is that functions have a holistic[33] character: Some thing only has a function if it is connected to many other things that also have functions and

765.

31 Weber, "Holism, Coherence, and the Dispositional Concept of Functions".

32 A lot hangs on the way in which the *explanadum* of such a network of functions is construed. It is tempting to suggest that it has to be "self-reproduction of the individual" (as I have done in my *Philosophy of Experimental Biology*), however, this notion suffers from a certain indeterminacy that is introduced by the reflexive term "self." What is that "self" that is being reproduced? And what does its "reproduction" or "maintenance" involve, in other words: what are its persistence conditions? Note that the answer "the individual" doesn't really help because of the notion of biological individual is notoriously difficult (see Jack Wilson, *Biological Individuality - The Identity and Persistence of Living Entities*. Cambridge: Cambridge University Press 1999). This could make some room for pluralism.

33 See Michael Esfeld, *Holism in Philosophy of Mind and Philosophy of Physics*. Dordrecht: Kluwer 2001.

that conspire to maintain the organism's form. Furthermore, what some thing's function is can depend on what other things do to which it is connected. However, this holism need not necessarily be an obstacle to reductionism, unless the requirements for successful reduction are made excessively strong. For instance, it might still be possible that Kim's requirements (see Section 1) can be satisfied. Of course, on the view of functions that I have mentioned, some thing's function may not only depend on how this thing interacts with its immediate interaction partners (Kim's "causal role") but also on what the role of that thing is in the whole organism. But once this role is known, there are no obstacles to then identifying the realizers of these functions.

In the final section, I shall critically discuss an altogether different challenge to reduction in biology.

4. Michael Thompson's Account of Biological Regularities

After much debate on *ceteris paribus* laws and various "outbreaks of lawlessness"[34] in biology, many philosophers of biology including myself have found Jim Woodward's account of causation and explanation[35] very helpful to come to terms with causal regularities in biology. However, there is something that this account does not quite capture, and this is the question of what makes a certain causal generalization a *biological* generalization as opposed to merely a physical or chemical one. I think the following answer is not really satisfactory: "A causal generalization is biological if it concerns living organisms or parts thereof." For there are endlessly many causal generalizations about any part of an organism that could just as well be described as physical or chemical, for example, "blood vessels with a high content of elastin expand as internal fluid pressure increases."[36]

An interesting answer to the question of what characterizes biological generalizations can be found in the work of Michael Thompson.[37] It comes from general metaphysics and has therefore rarely been noted by philosophers of science. Thompson writes for example:

> Now suppose I say, 'Bobcats breed in spring': it is obvious that this isn't going to happen in any particular case unless certain conditions are satisfied. Perhaps a special hormone must be released in late winter. And perhaps the hormone will not be released if the bobcat is too close to sea level, or if it fails to pass through the shade of a certain sort of tall pine. But now, to articulate *these* conditions is to advance one's teaching about bobcats. [...] The

34 Sober (1998), *op. cit.*

35 James Woodward, *Making Things Happen: A Theory of Causal Explanation*. Oxford: Oxford University Press 2003.

36 C. Kenneth Waters, "Causal Regularities in the Biological World of Contingent Distributions", *Biology and Philosophy* 13, 1998, pp. 5-36.

37 Thompson, *op. cit.*

> thought that *certain hormones are released*, or that *they live in such-and-such altitudes and amid such-and-such vegetation*, is a thought of the same kind as the thought that thy breed in the spring. […] These conditions are presupposed by the life-form itself.[38]

Thompson thinks that there is an important difference between biological generalizations such as 'bobcats breed in spring' and purely physical generalizations such as 'water boils at 100°C'. But the difference is not that one requires *ceteris paribus* clauses while the other doesn't. They both do. i.e., both generalizations are subject to certain conditions that must obtain for the generalizations to be manifested. In the first example, it is necessary that certain environmental cues that trigger mating behavior in bobcats occur (e.g., longer days, milder temperatures) and that nothing interferes (e.g., a shortage of prey). In the second example, it is necessary that normal atmospheric pressure obtains and that the water has not been salted. But according to Thompson, in the biological case it is itself a fact about this species *that* these conditions obtain. Bobcats will seek an environment where the conditions for breeding are favorable, such that the regularity will obtain. By contrast, there is no law about water that says that all water tends to occur under conditions such that the regularity "water boils at 100°C" or any other such regularity will obtain. In fact, the latter generalization has a purely *hypothetical* character: It only says, water boils *if* the temperature is 100°C or more. By contrast, the biological generalization is *categorical* in nature. It reads as it is written: bobcats breed in spring. *That* bobcats live in places where there is a seasonal change in temperature and day length that triggers their breeding is part of the nature of bobcats.

It is clear that Thompson has quite a different conception of regularities or laws than contemporary philosophy of science, in fact, it is closer to Aristotelian forms than to laws of nature in the modern sense. According to Thompson, each organism instantiates a certain "life-form" that is characterized by such caterogical laws as the ones about bobcats in his example. His notion of life-form seems to be one of a complex irreducible essence, much like Aristotle's concept of *eidos*. Of course, as such this conception is problematic, especially in light of all the arguments against biological essentialism that have been produced in recent years.[39] However, there might be some merit in Thompson's suggestion that what characterizes biological generalizations is in part the way in which different generalizations conspire to ensure each other's being manifested by individual organisms. There might perhaps even be an analogy to what some philosophers of science have said about natural kinds in biology, for instance, Richard Boyd's theory of homeostatic property clusters.[40]

---

38 *ibid.* p. 287.

39 John Dupré, *The Disorder of Things: Metaphysical Foundations of the Disunity of Science*. Cambridge Mass.: Harvard University Press 1993.

40 Richard Boyd, "Homeostasis, Species, and Higher Taxa", in: Robert A. Wilson (ed.), *Species: New Interdisciplinary Essays*. Cambridge, Mass.: MIT Press, pp. 141-185.

According to Thompson's account, what makes certain regularities biologically salient is that they ensure that *other* regularities are instantiated, regularities that are themselves important for the survival of the individual, and so on. This is quite reminiscent of my tentative answer to the question of what makes certain activities in an organism functionally relevant (see the preceding section). On this account of functions as well as on Thompson's account of biological laws, there exists a highly complex relation between the different parts of an organism, a relation that obtains exactly if the parts are organized such that the system sustains itself. This kind of focus on self-reproduction is what distinguishes biology from other natural sciences.

Where I must part with Thompson is here: I see no principled way of drawing a line between essential and non-essential parts of an organism. Which laws are associated with the form and which ones aren't? Furthermore, I see no reason why Thompson's account of biological laws should be inconsistent with an adequate form of reductionism. Even if it is their relation to the instantiation conditions of other laws that makes certain laws about biological entities salient, there is no reason why these relations cannot be fully understood and expressed in physical-chemical language.

One final point: It should also be noted that Thompson's categorical laws are only valid for the living state and if the organisms over which they range live in their normal environment. They contain no information what would happen, for instance, if North American bobcats were transferred to the Tropics. Would they still breed in spring? To answer this kind of question requires good old-fashioned causal laws that range not only over a set of actual states, but over counterfactual situations as well. Biologists can discover such causal laws as well, but they will be of the ordinary, hypothetical sort. In this respect, biology is no different from other natural sciences.

5. Conclusions

There have been many attempts to show that biology occupies some special place in the natural sciences, and most of them have attempted to show that biological theories (or laws) are irreducible to physical-chemical theories. This is obviously correct if "reduction" is understood in a strong, derivational sense, but far less obvious if a weaker sense of reduction such as Kim's is assumed. One of the most popular arguments against reduction, the argument from multiple realizability, is not convincing on such a weaker view. Additional arguments to the effect that some higher-level explanations do some explanatory work that cannot be recovered at the lower level rely strongly on intuitions as to what constitutes a good explanation and are not convincing to those who don't share these intuitions, for the intuitions of reductionists and anti-reductionists notoriously differ.

Further, I have considered Rosenberg's argument that (1) even though proximate biology needs evolutionary concepts (proper function) to individuate the parts of an organism, this (2) is no problem for the reductionist because the salient evolutionary principles are *fundamental physical laws*. The latter claim fails because natural selection theory is not a unified theory; it consists of a wide variety of specific models that deploy different fitness measures. Furthermore, there are alternative ways of how biologists can individuate the parts of an organism, for example, by causal role functions. I have discussed a rich version of causal role functions that might yield a natural system of functions for each type of organism. Even though functions are in a sense holistic properties on this account, reductionists need not worry about this.

Finally, I have critically examined M. Thompson's essentialistic account of biological laws according to which the latter develop an irreducible life-form for each species of organism. I argue that, while this account gives a good answer of what makes certain regularities biologically salient (or biological at all), it also provides no arguments against a suitably understood reductionism.

Universität Konstanz
Fachbereich Philosophie
Postfach D9
78457 Konstanz
Deutschland
Marcel.Weber@uni-konstanz.de

Claude Debru

Comments on Marcel Weber's "Life in a Physical World: The Place of the Life Sciences"

Marcel Weber's contribution is an extremely accurate and synthetic view of the state of some discussions which are typical of what is meant today by "philosophy of biology". Before commenting more closely his contribution, I would like to take some distance and to come closer to actual science (at least to some parts of the biological sciences). Looking at biology as a set of different sciences is more and more frequent. From mathematical biology to medical sciences and to biotechnologies, the field of biology became more and more diversified, although some common features remain at the most general level. Biology as a whole remains basically an empirical science, or a set of largely empirical researches. Philosophers dealing with biology should not underestimate biological empiricism and biological experimentalism. The extent to which this is the case, which is recognized by Marcel Weber, creates a real difficulty for philosophical thinking, because looking at biology as a largely empirical and basically experimental science, leads to the conclusion that biology is much less stabilized, at least in important parts, than it could be supposed. If you do not look at things from a purely conceptual point of view, if you avoid projecting on biology as a science in progress or rather as a set of closely related research disciplines, the idea of the structure of biological knowledge as the major problem of the philosophy "of" biology, then you get a much more realistic and richer picture of the biological sciences. Pluralism is now completely integrated in biological thinking, and experimentalism remains a most general feature. As far as I can see, Marcel Weber could agree entirely on these remarks, since he wrote a book on biological experimentalism.

Now let us go to the first item under discussion, reductionism. This subject has been discussed under precisely this term, reductionism, since at least one hundred and fifty years – and we still continue. In his Introduction to the study of experimental medicine (1865) Claude Bernard asked the question : can we "reduce" biology to physics and chemistry. His answer was partly yes and partly no, and was based on the state of the physiology of metabolism. The catabolic part of metabolism was understandable in terms of the molecular activity of enzymes, which were known as proteins. The biosynthetic part of metabolism was not understandable in terms of the chemical activity of enzymes. Later it became understandable, when it was realized that enzymes acted upon equilibrium reactions, which could be accelerated in both degradation and synthesis directions. At about the same time as Bernard's theoretical reflections, Ernst Mach asked the same question: can we reduce (zurückführen)? As a physicist he was extremely successful in his

F. Stadler (ed.), *The Present Situation in the Philosophy of Science*,
The Philosophy of Science in a European Perspective, DOI 10.1007/978-90-481-9115-4_13,

approaches of experimental psychology and psychophysiology. If we continue in this historical account, we can observe that Claude Bernard's theoretical view on the constancy of the internal environment became entirely understandable, regarding the chemical properties of blood, in terms of biophysical chemistry, thanks to the American physiologist Lawrence Henderson, who devised the equations governing blood equilibria, certainly one of the major theoretical results in classical biology. The same kind of demonstration could apply to many fields of biology.

So why do we continue asking the same question? Is it because of our deep reluctance to change our mental habits, to modify our cherished attitudes – as another great Viennese scientist, Erwin Schrödinger, once said in the concluding paragraphs of his Nobel Lecture in 1933 "Der Grundgedanke der Wellenmechanik": "mit solchen alten, lieben und unentbehrlich scheinenden Begriffen wie wirklich oder bloss möglich" etc. (with our old, cherished and apperently indispensable concepts like real and simply possible etc.)?[1] Schrödinger was certainly an extremely critical mind. One of the reasons why we continue asking the same question about reductionism is perhaps a cognitive one: this would be because we presently do not have any satisfactory mean to reconstruct, or to visualize simultaneously all aspects of a given biological reality, to go from a very partial view of the parts to a fully integrated view of the whole. Another reason would be more ontological : this would be the peculiar multilevel structure of organisms which requires a much more sophisticated account than the old, nineteenth century type of reduction.

Marcel Weber asks the question: "Why Rosenberg's defence of reductionism fails", in spite of strong arguments? The argument, that the principle of natural selection should be considered as a physical law should be considered very closely. Indeed, such a view has been discussed in classical contributions to biophysical chemistry. In 1979, Manfred Eigen and Peter Schuster published their famous memoir *The Hypercycle. A Principle of Natural Self-Organization.* In this memoir they devised the mathematical theory of a new kind of biochemical cycle, the hypercycle. True, biochemical as well as chemical cycles were well-known much before Eigen and Schuster's work. The complication introduced by Eigen and Schuster is that instead of dealing with catalytic systems, which are already cyclic, they dealt with autocatalytic, self-replicative systems, which they called hypercyclic. A hypercycle is a cyclical arrangement of cyclic units. A catalytic hypercycle is "a system which connects autocatalytic or self-replicative units through a cyclic linkage".[2] This means clearly a new level of organization. Hypercycles have particular, emerging properties. They share with self-replicative units the property of conserving a certain amount of information, which is a prerequisite for Darwinian selection and evolution. They have additional integrating properties, which enhance their selective power. Indeed, according to Eigen and Schuster, "they com-

1 Erwin Schrödinger, *Was ist ein Naturgesetz? Beiträge zum naturwissenschaftlichen Weltbild.* München–Wien: R. Oldenbourg 1967 p. 100.

2 Manfred Eigen and Peter Schuster, *The Hypercycle. A Principle of Natural Self-Organization*, Berlin–Heidelberg–New York: Springer Verlag 1979, p. 6.

pete even more violently than Darwinian species with any replicative entity not being part of their own. Furthermore, they have the ability of establishing global forms of organization as a consequence of their 'once-forever- selection behavior, which doesn't permit a coexistence with other hypercyclic systems, unless these are stabilized by higher order linkages".[3] These hypercycles are dynamical systems which are described by non-linear differential equations, with their typical behaviors in terms of limiting cycles and attractors.

According to Eigen and Schuster, hypercycles are necessary prerequisites of Darwinian systems. Does this mean that natural selection is a general law of nature? Certainly not. "What is the molecular basis of selection and evolution? Obviously, such a behavior is not a global attribute of any arbitrary form of matter but rather is the consequence of peculiar properties which have to be specified". These properties are dynamical ones. We can conclude that "systems of matter, in order to be eligible for selective self-organization, have to inherit physical properties which allow for metabolism and for self-reproduction. These requirements are indispensible".[4] It is perfectly true that Eigen and Schuster developed the concept of molecular Darwinism. It is certainly not true that they made natural selection a general law of nature. At the biochemical level, the systems which are described by Eigen and Schuster are highly elaborated, like enzymes etc. So I am perfectly in agreement with Weber's criticism of Rosenberg.

Now I would like to make an additional comment on the notion of function and on the problem of "individuating the parts". Biochemistry, physiology, and even more so neurophysiology are characterized not only by functionality, as you may define it in a very abstract way, but mainly by polyfunctionality and linkage between functions, higher order linkages (like hypercycles or other kinds of linkages like allosterism) which you may find already at the level of single molecules. These linkages may be perfectly well described in many ways, including fundamental thermodynamics (this was the work of Jeffries Wyman, a biophysical chemist, in the sixties and seventies).[5] Polyfunctionality is a kind of rule. Proteins of course are the most intensely studied examples of these behaviors. And it cannot be otherwise, because if no linkage between functions, no function at all (as we know already from the hypercycle model). This makes the question of defining functions even more complicated, and the necessity of a more global view more evident. I agree on this point with Marcel Weber rather than with Rosenberg.

If we look at the questions under discussion, reduction, function, and law, all these terms are typical of nineteenth century science. More recently, biologists started to discuss emerging properties of various kinds, levels of organization, polyfunctionality, models and simulations, computational complexity, all features (sometimes unsolved like the computational complexity of biological structures)

3 Ibid., p. 6.

4 Ibid., p. 8.

5 See Claude Debru, *L'esprit des protéines. Histoire et philosophie biochimiques*, Paris: Hermann 1983.

which are the result of a much deeper understanding in biological sciences. Philosophers should become more acquainted with that and go to the lab. Indeed, philosophers could be extremely useful if they would cope with real, particular biological problems (which they certainly can do in a creative fashion) – for instance (speaking of functions) they could cope with the problem of unknown functions. Indeed there are physiological processes whose functions are still unknown, and this is particularly the case in neurophysiology, a rapidly changing field of investigation. The case of paradoxical (or rapid eye movement) sleep, which corresponds to dreaming, is a good example of that.[6] Paradoxical sleep is certainly most important for physiological regulations in the brain. But its functions remain presently entirely unknown – although there are many speculations. Its complicated biochemical mechanisms are more or less entirely or almost entirely known at the molecular and cellular levels, but the output of these processes is still unknown, so that physiologists cannot define the functions of paradoxical sleep mechanisms. This remains a subject for future physiology. Philosophers could be extremely useful in this enquiry. They could collaborate with neuroscientists as they did already in the past with much success (the best known example is Popper and Eccles). But this is not a matter of "philosophy of". Rather this is philosophy pure and simple. Speculating about possible functions for paradoxical sleep is not doing "philosophy of" biology, but rather doing philosophy within physiology. We should put again philosophy pure and simple on the agenda – together with epistemology, which is more modest and conveys a sense of modifying questions and varying hypotheses and models. I don't think we can look at biology as an entirely unified science, even from a philosophical point of view. Natural selection is certainly a major part of the game. Chemistry is another part of the game, with its own rules and constraints. I once heard a Strasbourg chemist discussing François Jacob's idea of molecular tinkering and making precisely this point. Certainly nature cannot play any game in biology. We should try to combine several kinds of causes and look more closely at the stabilizing role of emerging properties.

Department of Philosophy
Ecole Normale Supérieure
45 rue d'Ulm
75005 Paris
France
claude.debru@ens.fr

6 See Claude Debru, *Neurophilosophie du rêve*, Paris: Hermann 2006.

Thomas A.C. Reydon

How Special Are the Life Sciences? A View from the Natural Kinds Debate

1. Introduction

Philosophers of the special sciences seem to find it important to ask whether or not particular groupings of things that feature in particular special sciences can be conceived of as natural kinds.[1] For example, a quick search of the philosophical literature of the past decades comes up with several dozens of papers targeting the question "Is … a natural kind?", many of these concerning kinds of emotions and the emotion category in psychology[2] and the category of concepts in psychology / cognitive science.[3]

Motivating these papers often are concerns about the *scientific status* of the field that studies the kind in question – i.e., concerns about issues such as the place that the field occupies within the whole of science, its independence from and relationships to other fields, its reducibility, its explanatory power and autonomy, its being a unified and self-contained field of work, etc. If it can be made plausible that the kind or kinds that constitute the specific objects of study of a particular field of investigation are *natural* kinds, the thought is, this can be taken as an indication that the field is a comparatively independent, autonomous and self-contained science that has its own proper place amidst the other sciences, that stud-

1 I use 'special sciences' in the sense of Jerry A. Fodor, 'Special sciences (or: the disunity of science as a working hypothesis)', in: *Synthese* 28, 1974, pp. 97-115.

2 E.g., Louis C. Charland, "The natural kind status of emotion", in: *British Journal for the Philosophy of Science* 53, 2002, pp. 511-537; Paul E. Griffiths, "Emotions as natural and normative kinds", in: *Philosophy of Science* 71, 2004, pp. 901-911; Paul E. Griffiths, "Is emotion a natural kind?", in: Robert C. Solomon (Ed.), *Thinking About Feeling: Philosophers on Emotions*. New York: Oxford University Press, 2004, pp. 233-249; Jesse J. Prinz, *Gut Reactions: A Perceptual Theory of Emotion*, New York: Oxford University Press, 2004, Chapter 4; Lisa F. Barrett, "Are emotions natural kinds?", in: *Perspectives on Psychological Science* 1, 2006, pp. 28-58; Alexandra Zinck & Albert Newen, "Classifying emotion: A developmental account", in: *Synthese* 161, 2008, pp. 1-25.

3 E.g., Edouard Machery, "Concepts are not a natural kind", in: *Philosophy of Science* 72, 2005, pp. 444-467; Edouard Machery, "How to split concepts: A reply to Piccinini and Scott", in: *Philosophy of Science* 73, 2006, pp. 410-418; Edouard Machery, "100 years of psychology of concepts: The theoretical notion of concept and its operationalization", in: *Studies in History and Philosophy of Biological and Biomedical Sciences* 38, 2007, pp. 63-84; Gualtiero Piccinini & Sam Scott, "Splitting concepts", in: *Philosophy of Science* 73, 2006, pp. 390-409.

F. Stadler (ed.), *The Present Situation in the Philosophy of Science*,
The Philosophy of Science in a European Perspective, DOI 10.1007/978-90-481-9115-4_14,

ies its own specific domain of phenomena and that devises its own explanatory theories to account for these phenomena. Consider the statement by psychologist Zenon Pylyshyn that the presence of natural kinds specific to cognitive science was one of the factors that could open up an "(…) exciting possibility: the prospect that cognitive science is *a genuine scientific domain* like the domains of chemistry, biology, economics, or geology".[4]

But there are clear problems with this way of analyzing the scientific status of fields, as the conclusions that one will draw about the scientific status of a field of work depend on the particular account of natural kindhood that one adopts. Even if there were a generally accepted theory of natural kinds, it would not be clear exactly in which ways the scientific status of a field of investigation is affected by its "having" natural kinds or not. And as present-day philosophy is still lacking a generally accepted theory of what it means for a group of things to be a natural kind, it seems that the notion of 'natural kind' cannot be of much use when assessing the status of a particular field or a cluster of sciences, such as the life sciences. Indeed, Hacking recently argued that the notion of 'natural kind' won't be of much use when addressing *any* issue of philosophical interest, because not only there is no general theory of natural kinds but there is no agreement on what the *problem* is that a theory of natural kinds is supposed to resolve.[5] Rather, Hacking contended, framing philosophical questions in terms of 'natural kinds' only serves to make things unnecessarily complicated.

Against this background, my question is whether the relations between the life sciences and other scientific fields – i.e., the place of the life sciences among the other sciences – can be meaningfully characterized using the notion of 'natural kinds'.[6] Contra Hacking (who provided a general argument and did not talk about

4 Zenon W. Pylyshyn, *Computation and Cognition: Toward a Foundation for Cognitive Science*. Cambridge (Mass.): MIT Press, 1984, p. xi; emphasis added. For more recent statements to this extent, involving 'emotion' and 'concept' as denoting natural kinds of psychology or cognitive science, see Charland, *loc. cit.*, p. 512, Griffiths, "Emotions as natural and normative kinds", *loc. cit.*, p. 901, or Machery, "100 years of psychology of concepts: The theoretical notion of concept and its operationalization", *loc. cit.*, p. 66.

5 Ian Hacking, "Natural kinds: rosy dawn, scholastic twilight", in: Anthony O'Hear (Ed.), *Philosophy of Science (Philosophy – Royal Institute of Philosophy Supplement 61)*, Cambridge: Cambridge University Press, 2007, pp. 203-239. Cf. Paul M. Churchland, "Conceptual progress and word/world relations: in search of the essence of natural kinds", in: *Canadian Journal of Philosophy* 15, 1985, pp. 1-17 (p. 1).

6 Talk about the scientific status of a field and its place in science is often thought of as referring to the same set of issues – see, e.g., Mayr's long discussion of the place of the biology among the sciences (Ernst Mayr, *The Growth of Biological Thought: Diversity, Evolution, and Inheritance*. Cambridge (Mass.): Harvard University Press, 1982, pp. 21-82). Here, this question was pre-given, as this paper was written for the first conference of the ESF research network *The Philosophy of Science in a European Perspective*, in the context of Team B's work on the theme "Approaches to the Foundations of Science: The Place of the Life Sciences". Work on this paper was supported by

natural kinds in specific relation to scientific status), I hold that natural kind theory *can* be a useful tool when attempting to characterize the place of a scientific field of work within the whole of science and its relations to other sciences. My aim, then, is programmatic: to achieve clarity about *how* natural kind theories could be used fruitfully in this context by critically reviewing the principal ways in which they have been used by others, and in so doing set the stage for future work.

I shall begin by examining two different ways in which philosophers have tried to draw consequences about the scientific status of fields of investigation from considerations about the kinds that feature in them (Sections 2 and 3). While there are quite a few theories of natural kinds available in the literature, the theories that I shall look at represent the two most basic views of what natural kinds are. The discussion will show that framing questions about the place of a science in terms of 'natural kinds' has failed to yield unequivocal conclusions, instead complicating the philosophical discussion on the topic. So, this part of my discussion supports Hacking's pessimistic view of the usefulness of natural kind theory. I shall, however, go on to argue that the situation is not as problematic as it looks and that there might be ways of making natural kind theory useful in the present context (Section 4). In Section 5, I shall conclude with some remarks on the "place question" for the case of the life sciences (broadly conceived) against the background of the view suggested in Section 4.

2. The essence-view of natural kinds

Perhaps the most widespread account of natural kinds is what I (for obvious reasons) shall call the "essence-view" of natural kinds. On this view, a clear distinction can be made between "good" natural kinds that are characterized by kind essences and other kinds of kinds that are not characterized by kind essences. This dichotomy of kinds, in turn, is often taken as underwriting a view of science according to which the realm of science divides into two distinct domains, those sciences that are built around "good" natural kinds and those that aren't. The former are often taken to be the sciences pitched at the most fundamental levels of organization (physics and possibly some fields of chemistry), while the latter are thought of as being those fields of work that study phenomena at higher levels of organization (chemistry, the life sciences, psychology, cognitive science, the social sciences, economics, etc.).

The dichotomy between natural kinds and other kinds of kinds is common with authors who defend some form of traditional essentialism about kinds. The tradition can roughly be traced as follows. It originated with Plato and Aristotle, is prominent in, among others, Locke's discussion of "real essences" in his *Essay Concerning Human Understanding* and Mill's discussion of kinds in the *System of*

grants (RE 2613/1–1, RE 2613/1-2) from the German Research Council (DFG).

Logic, was revived in contemporary philosophy in Kripke's and Putnam's works on the causal theory of reference and has its most recent manifestations in metaphysical positions such as Ellis's *Scientific Essentialism* and Oderberg's *Real Essentialism*.[7] According to this tradition, every natural kind is characterized by its own specific kind essence, a kind essence being conceived of as a set of properties (or possibly a single property) that all and only the members of the kind in question possess. Possession of each of the properties included in a kind's essence is separately necessary and possession of all of them is jointly sufficient for being a member of the kind. On this view, natural kinds are spatiotemporally unrestricted in the sense that a natural kind can in principle have members anywhere and at any time in the universe where the right conditions for their existence obtain.[8] However, because of their strict membership conditions natural kinds can only be found at those levels of organization where things *really are* neatly ordered into disjoint groups in such a way that for every group of things there is a kind essence. As one contemporary proponent of kind essentialism put it, in order for there to be natural kinds, the borders between kinds must be drawn by nature, not by us.[9] Even if in practice we might be unable to identify the kind essences of putative natural kinds, we should at least have good reasons to assume their existence.

On this view of what it means to be a natural kind, genuine natural kinds are hard to find. Given what the sciences tell us about what sorts of things exist in the world, we have reasons to believe that "proper" kind essences can be identified only for kinds on the most fundamental levels of organization, that is, presumably the organizational levels of elementary particles, atoms and comparatively simple molecules (for macromolecules it already becomes unclear whether there are sufficiently strict borders drawn by nature). In addition, many of the kinds that feature in scientifically important generalizations in the special sciences are such that

7 Brian Ellis, *Scientific Essentialism*. Cambridge: Cambridge University Press, 2001. David S. Oderberg, *Real Essentialism*. London & New York: Routledge, 2007. For a historical sketch, see e.g., John Dupré, "Natural kinds", in: William H. Newton-Smith (Ed.), *A Companion to the Philosophy of Science*. Oxford: Blackwell, 2000, pp. 311-319. I realize that some authors disagree with my rough historical sketch and trace the philosophical tradition of discussion about natural kinds through the history of philosophy in a different manner (e.g., Ian Hacking, "A tradition of natural kinds", in: *Philosophical Studies* 61, 1991, pp. 109-126; Richard N. Boyd, "Realism, anti-foundationalism and the enthusiasm for natural kinds", in: *Philosophical Studies* 61, 1991, pp. 127-148). Furthermore, I realize that there are important differences between the various positions that have been defended and are defended by authors who stand in the essentialist tradition. But these are issues that I cannot take up in the present paper.

8 This is why Millikan calls such natural kinds "eternal kinds", distinguishing them from kinds that cannot have members at any time or place (Ruth G. Millikan, "Historical kinds and the 'special sciences'", in: *Philosophical Studies* 95, 1999, pp. 45-65; Ruth G. Millikan, *On Clear and Confused Ideas: An Essay About Substance Concepts*. Cambridge: Cambridge University Press, 2000).

9 Ellis, *loc. cit.*, pp. 19-20.

their members cannot occur anywhere and at any time in the universe, but only at specific times and places. Biological species are a case in point. While species have often been considered one of the paradigm examples of natural kinds, the members of a species cannot occur at any time or place in the universe where the conditions for their existence happen to obtain, as they must stand in one continuous reproductive lineage in order to count as members of the same species. Moreover, biological theory gives us reasons to believe that there are no kind essences for biological species, as variation among the members of a species is crucial for evolution to occur. The fact that the organisms of a species can in principle vary in any of their traits, however, conflicts with the idea that there should be one or several essential properties that all organisms of a species have in common.

Although on this view many (or most, or all) of the kinds that the special sciences study seem not to be genuine natural kinds at all, assessments of the scientific status of fields based on whether they study essentialist natural kinds tend to remain inconclusive. Still, a consequence that is often drawn is that the special sciences are qualitatively different from those fundamental fields of science that center round natural kinds. For example, on the view that one of the primary aims of science is to provide us with an inventory of the natural kinds that make up the furniture of the world and to uncover their innermost natures,[10] one should conclude that the fields that do not study natural kinds are more marginal to realizing the aims of science than the fundamental sciences. If one thinks that citing the natural kind membership of things has a special explanatory force – for example, if one agrees with Ellis's position that the laws of nature must ultimately be grounded in natural kinds –, one should conclude that the sciences that do not study natural kinds lack this special explanatory force. For the life sciences, this is sometimes taken to imply that most of them are "sciences of case studies", i.e., fields of investigation that do not devise general explanatory theories, but rather account for individual cases by relying on the explanatory power of other sciences.[11]

However, traditional kind essentialism can be criticized on at least two counts. First, it rests on questionable a priori assumptions about what it is for a group of things to be a natural kind. Apart from being rooted in a venerable, long-standing philosophical tradition, the idea that the furniture of the world comes in natural

10 A view that used to be held more widely by past philosophers than it is today (cf. Dupré, *loc. cit.*, p. 311; Stathis Psillos, *Philosophy of Science A-Z*. Edinburgh: Edinburgh University Press, 2007, p. 156).

11 Rosenberg argued that most of biology consists of sciences of case studies that do not have their own explanatory power but use the explanatory power of the physical sciences, the principle of evolution and the laws of molecular biology (Alexander Rosenberg, *The Structure of Biological Science*. Cambridge: Cambridge University Press, 1984, pp. 202, 211, 219-225). For a similar point about ecology, see Kristin S. Shrader-Frechette & Earl D. McCoy, *Method in Ecology: Strategies for Conservation*. Cambridge: Cambridge University Press, 1993.

kinds, each of which individuated by its own kind essence (in the strict sense discussed above) does not seem to have much to support it. Ultimately, this idea remains an a priori assumption about what the world is like that can be rejected as easily as it can be accepted.

In addition, traditional kind essentialism fails to provide something an adequate theory of natural kinds should provide, namely an explanation of what it is that makes members of the same kind similar. Let me use a well-worn example to illustrate this. According to traditional essentialism, the kind essence of gold is the atomic property of having 79 nuclear protons. This property, it is held, explains the other properties that gold atoms exhibit: the various material properties of gold atoms and their characteristic behaviors in various circumstances causally flow from their having 79 nuclear protons. But this does not explain why different gold atoms have *similar* properties – rather, it shifts the question from why gold atoms have similar observable properties and behaviors to why gold atoms have a similar essential property that underlies their observable properties and behaviors. Presumably, this problem traces back to the roots of the tradition of kind essentialism. For both Plato and Aristotle, kind essences were given aspects of nature that were not themselves in need of deeper explanation. Kind essences were considered to be basic features of the world, either existing as Platonic ideas that are imperfectly instantiated by material things or as Aristotelian forms that are immanent in material things as one of their conditions of existence. However, as Millikan rightly pointed out, a task of natural kind theories is not to explain why things have the observable properties that they have (which is what kind essences explain) but why members of the same kind have *the same* or highly *similar* properties.[12] And here, explaining similarity in higher-level properties in terms of sameness of underlying essence amounts to begging the question.

In sum, the essence-view of natural kinds gives rise to a view of science as divided into "genuine" sciences built around natural kinds and other fields not built around natural kinds and hence not to be considered "genuine" sciences. But as the essence-view is problematic in a number of ways, the ensuing view of science should be doubted.

3. The law-view of natural kinds

A second widely held view of what natural kinds are links natural kinds to laws of nature. Perhaps the most prominent example of an argument that uses this view to derive conclusions about the scientific status of fields of investigation is Fodor's argument against the idea that science is unified.[13] In his argument, Fodor

12 Ruth G. Millikan, "Response to Boyd's commentary", in: *Philosophical Studies* 95, 1999, pp. 99-102 (p. 100).

13 Fodor, *loc. cit.*

starts from the basic assumption that every scientific field of work is organized around one or a number of natural kinds that are specific to the field in question. According to Fodor, a particular scientific field is individuated to a large extent by the predicate terms that feature in the field in question but not in other fields. Those predicate terms that feature in the laws that are proper to the field in question, Fodor holds, denote the field's own natural kinds: "roughly, the natural kind predicates of a science are the ones whose terms are the bound variables in its proper laws".[14] The notion of 'law' is taken here in a suitably broad sense: the laws of a field of work are its explanatory generalizations, where both non-universal generalizations and universal laws are counted.[15] Fodor's conception of natural kinds is widespread: it is often claimed that natural kinds and laws of nature are inseparably connected in that natural kinds are those kinds that are mentioned in laws of nature and, vice versa, laws of nature are those generalizations that reach over natural kinds. On this view, a science like biology would be individuated by the predicate terms that are typical for biological discourse, which include general kind terms such as 'species' or 'gene', as well as specific kind terms such as '*Pan troglodytes*' or '*Drosophila melanogaster antennapedia* gene'.

Fodor's argument hinges on the relations that obtain between the natural kinds of a particular special science and those of lower-level sciences, typically physics. A necessary requirement for a field of science to be reducible to physics, Fodor holds, is that each of the proper natural kinds of the field can be reduced to a natural kind of physics by way of a 1–1 mapping.[16] That is, each natural-kind denoting predicate term S_n of the field that is to be reduced must be connected with a natural-kind denoting predicate term P_n of the reducing field (here, physics) by means of a bridge law of the form $S_n x \leftrightarrow P_n x$.[17] But, Fodor argues, for many of the kinds that feature in special sciences this requirement is not satisfied: the special sciences commonly make important generalizations over kinds whose members are not all of the same physical kind. Fodor observes that many special sciences use *functionally* defined kinds as the bases of generalizations and argues that such kinds typically cannot be mapped in a 1–1 way onto physical kinds because functions tend to be multiply realizable. That is, in many cases the function that individuates a special-science kind can be performed by various entities with different material structures that – because of their structural differences – do not constitute a single natural kind of physics.[18]

14 *Ibid.*, pp. 98, 102.

15 Millikan, "Historical kinds and the 'special sciences'", *loc. cit.*, p. 55.

16 *Ibid.*, pp. 102, 104.

17 S_n and P_n are predicates specific to the special science and to physics, respectively, and $S_n x$ and $P_n x$ are sentences of the form "x is a S_n" and "x is a P_n". The bridge law $S_n x \leftrightarrow P_n x$ thus states that all things that are S_n's are also P_n's and vice versa, in this way mapping a natural kind of the special science under consideration onto a natural kind of physics in a 1-1 manner.

18 Functions are said to be *multiply realizable* in the sense that a particular function can

Consequently, the fields that focus on such kinds are not fully reducible to physics and are to be considered as to some degree self-contained fields of work that occupy an autonomous position among the other sciences. As many of the kinds that feature in the life sciences are functionally defined, this would have important consequences for the status of the life sciences: in contrast to the analysis presented in Section 2, Fodor's analysis would lead to the conclusion that the life sciences are autonomous sciences that do not depend for their explanations on the physical sciences. Even though many special-science kinds are defined by means of different sorts of properties than physical-science kinds (functional vs. non-functional properties), all are proper natural kinds that constitute the focal points of their own sciences. Hence – and this is Fodor's general conclusion –, science is not a unified phenomenon:

> there are special sciences [...] because of the way the world is put together: not all natural kinds (not all the classes of things and events about which there are important, counterfactual supporting generalizations to make) are, or correspond to, physical natural kinds.[19]

As a general critique of Fodor's argument I want to point to some assumptions that underlie it. Fodor simply posits that the bound variables in the predicative sentences specific for a particular field of science, importantly including those bound variables that refer to *functional* properties, denote that field's natural kinds.[20] But it is unclear whether this assumption holds: most philosophers agree that because functions are often multiply realizable, functionally defined kinds generally aren't natural kinds.[21] Thus, not all predicate terms that appear in scientific generalizations necessarily refer to natural kinds. If this consensus view is correct, the question is whether the (ir)reducibility of a particular science's functional kinds to

often be realized by entities with diverging material structures. Conversely, material entities tend to be *multiply functional* in that an entity with a particular material structure usually is able to perform different functions when placed in different contexts.

19 *Ibid.*, p. 113. In addition to this metaphysical argument, Fodor makes the epistemological argument that it is not even important whether or not the physical descriptions of the kind's members are the same, as this does not affect the epistemic importance of the generalizations that are made over the kind in the special science that uses it (*Ibid.*, p. 103).

20 For a similar assumption in the context of a similar argument, see Harold Kincaid, *Individualism and the Unity of Science: Essays on Reduction, Explanation, and the Special Sciences*. Lanham, MD: Rowman and Littlefield, 1997, p.75.

21 According to some authors, however, there is no *intrinsic* difference between functional and natural kinds and at least *some* functional kinds should be recognized as natural kinds (e.g., Richard N. Boyd, "Kinds, complexity and multiple realization", in: *Philosophical Studies* 95, 1999, pp. 67-98 (pp. 92-96); Ingo Brigandt, "Natural kinds in evolution and systematics: Metaphysical and epistemological considerations", in: *Acta Biotheoretica* 57, 2009, pp. 77-97; Thomas A.C. Reydon, "How to fix kind membership: A problem for HPC-theory and a solution", *Philosophy of Science*, forthcoming).

natural kinds of physics has any bearing on the relationship between that science and physics. For reducibility presumably is an issue that arises between the *same* kinds of elements of two different sciences: the question is whether the theories, laws, explanatory generalizations, natural kinds, etc. of one science are reducible to the theories, laws, explanatory generalizations, natural kinds, etc. of another science. If functional kinds are categorically different from natural kinds, trying to reduce the functional kinds of one science to the natural kinds of another science amounts to a category mistake – and nothing much can be concluded from the irreducibility of the *functional* kinds of one field to the *natural* kinds of another field. The upshot is that as long as it has not been established that Fodor is correct in admitting functionally defined kinds to the natural kind fold, his argument rests on a questionable assumption.

A second – and in the context of the sciences of life and mind probably more forceful – criticism of Fodor's argumentation is that it crucially depends on assumptions about the nature of the functions that individuate kinds in the special sciences. For the degree to which functions *actually are* multiply realizable seems to depend on what one takes functions to be. On one important notion of biological function, that of function as selected effect, functional kinds seem multiply realizable only in a very weak sense, as an entity's function is inseparably linked to the specific selectional history of *that particular* entity. An entity's selected-effect function is that activity that *its* ancestors were originally selected for – thus, a selected-effect function by definition is a function that can only be performed by evolutionarily related entities that share the same ancestors.[22] And evolutionarily related entities tend to have very similar (though not *exactly* the same) material structures precisely because they stand in one line of descent. This is different when one adopts a notion of function as activity or as causal role, where whether different entities can perform the same function is independent of their relatedness.[23]

22 What about convergent evolution, i.e., cases in which the same function has evolved independently in a number of lineages? According to Millikan ("Historical kinds and the 'special sciences'", *loc. cit.*, pp. 59-60) a kind individuated by a convergently selected function would not constitute a natural kind over which scientifically useful generalizations hold, as the mere fact that things perform the same function (causal role) does not make it the case that they are similar in any other respects. I would suggest that in cases of convergent evolution of the same causal role function there are multiple functional (as selected effects) kinds, each individuated by one particular selectional history of one particular lineage. Thus, 'vertebrate lens eye', 'insect facet eye' and 'trilobite crystalline eye' would denote different functionally defined natural kinds of biology that presumably would need to be further subdivided into more specific natural kinds.

23 For an in-depth discussion of these different notions of function – not considering multiple realizability, however –, see Arno G. Wouters, "Four notions of biological function", in: *Studies in History and Philosophy of Biological and Biomedical Sciences* 34, 2003, pp. 633-668; Arno G. Wouters, "The function debate in philosophy", in: *Acta*

So, let me again turn to the life sciences. What kinds of functions play central roles in the different fields of life science? According to some authors, the life sciences are primarily – or perhaps exclusively – concerned with selected-effect functions.[24] Others hold that the functions that individuate kinds in the life sciences oftentimes are causal-role functions or activity-functions, so that one needs to distinguish between different notions of function that biologists use, depending on the particular field that they work in.[25] It is unlikely that this debate can be resolved by means of a global argument about the notion of 'function' in the life sciences – and even less likely when the notion of 'function' throughout the whole of science is considered. Rather, a case-by-case analysis seems more appropriate, explicating the notion of function that is involved in every individual case in which life scientists talk about the functions of traits or entities and group things into functionally defined kinds. This means that an overall analysis à la Fodor of the reducibility of functionally defined life-science kinds to kinds of physics must fail.

In sum, the law-view of natural kinds gives rise to a view of science as encompassing many distinct fields, physical as well as special sciences, that all center around their proper natural kinds (featuring in their proper explanatory generalizations) that cannot be mapped onto one another in a 1–1 manner. All these different sciences thus have their own place within the whole of science, standing side by side as comparatively self-contained and autonomous fields of investigation. But as the law-view is problematic in a number of ways, the ensuing view of science is to be doubted.

4. A possible way out

I have outlined and criticized two basic views of what natural kinds are and pointed to their disparate implications for assessments of the place of a field of work within the whole of science. Apparently, the conclusion must be that answers to questions of the form "Is ... a natural kind?" and consequences drawn from them for the scientific status of a particular field of work are vacuous as long as there is no generally accepted theory of natural kinds. Different views of what it is to be a natural kind lead to different answers and hence to different assessments of the status of the field in question. Moreover, if different authors base their assessments on different accounts of natural kindhood this will yield incompatible results: while

Biotheoretica 53, 2005, pp. 123-151.

24 E.g., Karen Neander, "Functions as selected effects: The conceptual analyst's defense", in: *Philosophy of Science* 58, 1991, pp. 168-184; Millikan, "Response to Boyd's commentary", *loc. cit.*

25 It seems that at least two notions of 'function' play a role in biology (Kim Sterelny & Paul E. Griffiths, *Sex and Death: An Introduction to Philosophy of Biology*. Chicago & London: University of Chicago Press, 1999, pp. 223-224 and references therein), but probably more (Wouters, *loc. cit.*).

one author uses view A to argue that field F is a genuine, autonomous science, another author uses view B to conclude that field G merely consists in one case study after another. It is unlikely that this will yield much clarity about the places of different fields within the whole of science. Under these circumstances it seems that Hacking was right to state that framing philosophical questions (in this case the "place/status question") in terms of 'natural kinds' doesn't do much besides complicating the problem and making it harder to find good answers.

However, the two views discussed above are not as disparate as it might seem. Both attempt to do justice to widespread intuitions about what it means to be a natural kind: the idea that we group things into kinds in such a way that these groupings represent some aspects of the natural state of affairs and that *precisely because* of this, these groupings are suitable for generalizing about. The essence-view gives priority to the metaphysical intuition that natural kinds represent aspects of whatever order there is in the world, or that natural kinds in some sense exist "out there" in nature.[26] That we can make generalized statements about natural kinds is secondary, as it is a consequence of what makes a group of things into a natural kind in the first place. The law-view gives pride of place to the epistemological intuition that natural kinds are just those groupings of things that useful generalizations refer to.[27] The law-view encompasses the often assumed intimate link between laws of nature and natural kinds, according to which laws reach over kinds and kinds are referred to by the predicates that feature in laws, but goes further by conceiving of laws in a non-strict sense (recall that according to Fodor the laws of a science are just its explanatory generalizations). The "specialness" of natural kinds thus lies in the fact that we refer to them in important generalizations; what metaphysically supports this epistemic success is secondary, as there probably are a great many factors in nature that can do the job.

This difference, I want to suggest, is more a difference in *emphasis* than a deep incompatibility of positions.[28] Ultimately, the thought underlying the two

26 Cf. Dupré, *loc. cit.*, p. 311; Kathrin Koslicki, "Natural kinds and natural kind terms", in: *Philosophy Compass* 3, 2008, pp. 789-802 (p. 789).

27 For example: "natural kinds reflect a strategy of deferring to nature in the making of projectability judgments" (Boyd, "Realism, anti-foundationalism and the enthusiasm for natural kinds", *loc. cit.*, p. 139). "Natural kinds are scientific categories posited by our theories as *epistemological devices*; insofar as they have ontological status, it is as features of the ways in which causal structures in the world interact with our classificatory practices in such a way as to support reliable induction and explanation. The naturalness of natural kinds consists in their aptness for induction and explanation." (Roberto A. Keller, Richard N. Boyd & Quentin D. Wheeler, "The illogical basis of phylogenetic nomenclature", in: *Botanical Review* 69, 2003, pp. 93-110; p. 102). See also Philip Kitcher, *The Advancement of Science: Science without Legend, Objectivity without Illusions*. New York: Oxford University Press, 1993, p. 172; Boyd, 'Kinds, complexity and multiple realization', *loc. cit.*

28 See also Thomas A.C. Reydon, "Natural kind theory as a tool for philosophers of science", in: Mauricio Suárez, Mauro Dorato & Miklós Rédei (Eds): *EPSA07: Launch of*

views is the same. After all, if successful generalizations would not reflect aspects of whatever order there is in the world "out there", their success would remain a miracle. (One could accept that *occasionally* a generalization is successful by accident, but not that this happens all the time.) Conversely, if there *is* some order in the world, it would be surprising if this wouldn't be reflected to some extent in our generalizations. The question, then, is where emphasis *should* be placed when asking whether a particular grouping is a natural kind – on metaphysical or epistemological criteria for being a natural kind.

This question can be answered by examining the prospects for success of both approaches. Conceiving of natural kinds as those groupings of things for which a kind essence exists is, as explicated above, problematic for two reasons: it rests on a priori metaphysical assumptions about what it is to be a natural kind and it fails to adequately explain one of the main explananda of natural kind theory (namely, why the members of a kind are similar). I take these problems as fatal for any metaphysics-first approach. For beginning with setting up a metaphysics of natural kinds will *always* involve adopting some a priori criteria of what it is to be a natural kind that include some groupings and exclude others from being further considered. The metaphysics-first approach will, therefore, suffer from a certain degree of myopia about natural kindhood.

Taking natural kinds as simply those groupings over which we successfully generalize, however, entails exactly the opposite problem. On this view, any generalization that is sufficiently stable (i.e., has proven to hold without confronting too many exceptions) should be taken as referring to a natural kind and any group term featuring in a successful generalization thus should be taken as a natural kind term.[29] But this goes against the intuition that, as kinds represent whatever order there is in the world, any entity is at most a member of a limited number of kinds – that is, that the question "What kind of thing is this?" is susceptible only to a limited number correct answers.[30] And as there typically are innumerably many similarities between entities, counting every similarity that appears in a generalization as individuating a natural kind would allow every entity to be a member of innumerably many natural kinds. Thinking of natural kinds as simply those groups referred to in successful generalizations entails the risk of being too generous with

the European Philosophy of Science Association. Dordrecht: Springer, forthcoming.

29 That this is indeed what some advocates of this approach have in mind can be seen from Boyd's remarks that "we should *always* require the sort of semantic machinery indicated by the theory of natural kinds when our aim is induction and explanation […]. Kinds useful for induction or explanation must *always* "cut the world at its joints" in this sense: successful induction and explanation *always* require that we accommodate our categories to the causal structure of the world." (Boyd, "Realism, anti-foundationalism and the enthusiasm for natural kinds", *loc. cit.*, p. 139; emphasis added).

30 In the case of organisms, for example, "It's a *Drosophila melanogaster*.", "It's an insect.", etc., but not "It's a black thing.", "It's a flying object."

respect to allowing groups into the natural kind domain: if taken to its extreme, ultimately there will be no non-natural kinds left.[31]

Still, the epistemology-oriented view of natural kinds has better prospects for success than the metaphysics-oriented view. For one, as we do not have direct access to the natural kind structure of the world (if there is such a structure), the best we can do is examine those kinds that feature in successful attempts to generalize, systematize knowledge, explain phenomena, predict future events, etc. – i.e., in successful epistemic practices.[32] The problem is where to stop: should all useful generalizations be considered as being about natural kinds, or do only generalizations of a privileged sort refer to natural kinds and, if the latter, what makes these generalizations privileged? The question is how to truncate the domain of generalizations that refer to natural kinds in a non-arbitrary manner.

A useful heuristic might be the following. As Millikan pointed out, it surely is not the case that every single "good" generalization reaches over its own natural kind. If we take the intuition seriously that natural kinds in some way represent the furniture of the world, it is plausible that every natural kind should lie at the intersection of *many* generalizations.[33] The underlying thought is that kinds that come to feature in more and more distinct generalizations as time progresses will become established in the ontologies of the fields that use them. In contrast, kinds of which the usefulness persistently does not go beyond featuring in just one or two generalizations will ultimately become ontologically suspect. Although this does not hold exclusively for the sciences, but generally in all cases in which things are grouped into kinds, the kinds that feature in the sciences are typically of the sort over which many different generalizations can be made: "[a] science begins only when, at minimum, a number of generalizations can be made over instances of a single kind".[34] Millikan traced this insight back to John Stuart Mill, who wrote about kinds that

> a hundred generations have not exhausted the common properties of animals or plants … nor do we suppose them to be exhaustible, but proceed to new observations and experiments, in the full confidence of discovering new properties which were by no means implied in those we previously new.[35]

---

31 For a similar point, see Millikan (*On Clear and Confused Ideas: An Essay About Substance Concepts, loc. cit.*, p. 15). An example is Dupré's "promiscuous realism" about natural kinds (John Dupré, *The Disorder of Things: Metaphysical Foundations of the Disunity of Science*. Cambridge (Mass.): Harvard University Press, 1993; 'Natural kinds', *loc. cit.*, p. 318).

32 Boyd, "Kinds, complexity and multiple realization", *loc. cit.*, p. 69.

33 Millikan, "Historical kinds and the 'special sciences'", *loc. cit.*, p. 48; *On Clear and Confused Ideas: An Essay About Substance Concepts, loc. cit.*, p. 15, 17.

34 Millikan, "Historical kinds and the 'special sciences'", *loc. cit.*, p. 48.

35 Mill, quoted in Millikan, *On Clear and Confused Ideas: An Essay About Substance Concepts, loc. cit.*, p. 16. This is also the gist of Russell's assertion that a natural kind is a class of things that possess a number of properties that are not known to be logically

The suggestion, then, is to conceive of natural kinds as groups of entities that are similar in numerous respects, where these similarities are not accidental but due to causal factors in nature. That is, the members of a natural kind are similar *for a reason*.[36] The natural kinds of a particular field, then, are the kinds that lie at the intersections of *many* of that field's generalizations – not simply those that feature in *any* of the science's generalizations (as on Fodor's account), nor those that meet strict metaphysical criteria for being a natural kind. Natural kinds can thus be found by searching for kinds that feature in many generalizations and investigating which causal factors in nature underlie the similarities between the members of a putative kind. As whether a kind term features in many generalizations is a heuristic factor, the question how many generalizations count as "many" is not a serious problem. This way of thinking about natural kinds avoids the problems confronted by the essence-view (apriorism about the nature of natural kinds) and the law-view (counting all kinds referred to in generalizations as natural kinds) while doing justice to the central intuitions that underlie both these views.

5. How special are the life sciences?

So, how does this view of natural kinds work out when analyzing the place of the life sciences within the whole of science?

A first thing to note is that if featuring in many generalizations is taken as the principal indication of natural kindhood, it remains unclear how kinds and fields of work are related. As one kind can feature in generalizations in a number of fields of investigation, for every kind the question arises to which field it should be allocated. Conversely, for every field of investigation that is recognized, the question arises which are its proper natural kinds and which are more marginal because they actually are proper kinds of another field. For instance, does '*Drosophila melanogaster antennapedia* gene' denote a proper natural kind of genomics, genetics, developmental biology, or simply of biology? Presumably, the relation between fields and kinds needs to be assessed on a case-by-case basis. Such questions do not arise on Fodor's account, as Fodor assumes that a field is individuated by its own proper predicates. But if kinds typically feature in many generalizations, this cannot generally be the case, as there is no reason to assume that the generalizations that refer to a kind *K* will all feature only in one field.

---

interconnected (Bertrand Russell, *Human Knowledge: Its Scope and Limits*. London: George Allen & Unwin, 1948, p. 317). Millikan ("Historical kinds and the 'special sciences'", *loc. cit.*, p. 57) thinks of the "many generalizations" idea as spanning a continuum of clearly "good" natural kinds to clearly non-natural kinds. Note that Millikan (*ibid.*, pp. 47-48) presented her approach as a further elaboration of Fodor's.

36 Millikan speaks of a "real ground", "ontological ground", or a "ground in nature" that underlies the similarities between things of a kind; Millikan, *ibid.*, p. 50; *On Clear and Confused Ideas: An Essay About Substance Concepts*, *loc. cit.*, pp. 16-18, 25.

Before examining the kinds of the different fields of work that make up the life sciences, then, these fields themselves need to be individuated first. But now an obvious problem is that there is no standard list of all the various fields of life science. Simply using common sense and listing the fields that are mentioned in the standard textbooks won't do, as this might result in a too fine-grained or too coarse-grained list. Should our list read: ecology, physiology, genetics, ... – or should it be: organismal ecology, population ecology, community ecology, ecosystem ecology, evolutionary ecology, animal physiology, plant physiology, developmental genetics, comparative genomics, ...? Should philosophers examine the place of ecology among the sciences by looking at ecological kinds in general, or the places of organismal, population, community, etc. ecology by examining each of these fields' proper kinds separately? Or perhaps both? And what happens in cases in which one kind can be equally legitimately allocated to several fields? The answers depend on the criteria that one adopts for individuating scientific fields; and there do not seem to be any unequivocal criteria available. On one account, for example,

> a field is an area of science consisting of the following elements: a central problem, a domain consisting of items taken to be facts related to that problem, general explanatory factors and goals providing expectations as to how the problem is to be solved, techniques and methods, and, sometimes, but not always, concepts, laws and theories which are related to the problem and which attempt to realize the explanatory goals. A special vocabulary is often associated with the characteristic elements of a field. [...] Fields emerge in science, evolve, sometimes even cease to be. [...] Although any or all of the elements of the field may have existed separately in science, they must be brought together in a fruitful way for the field to emerge. Such an emergence is marked by the recognition of a promising way to solve an important problem and the initiation of a line of research in that direction.[37]

To be sure, this is a rather open-ended way of individuating fields that does not provide any strict criteria by means of which fields of investigation can be delimited or it could be determined which kinds are to be allocated to which fields.

The conclusion, then, must be that on the current state of affairs in philosophy the question "How special are the life sciences?" cannot be answered – at least not by relying on an analysis of the putative natural kinds that feature in the various fields of life science. The answer could emerge after a consensus emerges on which different fields of investigation make up the life sciences and the proper kinds of each field have been established. Quite likely, it will be different for different fields. There is, after all, no reason to assume that scientific status is an all-or-nothing affair (status often comes in degrees, after all) or that the place of a field within the whole of science will be the same for all fields of life science. To

37 Lindley Darden & Nancy L. Maull, "Interfield theories", in: *Philosophy of Science* 44, 1977, pp. 43-64 (pp. 44-45). For a sociological account of scientific fields, see Richard Whitley, *The Intellectual and Social Organization of the Sciences (second edition)*. Oxford: Oxford University Press, 2000.

be sure, this conclusion is unsatisfactory, as it means that using natural kind theory to analyze the place of the life sciences within the whole of science will involve a lot of work, most of which still remains to be done. In addition, the prospects of ultimately reaching an unequivocal answer are not clear. But I hope that at least some clarity has been achieved about what a project that attempts to use natural kind theory in this context should look like – and that it has become clear why asking "Is … a natural kind?" is not the right question to start from.

Center for Philosophy and Ethics of Science (ZEWW)
Leibniz Universität Hannover
Im Moore 21
D-30167 Hannover
Germany
reydon@ww.uni-hannover.de

Miles MacLeod

The Epistemology-only Approach to Natural Kinds: A Reply to Thomas Reydon

What I'm going to comment on here mostly are the issues of natural kinds that Thomas Reydon raises with respect to his 'epistemology-first' strategy, rather than specifically the problem of the special sciences, although is clear that it is with respect to the special sciences and the problem of field demarcation that any notion of natural kind has its most pressing application. I'm going to suggest a perhaps rather pragmatic extension of this strategy which would attempt to provide a degree of philosophical usefulness in this regard but avoid the traditional problems which I don't think we escape unless this strategy is pushed to the logical conclusion of an *epistemology-only* approach.

As Reydon mentions in his text, there is today a rather large shadow looming over those who trade in natural kinds. This has been cast by Ian Hacking, who has in his inimitable way declared the concept to be arbitrary in its many inconsistent formulations, to have failed in its chief ambitions of providing a general account of categorisation in science, and ultimately to be of no consequence to philosophy of science.[1]

But if we read Hacking closely what I think we find he is really objecting to is a unified notion of natural kind; namely, that some single unitary logical or metaphysical definition grounds the relations and properties of seemingly diverse class structures in science. "It is the idea of a well-defined class of natural kinds that has self-destructed. ..."[2] Yet it is precisely in the directions of more pluralistic and more naturalistic understandings of natural kinds away from essentialism that theorists have been heading. Hacking shows his sympathy to this. In the paper I am citing for instance he criticises almost everybody but seems willing to entertain Richard Boyd's Homeostatic Property Cluster conception for species and the like, although he thinks it not relevant for all potential natural kinds (which I don't believe was ever Boyd's claim).[3] He does however prefer to regard these more pluralistic approaches to natural kinds as simple evidence of the vacuousness of

1 Ian Hacking, "Natural Kinds: Rosy Dawn, Scholastic Twilight", in: *Royal Institute of Philosophy Supplement*, 82, 2007, pp. 203-239.

2 *Ibid.*, p.209.

3 The cluster theory goes back to papers such as Richard Boyd, "How to be a Moral Realist", in G. Sayre-McCord (Ed.), *Essays on Moral Realism*, Ithaca:Cornell University Press 1988, pp. 181-228; Richard Boyd, "What Realism Implies and What it Does Not", *Dialectica*, 43, 1989, pp.5-29; Richard Boyd, "Realism, Anti-Foundationalism, and the Enthusiasm for Natural Kinds", *Philosophical Studies*, 61, 1991, pp. 127-148.

F. Stadler (ed.), *The Present Situation in the Philosophy of Science*,
The Philosophy of Science in a European Perspective, DOI 10.1007/978-90-481-9115-4_15,

the natural kind concept. I don't think that's right. If the imperative is that natural kinds be shown to be philosophically useful in our accounts of science and solving the problems that emerge therein, then I think the best hope is an approach that is willing to accept that it is not confined to some unitary metaphysical account of what natural kinds are. While Hacking objects to the epithet 'natural' as applied to kinds, he does appreciate that kinds themselves have certain epistemic value in scientific practice in for instance induction and explanation, and it's exactly I believe along the lines of these observations that *natural kinds* can be given sense and seen to be useful conceptually, if not essential, to our understanding of scientific practice.

The approach to natural kinds which seems more promising in this respect, and I will give some good reasons for thinking this below, treats 'natural kinds' as first and foremost epistemic devices, as epistemically relevant groupings: that is, tools of inductive generalisation and explanation. It lets science decide in each instance through its own investigative processes and in its own language what the underlying reductive basis of this success might be. This is often a natural and central part of a field's investigative processes (investigating its own natural kinds), but it needn't be. Scientists might work at a level of explanation where reduction in this way is simply not the aim. The success itself justifies their belief in it and in turn, reliance on it. Reydon's epistemology-first approach fits him within this line of thought. So let me start by expressing why I think it makes sense to seek an answer to the problems of fields and the status of special sciences, and other problems too, through this epistemically based criterion for identifying kinds.

As Reydon points out a real problem with using natural kinds to resolve philosophical problem such as the status of fields in the special sciences, is the stark dependence this has on the notion of natural kind one chooses to support and the extremes that result. Essentialism includes almost nothing in the special sciences, a Fodorian approach almost everything.[4] Most would agree however that despite the inconclusiveness of the debate there is a sense to the concept that stems from its role and success in scientific practice, even if a precise ontological formulation of what it is to be a natural kind can't be given. This intuition gets lost perhaps by the attribution to natural kinds of content and structure that tries to classify or account for them in terms for instance of 'essences' or 'functions' which turn out ultimately to be problematic notions.

But the fact is there remains a practice of using kinds generally in science for epistemic purposes. And it is not just explanation and generalisation that gives the

4 See Jerry Fodor, "Special Sciences (or: the Disunity of Science as a Working Hypothesis)", *Synthese*, 28, 1974, pp. 97-115. On Fodor's approach "the natural kind predicates of a science are the ones whose terms are the bound variables in its proper laws" (p. 102). This of course leaves open the notion of what a 'proper law' is which Fodor admits he can't say, except that later in the text he argues that 'lawlikeness' itself depends on its relata being natural kind terms. It appears he has some kind of equivalence relation in mind between laws and natural kinds.

use of these kinds their epistemic character. They form the very conceptual basis of scientific systems of thought and reasoning and the basis upon which research is organised and directed, since it is in terms of kinds that research questions and issues of investigations are often realised and posed, and at the same time resolved by assigning new properties to them and forming new relations between them. They are *epistemic* in so far as they are a basis upon which scientific knowledge is sought for, obtained and formulated. Now if ultimately our intention is to understand the basis of various elements of scientific practice, such as the practice of organising science in fields, including what makes them distinct from one another, then I think it is natural to perceive that *natural kinds* when evaluated and identified in these epistemic terms are critical to this. After all a scientific field is in practice itself 'epistemic'. One way of looking at a field is as a grouping of phenomena on the basis of connections, patterns and relations that suggest there exist revealing generalisations and underlying explanations to this order. A persisting field is one in which such generalisation occurs and successful explanatory frameworks have developed. Discovering what defines or makes a field then depends on what underlies these generalisations and explanatory successes, and here of course is where we would anticipate the primary and active epistemic role of natural kinds as the bases of these.

I take it that this is a much more epistemic approach to fields and natural kind. But I think this is the way 'natural kind' emerges as a relevant concept. The governing presupposition is that the status of a field is not *directly* underwritten by metaphysical considerations, but by epistemic ones, such as its investigative and explanatory activity, and success in this regard.

This is a perspective which adds support to what Reydon is doing. On this basis I think he is right to think that there are good principles at least to favour the epistemological approach, as he calls it, in the attempt to discover a more astute and useful notion of 'natural kind'. His novelty in this regard is to pose heuristically, what we might plausibly think is also a standard for scientific practice for identifying 'natural kinds', which is the success of a kind in multiple different generalisations. Let me give some broader reasons for thinking that kinds for which there are multiple generalisations should be exactly the basis which is relied on in practice to identify 'natural kinds' since it underwrites the kinds of special epistemic roles they have, which includes the organisation of fields. After all if a kind can be associated with various different generalisations then there is an obvious established usefulness to the concept as a systemising element for which the role as a unifying principle no doubt gives it explanatory value … think of the kind 'acid' for instance which is explanatorily useful in many different circumstances because of the number of generalisations that are made about it. As Reydon puts it natural kinds are those kinds that lie at the 'intersection' of many of the field's generalisations, which suggests a centrality to those kinds in just this regard. At the same time, a point noticed by Millikan but not quite put in the epistemic terms I'm putting it, the success of many generalisations underwrites the expectation that the kind is a subject of investigation (by having a reality) and thus has further discoverable properties, relations, ultimately generalisations ... and thereby further explanatory applications.[5] In fact it's on the basis that we can explicate what the sense of 'natural' contributes to the concept in epistemic terms. Part of what 'natural' implies of a kind like phosphorous for instance is that there is always more to say of it and discover of it: that it is not exhausted by any set of properties or one description. As Millikan puts it, "science begins only when, at minimum, a number of generalisations can be made over instances of a single kind".[6] This identification of 'natural' with 'many generalisations' thus reflects, I would postulate, the decisions scientists themselves make in many cases about what counts as a natural kind which in turn governs how they employ it and rely upon it (particularly if, like Fodor, we think reductionism is not necessarily a regulative constraint on the establishment of natural kinds in scientific practice). If so then the 'multiple generalisations' test for identifying natural kinds picks out a significant aspect of the epistemic basis of such kinds, and by virtue picks out the groupings from which fields themselves acquire epistemic value and success, and present as productive scientifically. The meaning of 'natural' here then plays out in terms of the central belief scientists have in the kind and the corresponding application that is made of it as a result of those beliefs.

Of course all this is mostly supportive, but I think it compels a different but nonetheless logical outcome from the one Reydon aspires to. I suggest that to take the kind of approach I'm expressing one really has to be willing to concern oneself solely with the practice of natural kinds as so defined. Attempting to prescribe any kind of ontological criterion I think compromises this usefulness, because it will inevitably cut across this practical dimension, especially with respect to the special sciences.

Reydon expresses that the 'many generalisations' viewpoint is to be treated as a heuristic strategy for finding natural kind candidates, but not itself as a way of demarcating natural kinds. To find 'real' natural kinds we still must turn ourselves to discovering whether or not there is a *causal structure* underlying those kinds. There must be in other words some naturalistic explanation for the success of a potential natural kind term, i.e. some basis for treating its members as a group with which one can project and explain, to consider it a natural kind. I want to put however that if our aim is a useful notion of natural kind for fathoming scientific practice then this added condition is counterproductive. Firstly in many instances fields employ terms as natural kinds, relying on them in these sense above, without being able to reduce them or explain them at a different level their success. It is often taken for granted that there is an underlying basis which might be complex or multiply-realisable but has the coherence nonetheless to provide this success

5 Ruth Millikan, "Historical Kinds and the Special Sciences", *Philosophical Studies*, 95, 1999, pp. 45-65.

6 *Ibid.*, p. 48.

and make it a concept that can be relied upon further. Thus setting these kinds of conditions threatens our chance of accounting for one highly important aspect of practice; using kinds as natural kinds without reduction. Secondly I would take it however that there is still the problem of what to do about multiple-realisability. How do we treat kinds that are discovered to have more than one causal basis? Are they natural kinds or not? Eventually this is the kind of question one presumably has to resolve if one wishes to demarcate the special sciences 'metaphysically' by natural kinds. Which means we are right back somewhat where we started with these old familiar metaphysical problems. These however run adverse to the conception of a useful notion of natural kind. Certainly if your goal is describing the boundaries of a field, as opposed to say the scientific recognition of a field, then taking a position seems to predetermine the answer. Yet in terms of their epistemic roles generating inductions and explanations for instance the natural kinds function in similar ways in those deemed to be fields and those deemed not by a metaphysical standard. Surely the answer to what demarcates fields is a question of practice in respect of the use of concepts and the beliefs involved, but not one of metaphysics.

As Reydon points out Fodor's viewpoint, which treats functionally-defined kinds as natural kinds, is criticised because a functional kind seems like a different kind of thing from a natural kind and because 'functional' is not interpretable in any one precise way. But of course Reydon's heuristic itself would seem to accept that at least at the outset functional kinds, however one defines functional, as potential natural kinds because functional kinds might well also be kinds that are successfully involved in multiple generalisations. But if we attempt to pare kinds down by their causal basis in such a way then presumably we lose the connection we might otherwise be able to establish in a general way in scientific practice between the epistemic value of natural kinds and the demarcation and status of a field.

So I want to put I suppose the following question. If our aim is *contra* Hacking to have a useful concept of natural kind, why not just employ it as an epistemic category for which we cache out 'natural' in terms of the way in which scientists place their beliefs and use the concept by virtue of those beliefs. Natural kinds in this way are seen as tools of practice and are explanatory for philosophers as devices that explain how scientific practice functions, including its division into fields. Do we really need to more than this? Is it just counterproductive to expect more? We note this view isn't as wide-open as Fodor's because the epistemic criterion of many-generalisations is stricter and a more compelling basis for beliefs in an underlying reality to the kind. But we don't need to specify what this reality needs to consist in, or to put it another way, it's not our interests to do so.

Yet the rather obvious observation should be made that even though I think there's good reason to pursue this epistemological approach, as Reydon does, it seems to fail in the task set for it in his proposals here. Relying on a 'many generalisations' heuristic won't in fact *alone* help with the task of demarcating fields and giving them status thereby. It doesn't seem like fields are themselves simply built atop a set of refined natural kinds. After all identifying natural kinds by 'many generalisations of a field' requires first picking out what the *generalisations of a field* are. Let me say that I think the impression that a field in the special sciences is more complex than a set of natural kinds is surely true. One can't define a field simply in terms of them. But this doesn't stop us maintaining as I have above, that natural kinds are an essential part of the way a field operates as an epistemic unit and essential to any understanding of this. This is where the useful of the natural kind concept lies. And however we choose to define fields, natural kinds will be integrated essentially into this definition. After all a field may well have central problems but those problems themselves may well be problems of the natural kind structure, or at least expressed in the vocabulary of these kinds. The methodology of the field will itself organise itself around the kinds it considers fundamental and so on. Natural kinds represent a central part of the categorical structure with and through which a field's scientists interpret the world, organise and understand the phenomena. They guide how the world is further investigated. Obviously the theory we need must be sensitive to the complexity involved, but natural kinds as part of the basic epistemic structure of a field will surely be part of the tools of philosophers for understanding the 'field' as a unit of scientific practice and finding what gives one status, at least for the scientists involved.

Initiativkolleg ‚Naturwissenschaft im historischen Kontext'
Universität Wien
Rooseveltplatz 10/9
1090 Wien
Austria
miles.macleod@univie.ac.at

MEHMET ELGIN

REDUCTIONISM IN BIOLOGY: AN EXAMPLE OF BIOCHEMISTRY

ABSTRACT

In this paper, I argue that the multiple realizability argument against reductionism does not work in biochemistry and that biochemistry as a reductionist project is a progressive research program. Since the anti-reductionist argument that appeals to the multiple realizability thesis doesn't work and since biochemistry that incorporates the principle that biological functions of biomolecules in living cells can be understood in terms of chemical and physical properties of those molecules is a progressive research program, I conclude that plausibility of reductionism is still worthy of further study.

I

Reductionism in biology is concerned with the relation between biological knowledge and chemical or physical knowledge.[1] There is the idea of theory-reduction, which concerns with whether and how a biological theory can be reduced to a chemical or a physical theory. There is the idea of explanatory reduction, which concerns with whether or how biological representations can be explained by chemical or physical representations.[2] In this paper, I will only focus on biochemistry and I will not discuss the nature of reduction relation. Instead, I will argue that the multiple realizability thesis does not show that type-type reduction is not possible at least in biochemistry. Second, I want to address the issue of reductionism in a different way by looking at a science (biochemistry) that forms a reductionist research program. To do so, I would like to answer the question of whether this reductionist research program leads to new empirical knowledge about the biological systems or whether it distorts our understanding of those systems. To answer this question, I would like to use Imre Lakatos' Theory of Scientific Research Programs. The reason I choose Lakatos' Theory is: 1. It specifically addresses the issue of whether research traditions are progressive i.e., whether they lead to new empirical knowledge. 2. Other theories of science such as falsificationism or inductivism (with the exception of Thomas Kuhn – I could as well use Kuhn's Theory and make similar points about this science since Kuhn's notion

1 Alan Love and Ingo Brigandt, "Reductionism in Biology", Entry in *Stanford Encyclopedia of Philosophy.*

2 Ibid.

F. Stadler (ed.), *The Present Situation in the Philosophy of Science*,
The Philosophy of Science in a European Perspective, DOI 10.1007/978-90-481-9115-4_16,

of progress in terms of increase in effectiveness of puzzle solving would do the same job) are not fitted to evaluate research traditions as integrated wholes. I acknowledge that conclusions I reach about Biochemistry Research Program on the basis of Lakatos' Theory will be sensitive to the appropriateness of that theory in understanding science. However, I will not address the broader issues about which theory of science is better.

II

The multiple realizability thesis has been introduced into philosophy by Putnam and Fodor.[3] The idea is this: the *same* kinds of higher level properties can be realized by *diverse* kinds of physical properties. If there is a higher level generalization of the form "If M then B" where M and B are higher level kinds, there are multiple physical properties P_1 to P_n that would realize M and there are multiple physical properties P'_1 to P'_n that would realize B, where Ps not only need not but also typically are not equal to $P's$. The relation of this thesis to reductionism is as follows: if a higher level type can be realized by diverse physical types at a lower level, the type-type reduction is not possible since there is no unique lower level type to which a higher level type can be related to. In philosophy of biology and philosophy of mind, the correctness of the thesis has been taken for granted because it has been thought that examples are everywhere (although its implications have been debated, see Sober[4]): just like one can make an automobile from very different physical materials and yet realize the function of being a car, in the same way minds and living organisms can be built up of diverse physical properties and yet do not lack anything in terms of their functions at higher levels.

The claim that multiple realizability is coherent was criticized by Larry Shapiro.[5]

> To say that a kind is multiply realizable is to say that there are *different* ways to bring about the function that defines the kind. But, if two particulars differ only in properties that do not in any way affect the achievement of the defining capacity of a kind then there is no reason to say that they are tokens of different realizations of the kind. Differently colored cork-

3 Hilary Putnam, "Psychological Predicates", in: W. Capitan and D. Merril (Eds.), *Art, Mind and Religion*. Pittsburgh: University of Pittsburgh Press 1967, pp. 37-48. Hilary Putnam, "Philosophy and Our Mental Life," in: *Mind, Language and Reality*. Cambridge: Cambridge University Press 1975, 291-303. Jerry Fodor, *Psychological Explanations*. Cambridge, MA: MIT Press 1968. Jerry Fodor, *The Language of Thought*. New York: Thomas Crowell 1975.

4 Elliott Sober, "The Multiple Realizability Argument against Reductionism", in: *Philosophy of Science* 66, 1999, pp. 542-564.

5 Larry A. Shapiro, "Multiple Realizations", in: *The Journal of Philosophy* 97, 12, 2000, pp. 635-654.

screws, alike in every other aspect, are not tokens of different realizations of a corkscrew because differences in color make no difference to their performance as a corkscrew.[6]

We can extract the following criterion from this passage:

> If there are two *different kinds of realizers* of *the same higher level kind*, then they must differ in their causal powers that are relevant to the function of a multiply realized state.

Contrapositive of this conditional statement runs as follows:

> If *different realizers* of *the same higher level kind* don't differ in their causal powers that are relevant to the function of a multiply realized state, then those realizers *are not different kinds of realizers* of *this higher level kind.*

This criterion implies that no matter how diverse realizers of a given state may seem, as long as they share a common causal power related to the function at a higher level, with respect to that higher level type, realizers don't fall under *distinct kinds* at the lower level. They may differ in other physical or chemical characteristics. It is highly dubious, according to Shapiro, that realizers at a lower level would differ in their relevant causal capacities in bringing about the higher level properties and we would still call them the realizers of *the same higher level type*.

Realizers of a higher level type may fall under distinct kinds with respect to many properties. For example, we can build pendulums from many different kinds of physical material but if all these realizers obey pendulum law, they do not constitute *different kinds* of realizers *with respect to the function of pendulums.*[7] In the same way, physical or chemical properties that realize biological properties of a living cell may or may not fall under distinct kinds depending on which characteristics we are interested in them. If we are interested in their relation to biological property in question (their ability to bring about biological properties), they may not fall under different kinds. If on the other hand, we are interested in classifying their other physical properties that are not relevant to their ability to bring about those biological properties, they may fall under distinct kinds. If this is true, it is then at least possibility that realizers of a given biological function may have at least some chemical or physical characteristics that are relevant to their ability to bring about that function. Then, the type-type reduction is at least a possibility. There is no implication in this thesis that all biological knowledge can be reduced to chemical and physical knowledge. It may be that some parts of biology resist this. However, it entails that the claim that reductionism is in principle not possible due to multiple realizability is false. It follows that the fact that the state is "multiply realized" does not entail that there cannot be kind generalizations about lower level physical or chemical properties relevant to the realization of a higher

6 Ibid., p. 644.

7 Robert Batterman, "Multiple Realizability and Universality", in: *British Journal for the Philosophy of Science* 51, 2000, pp. 115-145.

level type because (again) distinct realizers of a system may exhibit some features that are universal with respect to the behavior under consideration.

Consider the relation between protein function and its structure and sequence. To make sense of this relation, we should specify exactly what we call a higher level type (in this case about the specific function of a protein). Then, we must be clear about the relevant lower level property that we may say responsible for that specific function. For example, the same protein may be responsible for several functions or a protein having different sequences may realize the same function. Does it follow from this that multiple realizability thesis is right and consequently reductionism is false? No. If the two sequences are different and yet still realize the same function, then the relevant question is what part of the sequence is responsible for the function. What is a kind? Should we take two sequences as different lower level kinds because they differ only in one place or many? Sometimes only one change in the sequence may be enough to call them different kinds but sometimes even if they differ in many places it may not be. This depends on what function we are investigating. So whether something counts as a relevant kind at the lower level depends on the function we are interested in. If there is a part in the sequence that makes a difference in the realization of the function that is the relevant property we should be focusing on as a lower level kind. If we can identify such a part and if that part is common in all of the realizers of the function (in all the different sequences that realize the function) then we have identified a kind at the lower level with respect to the function in question. If a higher level generalization is "all proteins have some biological functions" and we want to reduce this to a lower level kind, then we should direct our attention to their chemical or physical attributes that enable all of them to do some biological functions. If a higher level generalization is "Protein X has a biological function Y" and we want to reduce this to a lower level kind, then we should look at the chemical or physical features of this protein that we can assign responsibility for that specific function. So a higher level type can be generalizations about a single protein or all proteins; but depending on a specific kind of higher level generalization, the lower level physical or chemical generalization may be different.

> In both closely and distantly related proteins the general response to mutation is conformational change. Variations in conformation in families of homologous proteins that retain a common function reveal how the structures accommodate changes in amino acid sequence. *Residues active in function, such as the proximal histidine of the globins or the catalytic serine, histidine and aspartate of the serine proteinases, are resistant to mutation because changing them would interfere, explicitly and directly, with function*[8] (Italics are mine).

Whether this situation is widespread or rear is irrelevant. However, this example illustrates how despite many differences in the sequence we can still call all of

8 Arthur M. Lesk, *Introduction to Protein Architecture*. Oxford: Oxford University Press 2001, p. 172.

them the same kind of sequence with respect to a common property that is relevant to the specific function. The above claim by Lesk says that there is a part in the sequence that is preserved despite the fact that other parts show variations. The explanation is that, the part that is preserved plays a vital role in the realization of the function. So with respect to this specific function, there is a common lower level kind to which that function can be reduced. This is what Shapiro's criterion of multiple realizations predicts.

III

The failure of a priori arguments against reductionism paves the way for a defense of methodological reductionism. The basic idea here is that if a research program with reductionist tenets such as biochemistry leads to new empirical knowledge about biological phenomena, the issue of whether reductionism is justified can be addressed on these methodological grounds. Here the issue does not concern with the truth of reductionist thesis; it mainly concerns with the heuristic value of it – i.e., whether sciences such as biochemistry, biophysics etc. are justified in following reductionist tenets. It is important, however, to note that if a reductionist research program constantly succeeds in the discovery of new empirical knowledge, this will provide plausibility for the reductionist thesis even if it does not justify its truth.

According to Imre Lakatos,[9] the basic unit of science is a research program. Scientific Research Programs (SRP) consist of negative and positive heuristics: negative heuristics determine what is not allowed in SRP and positive heuristics determine what is permitted. SRP also consists of two sets of assumptions: hard core and protective belt. Hard core assumptions are the fundamental principles of SRP (for example, in Newtonian physics they would be three laws of motion plus gravitational law or in evolutionary biology they would be formulations of principles that define how evolutionary forces affect genetic structure of a population) and protective belt assumptions are anything that may be needed to relate these hard core assumptions to the world. When there is a mismatch between theoretical results and the actual observations, negative heuristics say that no change in hard core assumptions is allowed. Positive heuristics say that only non-ad hoc changes are allowed. In the case of a gap between theoretical results and actual measurements, there is no recipe about what kind of changes can close the gap. However, sometimes methodological principles may lead us to make changes in certain directions. For example, in Newtonian physics, commitment to the idea

9 Imre Lakatos, "Falsification and the Methodology of Scientific Research Programmes", in: Imre Lakatos and Alan Musgrave (Eds.), *Criticism and the Growth of Knowledge*. New York: Cambridge University Press 1965. Imre Lakatos, *The Methodology of Scientific Research Programmes*. Philosophical Papers Volume 1. New York: Cambridge University Press 1978.

that nature should be explained in mechanical terms will lead scientists to look for new models that will not violate this maxim (so when one mechanical model fails they will look for another mechanical model that will do the job). In this sense, when we assess an SRP, we are also assessing these methodological maxims as well because such principles may sometimes be responsible for the failure of an SRP. According to Lakatos, failure or success of SRP cannot be put in absolute terms. For him, successful SRP makes both empirical and theoretical progress. Failed SRP is degenerative in the sense that it fails to make empirical progress. Since there is no recipe how to make appropriate changes in SRP when there is a gap between theoretical results and actual measurements, in most part creative and imaginative scientists determine the faith of SRP. In this sense, sometimes even the degenerative SRP may make a comeback.

In order to define theoretical and empirical progress, following Malcolm Forster,[10] let me introduce the term 'model'. A model is basically the combination of hard core assumptions and protective belt assumptions from which we obtain theoretical results that can be related to the world. When a theoretical result we obtain from a model does not match to the actual observations, then another model is called for. According to Lakatos, when we make changes in the old model, these changes should not be ad hoc; i.e., such changes should lead to new predictions and they should be independently testable. If these new predictions are empirically confirmed, then SRP is making empirical progress. If we construct more and more models and they make empirical progress, then SRP that these models belong to is progressive. However, if more and more models belonging to SRP start failing, then it is degenerative.

Does biochemistry contain reductionist tenets? Is it a progressive research program in Lakatos' sense? Biochemistry studies chemical processes and reactions that take place in living cells. There are varieties of different molecules in living cells. Molecules that Biochemistry studies are carbohydrates, proteins, enzymes, lipids and nucleic acids. The guiding idea of biochemistry is that processes of living cells can be understood in terms of the chemical properties of these molecules that form a living cell. One textbook defines the goal of biochemistry as follows:

> The overall goal of biochemistry is to describe life's processes using the language of molecules, that is, applying the principles and methods of chemistry to determine molecular structure from which it is often possible to explain biological function.[11]

---

10 Malcolm R. Forster, „The Hard Problems in the Philosophy of Science", in: R. Nola and H. Sankey (Eds.), *After Popper, Kuhn & Feyerabend: Recent Issues in Theories of Scientific Method*, Australasian Studies in History and Philosophy of Science, Kluwer Academic Publishers 2000, pp. 231-251.

11 Rodney F. Boyer, *Concepts in Biochemistry*. Hoboken, NJ: Wiley & Sons Inc. 2006, p. 2.

It is important to distinguish between two sets of theories that we may call hard core assumptions of biochemistry research program: 1. There are background theories, such as chemical, physical and biological theories. 2. There are set of principles about the nature, function and interactions of biomolecules that are building blocks of life. It is the second one that is distinctive about biochemistry and the reductionist nature of this research program lies in these second kinds of principles. Since the second claim is distinctive of biochemistry research program, it plays a vital role whether this science succeeds in realizing its goals. Lehninger, Nelson and Cox[12] write:

> The molecules of which living organisms are composed conform to all the familiar laws of chemistry, but they also interact with each other in accordance with another set of principles, which we shall refer to collectively as *the molecular logic of life*. These principles do not involve new or yet undiscovered physical laws or forces. Instead, they are set of relationships characterizing the nature, function, and the interactions of biomolecules.

The list of the principles concerning the molecular logic of life are:[13]

> A living cell is a self-contained, self-assembling, self-adjusting, self-perpetuating isothermal system of molecules that extracts free energy and raw materials from its environment.
> The cell carries out many consecutive reactions promoted by specific catalysts, called enzymes, which it produces itself.
> The cell maintains itself in a dynamic stady state, far from equilibrium with its surroundings. There is great economy of parts and processes, achieved by regulation of the catalytic activity of key enzymes.
> Self-replication through many generations is ensured by the self-repairing, linear information-coding system. Genetic information encoded as sequences of nucleotide subunits in DNA and RNA specifies the sequence of amino acids in each distinct protein which ultimaltely determines the three-dimensional structure and function of each protein.
> Many weak (noncovalent) interactions, acting cooperatively, stabilize the three-dimensional structures of biomolecules and supramolecular complexes.

The common theme in all these principles is the idea that life can be understood in terms of chemical or physical properties of biomolecules and their interactions. It is because of this common theme that I claim biochemistry is a reductionist research program. Biochemistry research program also includes methods and techniques about how to identify structure and how to relate this structure to a specific function. This involves the use of instruments, for example, NMR spectroscopy, X-ray crystallography, Cryoelectron microscopy and electron cystallography. It also involves certain heuristics about relating structure to function and interpretation of data provided by these instruments. Thus, biochemistry has features of a research program in Lakatos' sense.

---

12 Albert L. Lehninger, David L. Nelson and Michael M. Cox, *Principles of Biochemistry*. New York: Worth Publisher 1993, p. 4.

13 Ibid., p. 19.

The question now is whether biochemistry as a research program is empirically progressive. To answer this question, we have to look at the historical record of biochemistry whether its models constructed from hard core assumptions together with protective belt assumptions have led to new empirical knowledge about living cells and whether there are cues that point to expectations about further new empirical knowledge. To show that biochemistry research program has realized its goals to some extent, it suffices to list just some major discoveries about the structure of DNA, the structure and function of proteins, discoveries about the causes of many diseases, developments of new techniques and instruments in solving problems in biochemistry research program. There are more discoveries in the field than the number of Nobel prizes awarded but the selected list of noble prize awards will give some idea about its progress toward providing new empirical knowledge about biological functions: Fisher for enzyme action, Buchner for description of fermentation, Summer for crystallization of urease, Krebs for description of citric acid cycle, Watson and Crick for DNA double helix, Perutz for X-ray of protein crystals, Smith for restriction enzymes, Cech and Altman for catalytic RNA, Mullis for polymerase chain reaction, Horvitz for biochemistry of programmed cell death, Wüthrich, Fenn and Tanaka for NMR, MS structure of proteins, Mackinnon and Agre for Aquaporins and membrane channels and Hershko, Rose and Ciechanover for ubiquitin-mediated protein breakdown.[14]

Furthermore, just looking through paper publications related to biochemistry will show that more knowledge is being produced and on the way. 2005 JCR Science Edition reports that there are 261 journals listed under the category of "biochemistry and molecular biology" between 2003 and 2005. In these journals 236,517 papers published between 2000 and 2005 and the total number of citations these papers produced was 511,212.[15] Between 1971 and 1990, the percentage of biochemistry articles in chemistry papers published in the journal *Nature* is found to be 83 but the percentage drops to 73 in 1990s.[16] Around the same years the percentage of chemistry articles are 13 and the percentage of biology and medicine articles are 49.[17] More information about the performance of scientific fields is available in journals related to scientometric and bibliometric studies of performance evaluation of these fields. The figures I cited above point to some rough and ready ideas about the progress of biochemistry. These figures may include repetitive publications and not significant discoveries. However, even small percentage of these figures will show that this research program is empirically progressive in Lakatos' sense since it is not too harmful to assume that top journals in the field do not publish papers that are not original contribution to the field. There is also inter-

14 Boyer, *Concepts in Biochemistry*, p. 6.

15 Nan Ma, Jiancheng Guan, and Yi Zhao, "Bringing PageRank to the Citation Analysis", in: *Information Processing and Management* 44, 2007, p. 802.

16 D.B. Arkhipov, "Scientometric Analysis of Nature, The Journal", in: *Scientometrics* 46, 1, 1999, p. 62.

17 Ibid., p. 59.

esting statistics about the number of publications related to the subfields of biology. Between 1991 and 1998, in terms of average annual number of papers among the subfields of biology, with 320 papers molecular biology ranks first, with 155 papers medicine comes next, with 109 papers brain ranks third, with 21 papers natural history ranks fourth, and with 20 papers agriculture ranks fifth.[18] Even these figures give some approximate idea about the direction biological sciences heading. These figures should, of course, be detailed and should be subjected to serious analysis to provide more detailed answer to the question of whether sciences are heading in the direction of reductionism; but, as a starting hypothesis from these figures, we can say that biochemistry is progressive research program.

It must be noted, however, that Lakatos' SRP does not allow us to make judgments about the final faith of a research program. For, SRP can be progressive at one time and then may become degenerative later; it may be degenerative at one time but then with the imaginative and creative abilities of researchers working in the field it may become progressive again. So our evaluations of the overall success of a research program will always be relative to the information available to us at a given time period. Accordingly, my claim here is that the best evidence available to us now leads to the conclusion that biochemistry research program is empirically progressive.

IV

The fundamental question of biochemistry is defined as follows in one of the most influential textbook in the field: "Biochemistry asks how the thousands of different biomolecules formed from these elements interact with each other to confer the remarkable properties of living organisms."[19] Through the applications of techniques and concepts from chemistry, biochemists hope to understand this fundamental question. Given their track record in a short time, we should be optimistic about their possible success in answering this question. So we should be optimistic about this reductionist project. I believe we will make more progress in our philosophical projects about reductionism by studying in detail a science whose project is to understand biological phenomena in terms of chemical and physical concepts.

Department of Philosophy
Muğla Üniversitesi
Türkiye
melgin@mu.edu.tr

18 Ibid., p. 67.

19 Lehninger, Nelson and Cox, *Principles of Biochemistry*, p. 1.

Raffaella Campaner

Reductionist and Antireductionist Stances in the Health Sciences

Reductionism and antireductionism are among the most largely and hotly debated topics in philosophy of biology today. In this section of the volume, aiming to convey the current situation in the philosophy of the natural and life sciences, these topics are specifically addressed in Mehmet Elgin's paper, focusing on biochemistry. Elgin strongly supports reductionism, first by claiming that the now classical argument based on multiple realizability does not entail anti-reductionism and secondly highlighting how the version of methodological reductionism that biochemistry has been adopting – centered on "the principle that biological functions of biomolecules in living cells can be understood in terms of chemical and physical properties of those molecules"[1] – has proved largely successful, teaching "us new knowledge about the biological systems". Taken together, these two arguments are deemed to provide good grounds for a thorough defence of reductionism. While Elgin chooses biochemistry as his privileged standpoint on the issue, I shall dwell on the stances emerging in the health sciences. Although the health sciences are closely intertwined with – amongst others – biology and biochemistry, they also have their own peculiar features. Referring to some examples taken from different medical disciplines, I will question whether reductionism can be regarded as not only a viable, option, but as the best solution to gain new knowledge about biomedical systems. I shall suggest that considerations arising from current medical research and practice support a pluralistic approach to the topic, in which both reductionist and antireductionist stances can be accommodated in different ways.

1. Reductionism and explanation

The controversy over reductionism and antireductionism is multifaceted, and the development of a number of versions of each horn of the debate in the philosophical landscape makes it particularly intricate. In general terms, I shall take here the reductionist program of molecular biology to claim that all biological phenomena must in principle be fully reducible to physicochemical entities and their organization, and to the laws identified by physics and chemistry. Hence it is held that to understand something one has to look at the set of physical and chemical entities

1 Mehmet Elgin, "Reductionism in Biology: An Example of Biochemistry", this volume, p. x.

F. Stadler (ed.), *The Present Situation in the Philosophy of Science*,
The Philosophy of Science in a European Perspective, DOI 10.1007/978-90-481-9115-4_17,

of which it is composed, at their properties and arrangement: the level described by physics and chemistry is regarded as the lowest level to which everything that exists can be reduced by virtue of the fact that everything that exists is ultimately composed of physicochemical entities.

The reductionist mindset has largely pervaded molecular biology in the conviction that "because biological systems are composed solely of atoms and molecules [...], it should be possible to explain them using the physicochemical properties of their individual components, down to the atomic level"[2]. One of the versions in which reductionism has been developed is thus explanatory reductionism, with reductionism being considered the most appropriate key to explanation. "If it is a fact that the entities and properties of a realm can be identified with certain arrangements of basic types of components, then our understanding of this realm should improve, to some degree, by our becoming aware of the existence and nature of these components."[3] This target will be reached by appealing to physics and chemistry, which have been proved to explain a vast range of things by a small set of laws. One of the most forceful supporters of explanatory reductionism with respect to biology – more specifically, to experimental biology – has recently been Marcel Weber. Maintaining that modern experimental biology is thoroughly reductionist, he emphasizes that, while physics and chemistry are interested in identifying and describing natural laws, experimental biology just aims at *applying* them to biological phenomena for explanatory purposes. If, on the one hand, biology can be recognized as an autonomous science as far as its specific concepts, methodological standards and research methods are concerned, Weber stresses that, on the other hand, it depends on physics and chemistry as far as explanation is concerned. Biological terms such as "axon" or "synapse" play a descriptive role, whereas all terms figuring in genuine explanations are physical and chemical terms, referring, for instance, to molecules, macromolecular aggregates and purely physical entities. Biological terms are adopted simply to identify the kind of systems to which physical and chemical terms are to be applied.

Does this hold for the health sciences as well? Does such a perspective exhaust the stances emerging in the health sciences? The patient's body can be seen as

> composed of different anatomical systems, such as the respiratory or cardiovascular systems. These systems are, in turn, composed of various organs, such as lungs and hearts, which are made up of epithelial, muscular, nervous and glandular tissues. Finally, to complete the reduction, these tissues are composed of diverse cellular types that are made up of a variety of molecules.[4]

2 Marc H.V. Van Regenmortel, "Reductionism and Complexity in Molecular Biology", in: *European Molecular Biology Organization Reports* 5, 2004, pp. 1016-1020, quot. ex p. 1016.

3 Todd Jones, "Reductionism and Antireductionism: Rights and Wrongs", in: *Metaphilosophy* 35, 5, 2004, pp. 614-647, quot. ex p. 616.

4 James Marcum, "Biomechanical and Phenomenological Models of the Body, the Meaning of Illness and Quality of Care", in: *Medicine, Health Care and Philosophy* 7,

It is with respect to their affecting such systems that diseases are investigated, and their fundamental physicochemical features identified[5]. According to reductionists, biological systems – in the cases of the health sciences, bodies affected by diseases – are composed uniquely of atoms and molecules, and can therefore be fully described and explained in terms of the physicochemical properties of their constituent parts. These parts are usually combined in articulated systems, which have to be broken down into simpler pieces to determine the properties and connections between the parts.

> After describing a complex system in terms of its constituents, a biologist may be led to believe that he has "reduced" something complex to its simpler components [...] When cells and organelles are described in terms of their molecular constituents, it may, indeed, seem plausible that biological entities are nothing but physicochemical systems and that biology should be reducible to chemistry and physics.[6]

Much of the recent philosophical literature has emphasized the complexity of biological systems and the multi-level character of their analyses. The possible relations between reductionist tenets and a mechanistic perspective are being considered in different ways. On the one hand, mechanistic investigations have been opposed to reductionism. Carl Craver, a neo-mechanist, for instance, has pointed out that adequate mechanistic explanations of biological systems must have an interfield character. He stresses how typically reductionists reduce

> theories about phenomena at a higher level (e.g., gases, lightning, and life) [...] to theories about phenomena at lower levels (e.g., molecules, electrons, and physiological systems). [...] The mechanistic approach [he] develop[s] [instead] suggests a reasonable way to understand the level relationship and insists on recognizing interfield relations that oscillate upward and downward in a hierarchy of levels.[7]

3, 2004, pp. 311-320, quot. ex p. 313.

5 For some examples, see Iain H. McKillop, Diarmuid M. Moran, et al., "Molecular Pathogenesis of Hepatocellular Carcinoma", in: *Journal of Surgical Research* 136, 1, 2006, pp. 125-135; Maryann E. Smela, Sophie S. Currier et al., "The Chemistry and Biology of Aflatoxin B_1: from Mutational Spectrometry to Carcinogenesis", in: *Carcinogenesis* 22, 4, 2001, pp. 535-545.

6 Marc H.V. Van Regenmortel, "Pitfalls of Reductionism in Immunology", in: Marc H.V. Van Regenmortel and David Hull (Eds), *Promises and Limits of Reductionism in the Health Sciences*, Chichester: Wiley & Sons 2002, pp. 47-66, quot. ex p. 47.

7 Carl Craver, "Beyond Reduction: Mechanisms, Multifield Integration and the Unity of Neuroscience", in: *Studies in History and Philosophy of Biological and Biomedical Sciences* 36, 2, 2005, pp. 373–395, quot. ex p. 376. Craver dwells on the case of long-term potentiation and memory consolidation, concluding that the goal of reduction, especially pursued in the 1950s and 1960s, has been replaced by that of identifying multilevel mechanisms with an explanatory purpose.

On the other hand, William Wimsatt claims that mechanicism and reductionism can actually go hand in hand, stressing that much of the explanatory weight in explanatory reductions is borne by the *organization* of smallest parts *into mechanisms* accounting for the behaviour of the system at stake.[8] Clarifying such an organization is actually regarded as the major task reductions are supposed to do:

> a reductionist analysis offers a lower level mechanistic account of a higher-level phenomenon, entity, or regularity. To do so, one commonly decomposes a complex system into its parts, analyzes them in isolation, and then re-synthesizes these parts and the explanations of their behaviour into a composite explanation of some aspect of the behaviour of the system.[9]

An author that, as we shall see below, takes part in the debate with a specific focus on medicine, Marc H.V. Van Regenmortel, criticises reductionist thinking for analysing "complex network interactions in terms of *simple causal chains* and *mechanistic models*" and for favouring "causal explanations that give undue explanatory weight *to a single factor*".[10] This is taken to overlook the fact that clinical states are always the outcomes of many biochemical pathways and networks, and that diseases are the results of alterations to complex systems. Contemporary mechanistic views, which devote a particular attention to biological and biomedical systems, strongly emphasize the complexity of the phenomena investigated, such as the functioning of the human body and the development of diseases. Far from searching for "simple causal chains" or for "a single [explanatory] factor", they insist on outcomes being brought about by a plurality of causes, combined as different causal nets. Following Wimsatt's suggestion, a reductionist can therefore be seen as understanding "the character, properties, and behaviour of the studied system in terms of the properties of its parts and their interrelations and interactions",[11] where such parts can be conceived of as components in a mechanism. The

8 For a position that associates reduction with the search for mechanisms see also Sahotra Sarkar, "Models of Reduction and Categories of Reductionism", in: *Synthese* 91, 5, 1992, pp. 167-194.

9 William C. Wimsatt, "Reductionism and Its Heuristics: Making Methodological Reductionism Honest", in: *Synthese* 151, 3, 2006, pp. 445-475, quot. ex p. 466.

10 Van Regenmortel, "Reductionism and Complexity in Molecular Biology", *loc. cit.*, p. 1018, italics added. See also Marc H.V. Van Regenmortel, "Pitfalls of Reductionism in the Design of Peptide-Based Vaccines", in: *Vaccine* 19, 17-19, 2001, pp. 2369-2374: "Causal explanations are reductive in the sense that one factor is singled out for attention and given excessive explanatory weight" (p. 2370).

11 Wimsatt, "Reductionism and Its Heuristics: Making Methodological Reductionism Honest", *loc. cit.*, p. 467. On this, see also William C. Wimsatt, "Aggregate, Composed and Evolved Systems: Reductionist Heuristics as Means to More Holistic Theories", in: *Biology and Philosophy* 21, 5, 2006, pp. 667-702: "Such mechanistic explanations are also reductionist explanations of the behaviour or a property of a system in terms of the interactions of its parts and properties. Such a reduction need not deny the causal importance of higher-level phenomena, regularities, entities, structures, and mecha-

mechanistic account emphasizes how such relevant components work together, how they interact and are organized in productive continuity from some beginning to some final conditions. At the same time, neo-mechanicism counts among its advantages the capacity to provide insights into "interlevel forms of interfield integration", largely "neglected in much of the literature on reduction".[12]

2. Antireductionist Stances and the Health Sciences

Do reductionist tenets suffice for a full understanding of the behaviour of the systems investigated by the health sciences, or is something different or something more required? Some limitations arise from our inability to know the exact details the physicochemical entities involved in the development of a disease, or – if we are dealing with health – in the standard functioning of the human body. The acknowledgment that we cannot identify and keep track of the behaviours of the fundamental molecules that collectively act and interact into an organism does not exclude explanatory reductionism: a purely physicochemical explanation can eventually be reached, and reductive explicability can still be maintained in principle. Is a reductionist explanation, though, the only goal to be reached, in practice or just in principle, in the biomedical sciences?

Criticisms of a reductionist approach have been formulated in various fields of the health sciences on different matters. Some of them are targeted against the general conception of the human body and doctor-patient relationship that is taken to usually underlie the reductionist perspective. James Marcum criticises the "biomechanical model" of the patient's body, considered the predominant model in modern Western medicine. Advocates of this model, which basically construes the human body as a machine, "reduce the patient to separate, individual body parts in order to diagnose and treat disease", and fail to consider it as an embodied person: once thought of as governed by physical and chemical laws, the body is reduced to an abstract thing and totally "stripped of its lived context". The body is conceived as a mechanized material object, reducible to a collection of physical parts, that, when broken or malfunctioning, can be repaired or substituted. According to this model, illness is construed in terms of "diseased or dysfunctional body

nisms built upon them" (p. 669); "a reductive explanation of a behaviour or a property of a system is one that shows it to be mechanistically explicable in terms of the properties of and inter-actions among the parts of the system" (pp. 670-671). Wimsatt suggests that a "multilevel reductionistic analysis" can pick out "the appropriate levels for objects, processes, and phenomena, and articulates and explicates their relations to complete the explanatory task with no further mystery. It is to be distinguished from apocalyptic reductionism through the recognition of relevant dynamics at multiple levels (*ibid.*, p. 672).

12 Craver, "Beyond Reduction: Mechanisms, Multifield Integration and the Unity of Neuroscience", *loc. cit.*, p. 388.

parts separate from the overall integrity of the patient's body and lived context",[13] and the physician's role is comparable to that of a technician or a mechanic. As a consequence, Marcum stresses, the quality of medical care has been undergoing a crisis.

The need for a proper consideration of the patient as such in non-reductionist terms, with the "restitution of the intact person to his or her full personhood",[14] is also advocated by Alfred Tauber, who emphasizes the socio-psychological aspects of the healing process, and by Elizabeth Lloyd, who discusses cross-populational correlations between socioeconomic status and morbidity and mortality. Lloyd reviews a number of surveys aimed at demonstrating that a major income inequality in a given society is strongly related to that society's level of mortality: the more severly the poor are poorer than the rich in a certain society, the lower the average life expectancy everybody – not just the poor – has: "socioeconomic factors turn out to be powerful predictors of health outcomes [...], and these factors cannot be investigated if all research funds are concentrated at problems conceived at the molecular level."[15] She takes this to show that not all medical research can be exhausted by an analysis at the molecular level.

Further antireductionist stances can be found in contemporary medical research, with quite an impact on the methodology adopted and its practical implications. Some such stances have been expressed with respect to the discovery of new drugs and the development of vaccines. It has been claimed that the reductionist mindset, which is largely adopted in these contexts, actually turns out to have severe limitations. More specifically, as far as drug-discovery programmes are concerned, Van Regenmortel traces the decline in the approval of new drugs per year in the last decade back to unmitigated reductionism. He claims that dissecting the human body into its components underrates the importance of regarding organisms and patients as wholes, and loses a wealth of useful information on their workings. With regard to vaccine research, the idea that a biological phenomenon such as protection against infection can be reduced to the level of chemistry, although fashionable, is criticised and accused of serious shortcomings. Vaccination and protective immunity have a meaning only at the level of the whole biological organism, given that "molecules, tissues and organs cannot be vaccinated", and are anchored in the biological realm:

13 Marcum, "Biomechanical and Phenomenological Models of the Body, the Meaning of Illness and Quality of Care", *loc. cit.*, pp. 311-313.

14 Alfred I. Tauber, "The Ethical Imperative of Holism in Medicine", in: Marc H.V. Van Regenmortel and D. Hull (Eds.), *Promises and Limits of Reductionism in the Health Sciences*, Chichester: Wiley & Sons 2002, pp. 261-27, quot. ex p. 268. Tauber goes as far as to argue that even the moral and spiritual aspects of human beings are to be taken into account.

15 Elisabeth A. Lloyd, "Reductionism in Medicine: Social Aspects of Health", in: Marc H.V. Van Regenmortel and D. Hull (Eds.), *Promises and Limits of Reductionism in the Health Sciences*, Chichester: Wiley & Sons 2002, pp. 67-82, quot. ex p. 66.

immunogenicity depends on the biological potential of the host that is being immunised; [...] antibodies act in a collective manner and [...] the neutralizing synergy between various antibodies cannot be reduced to the simple additive effects of individual molecules.[16]

Hence it is stressed that vaccination and immune response can be elucidated only by studying the dynamics of the entire system of pathogen, antibody and host cell. Reductionist analyses in terms of the physicochemical principles underlying immunological recognition are claimed to be insufficient to design new vaccines, and attempts to reduce both vaccination and autoimmune phenomena to discrete molecular features of individual components of the immune system are bound to fail. Furthermore,

if all of the details of the incredibly complex workings of the immune system are spelled out but no mention is made to the effect that all these mechanisms have on the health of the organism, something desperately important has been left out.[17]

Another example of antireductionist thinking is provided by some reflections on the history of cancer research. It has been noticed that the established reductionist approach known as the "somatic mutation theory" (SMT) expounded by Robert Weinberg and colleagues, aiming to disentangle the genetic and molecular circuitry of carcinogenesis, has been challenged in the past few years by an organicist approach, the "tissue organization field theory" (TOFT) developed by Ana Soto, Carlos Sonnenschein and colleagues, focusing on tissues and the complex arrays of relationships ranging over several levels of entities.

From a reductionist perspective, cancer is considered as the product of mutated genes. From an organicist perspective, according to Sonnenschein and Soto, the architecture of the tissue cannot be reduced simply to the underlying individual cells and biomacromolecules that compose them. Rather, tissue architecture is an emergent property of 'the society of cells', i.e., it is not a function simply of the collective properties of the cells that make up the tissue.[18]

---

16 Van Regenmortel, "Reductionism and Complexity in Moelcular Biology", *loc. cit.*, p. 1019. In the field of vaccinology, the reductionist claim that "it will soon be possible to rationally design effective synthetic vaccines on the basis of our considerable knowledge of the molecular constituents involved in immunological interactions. [...] arises from an unwarranted faith in the powers of reductionism and [...] overlooks the fact that protection against disease achievable by vaccination is a [...] concept that is meaningless when expressed only in molecular or chemical terms" (Van Regenmortel, "Pitfalls of Reductionism in the Design of Peptide-Based Vaccines", *loc. cit.*, p. 2369).

17 David Hull and Marc H.C. Van Regenmortel, "Introduction", in: David Hull and Marc H.C. Van Regenmortel (Eds), *Promises and Limits of Reductionism in the Health Sciences*, Chichester: Wiley & Sons 2002, pp. 1-13, quot. ex p.11.

18 James Marcum, "Metaphysical Presuppositions and Scientific Practices: Reductionism and Organicism in Cancer Research", in: *International Studies in the Philosophy of*

While the reductionist perspective focuses on molecular entities and expressions of faulty genes, the organicist perspective focuses on the disruption of tissue organization, without reaching the lowest physicochemical level. TOFT sees cancer "as a problem akin to histogenesis or organogenesis gone awry, and thus to a problem of developmental biology"[19]. The strictly reductionist approach to cancer research – although largely successful and far from being defeated – has thus been criticised as insufficient. Even if a final triumph is a long way off, and judgements are therefore hard to formulate, it has been acknowledged that the organicist perspective has yielded some important observations on the different levels involved in carcinogenesis (e.g. social and environmental, organismal, tissue level). Different approaches to the topic affect research by orienting it in different directions.

Considerations on the adequacy of reductionism and antireductionism as general approaches in the achievement of knowledge shall also take into account which kind of knowledge we are thinking of. To start with, different conceptions may underlie the search for explanations. If the variety and richness of biomedical scientific activity is taken into account, it can be revealed how

> for a while certain levels will not be investigated, but advances at one level (and not always a lower level) will sometimes open up another level for investigation, so that through time the thrust of scientific investigation will wander from level to level in the organizational hierarchy. Nor is it the case that the lower-level explanations are always more useful than higher-level explanations. It depends on what you want to do,[20]

and on the conception of scientific explanation you are embracing. Sometimes we cannot physically explain why things are happening a certain way because we are ignorant of which laws or boundary conditions are involved. "Perhaps the correct physical explanation is computationally intractable for us [...]. There may even be physical laws, particles or properties that we don't yet know about."[21] If these cases can well occur, the adoption of a non-reductionist approach can also be dictated by a different conception of scientific explanation, since the very topic of explanation can be addressed in different ways. If, in general, it can be claimed that most scientific efforts in biomedical research are aimed at the construction of mostly detailed causal explanations in terms of fundamental physicochemical entities and laws, different conceptions of explanation can lead us to opt for a different approach. As highlighted by some of the latest philosophical literature on

Science 19, 1, 2005, pp. 31-45, quot. ex p. 37.

19 Christopher Malaterre, "Organicism and Reductionism in Cancer Research: Towards a Systemic Approach", in: *International Studies in the Philosophy of Science* 21, 1, pp. 57-73, quot. ex p. 60.

20 David Hull, "Varieties of Reductionism: Derivation and Gene Selection", in: David Hull and Marc H.C. Van Regenmortel (Eds), *Promises and Limits of Reductionism in the Health Sciences*, Chichester: Wiley & Sons 2002, pp. 161-173, quot. ex p. 171.

21 Jones, "Reductionism and Antireductionism: Rights and Wrongs", *loc. cit.*, pp. 618-619.

scientific explanation, higher-level explanations with – for instance – a pragmatic or unificatory character can be brought forth and favoured: if to explain is taken to be the search for very general principles governing a system – as the unificatory approach holds – or to include mostly contextual factors – as suggested by the pragmatic approach – outlining fundamental physicochemical properties and laws may not be the best option. The choice can depend on "what counts as a relevant question, an adequate explanation and a sufficient degree of understanding. Reductionists and antireductionists tend to disagree about what the relevant questions are and about what constitutes an adequate answer to these questions."[22] Even when lower-level explanations are available, appeal to more general biomedical features can prove essential and/or more adequate in the acquisition and/or transmission of medical knowledge. If an explanation of a disease is to be provided, for example, to a patient's relatives, or an explanation of the contributions of risk factors in the emergence of a pathology are to be outlined to a population for a prevention campaign, an antireductionist appeal to general biomedical or environmental principles can work perfectly. Explanations can be both downward-looking and upward-looking, at different times and in different contexts, and pluralism as far as the reductionism/antireductionism issue is concerned can therefore go together with some form of explanatory pluralism.

Furthermore, explanation is not necessarily the final goal of all medical enquiry. Explanatory knowledge is not the only kind of knowledge we may achieve: descriptive knowledge, unifying knowledge, classificatory knowledge, predictive knowledge, or sketchy knowledge aimed at practical interventions (for example, targeted at cures and therapeutic strategies) belong to scientific activity and contribute to scientific progress, and may well not be formulated in physicochemical terms. These kinds of knowledge may be due not only to temporary limitations in our ability to carry our reductions, but to specific interests and aims in a given context. In both cases, they provide some representations of biological systems and hence new knowledge about them. New problems may call for a number of different answers and need a number of different epistemic strategies. The adoption of a non-reductionist perspective may be driven by structural features of the kinds of knowledge sought for, and/or by the stage we have reached in the elaboration of scientific theorizing. If the explanations reached by biological and biomedical research, which are no doubt one of the main targets of such research, mostly follow a reductionist strategy, and their success can be taken to provide good grounds for supporting reductionism, not all progressive scientific knowledge is explanatory, nor does it start off with complete explanations. We often do not know how to explain a living system straight away in terms of its physicochemical microcom-

22 Van Regenmortel, "Pitfalls of Reductionism in the Design of Peptide-Based Vaccines", *loc. cit.*, p. 2370. See also Claude Debru, "From Nineteenth Century Ideas on Reduction in Physiology to Non-Reductive Explanations in Twentieth-Century Biochemistry", in: Marc H.V. Van Regenmortel and David Hull (Eds.) *Promises and Limits of Reductionism in the Health Sciences*, Chichester: Wiley & Sons 2002, pp. 35-46.

ponents, or may not be interested in doing so. Antireductionist tenets may orient scientific research, identify, define and circumscribe the system under inquiry, the properties we aim to discover and the relevant boundary conditions, as well as provide first stage descriptions, generalizations and crucial questions.

The differences between explanatory on the one hand, and unifying, descriptive, predictive strategies on the other, may underlie the fact that molecular research methodologies flank the exploration of non-reductionist research strategies. If we take a look at epidemiology, for instance, we can see how the main interest there is prevention: it is through manipulation of known variables and correlations that the appearance and spread of pathologies is avoided. This can be achieved even in the absence of explanatory knowledge. In many cases adequate explanatory knowledge is actually reached only long after knowledge aimed at effective prevention. Descriptions of the system at stake and means to exercise some form of control over it are sought, and enquiries concerning its relations with the environment are largely pursued for a risk assessment. As highlighted by Paolo Vineis and Micaela Ghisleni, the history of epidemiology shows a number of cases in which effective measures were established even many decades before the achievement of a detailed understanding of the functioning of the disease. Guidelines to start preventive actions were devised, for instance, for pathologies such as scurvy, pellagra, cancer of the scrotum, smallpox, cholera, yellow fever, and others. Awaiting a complete explanation of the disease would have entailed serious delays in the elaboration of the preventive measures. Instead of being primarily interested in the fundamental entities constituting the biomedical system at stake and in relations governing its internal working, epidemiology is hence strongly oriented towards the implementation of such preventive strategies. It is targeted at promoting measures such as – for example – a balanced diet and a better quality of the environment even in the absence of physicochemical satisfying explanations.[23]

23 See Paolo Vineis and Micaela Ghisleni, "Risks, Causality and the Precautionary Principle", in: *Topoi* 23, 2, 2004, pp. 203-210; Paolo Vineis, "La confusione tra cause e meccanismi nell'insorgenza delle malattie", in: *La Nuova Civiltà delle Macchine* XXIII, 3, 2005, pp. 113-118. As pointed out by Vineis and Ghisleni, things are different for fields such as toxicology, which is strongly oriented towards the evaluation of chemical substances. "Traditional toxicology is essentially characterized by an analytical approach (each chemical substance is evaluated in isolation) and based on strong theoretical premises (in particular, a threshold of toxicity)" (Vineis and Ghisleni, "Risks, Causality and the Precautionary Principle", *loc. cit.*, p. 207). Epidemiological observations have proved that the latter can be seriously misleading. See *ibid.*, pp. 207-208.

3. Reductionist and antireductionist stances: for a pluralistic approach

How do reductionist tenets fare with respect to such fields and lines of research as presented in section two? Is it really necessary to lodge oneself firmly at one extreme of the controversy between reductionists and antireductionists, or are there ways to preserve both the stances they express? In the light of the examples mentioned above, we believe that with respect to the health sciences the dispute can be largely seen as one over emphasis, conditional on the context and targets of the enquiry. On the one hand, the dissection of biomedical systems

> into [their] components has given us considerable insight into the workings of the system[s], a fact that even staunch anti-reductionists do not deny. What is controversial, however, is the extent to which descriptions of the isolated components in molecular terms are able to provide the type of explanation, the level of understanding and the predictive ability[24]

that scientists such as – as we have seen – immunologists and epidemiologists would like to have. We shall conclude that reductionist and non-reductionist stances, and, more precisely, reductionist explanations and investigative non-reductionism, can be adopted jointly for a tentatively exhaustive understanding of the phenomena at stake in the health sciences.

"The reduction of vital phenomena to irreducible physical and chemical elements and/or relationships is both a necessary and an insufficient condition"[25] in the health sciences. Explanatory reductionism can be admitted together with an antireductionist approach to both the general advancement of medical knowledge and the consideration of the embodied disease, i.e. the sick person. Explanatory reductionism can provide an essential tool in the understanding of diseases and the elaboration of general models of their functioning, but, given the peculiarities of the health sciences, the reductionist mindset does not seem to suffice. With regard to clinical practice, viewing and treating the patient as a biological entirety, not simply in molecular terms, and thinking in terms of a concrete illness, not an abstract disease, is a fundamental demand, to be met in order to establish a correct diagnosis and prognosis. The adoption of reductionism in an explanatory perspective does not imply a reification of the medical object as involved in clinical practice, nor a denial of other forms of knowledge in addition to explanatory knowledge. As we have seen, antireductionism, or, better, *non*-reductionism, can prove a very important epistemic strategy in other fields too, such as immunology and epidemiology. However, this is not to say that extreme holism has to be embraced. It certainly does not provide a viable methodological alternative. That a

---

24 Van Regenmortel, "Pitfalls of Reductionism in the Design of Peptide-Based Vaccines", *loc. cit.*, p. 2370.

25 Debru, "From Nineteenth Century Ideas on Reduction in Physiology to Non-Reductive Explanations in Twentieth-Century Biochemistry", *loc. cit.*, p. 40.

reductionist perspective has been and continues to be very successful in attaining explanatory knowledge can also be recognised by non-reductionists, who admit that reductionism is effective in explaining the chemical basis of a large number of living processes. Furthermore, the importance of medicine as a science largely rests on its capacity of healing, which in turn largely rests on a physicochemical basis. The effectiveness of pharmaceutical drugs, various kinds of cures and diagnostic techniques (see, for instance, electronic microscopes, chemotherapy and radiotherapy, positron emission tomography, to mention but a few) undeniably depends on physicochemical investigations.

Without denying that reductionism should be given a chance – as Elgin claims – and, rather, acknowledging that it is the most successful explanatory strategy, I suggest that examining the "present situation" in the biomedical sciences – which is the aim of this volume and the conference it originates from – we cannot ignore that some important non-reductionist stances are also present. Referring to various quotations from biochemistry textbooks, Elgin concludes "the common theme in all these principles is the idea that life can be understood in terms of chemical or physical properties of biomolecules and their interactions".[26] It is claimed that the success of biochemistry is largely due to this common theme, and that it is on these grounds that explanatory reductionism can be supported, holding that to explain a biological system is to provide an account of the combination and arrangement of more basic, physiochemical entities and properties. This does not clash with methods inspired by non-reductionism, indicating, for instance, how to operate the first cuts when one is investigating a very complex system, and how to select the features that will prove relevant. Without neglecting the popularity of reduction as a scientific strategy and without advocating anything like a "phenomenological" model of the body,[27] a reciprocal integration of both reduction and non-reductionist tenets can be hypothesized, which could ultimately be accepted in the kinds of biomedical examples mentioned above. Suggesting that reductionism as a crucial, successful strategy can be accompanied by, and benefit from, an antireductionist epistemology might help avoiding the limitations due to assuming just reductionism, which looks like "an insufficient presupposition for medical knowledge and practice".[28] While sympathetic with Elgin's reductionism as far as explanation is concerned, I find it plausible that both reductionist and antireductionist tenets can point to expectations about further knowledge, and thus contribute, in different ways, to progress in the life sciences and to a deeper consideration of the different disciplinary fields belonging to the health sciences.

26 Elgin, "Reductionism in Biology: An Example of Biochemistry", *loc. cit.*, p. x.

27 See Marcum, "Biomechanical and Phenomenological Models of the Body, the Meaning of Illness and Quality of Care".

28 James Marcum, "Reflections on Humanizing Biomedicine", *Perspectives in Biology and Medicine* 51, 3, 2008, pp. 392-405, quot. ex p. 395.

Neither reductionism nor anti-reductionism should be seen as the only game in town, and pluralism is not to be arrived at as "the last resort of losers".[29] A careful look at actual scientific practice reveals a high sensitivity of methodologies to contexts, which reductionism generally tends to de-emphasize. In evaluating whether reductionism should be privileged in our toolbox to enhance scientific knowledge, what field of biomedical research we are thinking of ought to be clarified. Medicine has the status of a strongly multidimensional science, which exploits the support of multiple disciplines, and presents multiple attitudes towards the reductionism/antireductionism issue. Instead of being a terrain for conflict between reductionist and antireductionist stances, the health sciences can be regarded as a stage in which different emphasis can be put on different knowledge processes *in different contexts*. Not *any* pluralism will do, but both reductionist explanatory methods and non-reductionist attitudes to the disease and to the sick patient merit consideration. Not only are reductionist and antireductionist tenets compatible once they are properly understood and contextualised, but they can be taken as a tool to better clarify and highlight *specific* features of *single* medical fields. To embrace a pluralistic approach in this sense is not to fragment the analysis, but rather to acknowledge and take into due account the methodological peculiarities of the different fields of medical research and their final aims, which include understanding the disease process, preventing it and taking care of the suffering patient. The multifarious adoption of reductionist and antireductionist stances in fields such as biomedical research and physiopathology, epidemiology, immunology and clinical medicine can hence be taken as hints towards the need to develop different *philosophies* of medicine, each to account for the foundational and methodological peculiarities of these fields.

Further bibliographic references

Philip Kitcher, "Explanatory Unification and the Causal Structure of the World", in: Philip Kitcher and Wesley Salmon (Eds), *Scientific Explanation*, Minneapolis: Univ. of Minneapolis Press 1989, pp. 410-505.

Peter Machamer, Lindley Darden and Carl Craver, "Thinking about Mechanisms", in: *Philosophy of Science* 67, 1, 2000, pp. 1-25.

Wesley Salmon, *Scientific Explanation and the Causal Structure of the World*, Princeton: Princeton University Press 1989.

Wesley Salmon, "Four Decades of Scientific Explanation", in: Philip Kitcher and Wesley Salmon (Eds), *Scientific Explanation*, Minneapolis: Univ. of Minneapolis Prcss 1989, pp. 3-219.

Wesley Salmon, *Causality and Explanation*, Oxford: Oxford University Press 1998.

29 Van Regenmortel and Hull, "Introduction", *loc. cit*., p. 1.

Sahotra Sarkar, "Genes versus Molecules: How To, and How Not To, Be a Reductionist", in: Marc H.V. Van Regenmortel and David Hull (Eds), *Promises and Limits of Reductionism in the Health Sciences*, Chichester: Wiley & Sons 2002, pp. 191-206.

Kenneth Schaffner, *Discovering and Explanation in Biology and Medicine*, Chicago: The University of Chicago Press 1993.

Kenneth Schaffner, "Reduction: the Cheshire Cat Problem and a Return to Roots", in: *Synthese*, 151, 3, 2006, pp. 377-402.

Bas van Fraassen, *The Scientific Image*, Oxford: Oxford University Press 1989.

Marcel Weber, *Philosophy of Experimental Biology*, Cambridge: Cambridge University Press 2005.

Deptartment of Philosophy
University of Bologna
Via Zamboni 28
40126 Bologna
Italy
Raffaella.Campaner@unibo.it

Part IV
Philosophy of the Cultural and Social Sciences

Wenceslao J. González

Trends and Problems in Philosophy of Social and Cultural Sciences: A European Perspective

1. The Levels of Analysis Dealing with the Cultural and Social Sciences
2. The Kind of Approach: "A European Perspective"
3. The "Naturalist Turn", the "Social Turn", and the Discussion on Scientific Realism
4. Explanation, Prediction, and Confirmation: Realm and Limits
5. The Debate on Mathematical Modeling in the Social Sciences and Consequences for Experimentation
6. The Sciences that Philosophy has hitherto Ignored: The Sciences of Design
7. New Directions in the Philosophy of Science
8. Final Remarks

For this initial conference of the program on *The Philosophy of Science in a European perspective*, the Steering Committee suggested an overview of the topics of Team C to be discussed during the five years of this project. The broad title – "Trends and Problems in Philosophy of Social and Cultural Sciences: A European Perspective" – was a way of meeting the aim suggested. Thus, the original focus was to offer some philosophical remarks in terms of possible lines of discussion in this common endeavor.[1] But this task requires previous reflection on the general framework: the kind of analysis to be developed and on what might be considered as "a European perspective".

The twofold consideration takes into account the elements of the proposal sent to the European Science Foundation.[2] The paper tries to make them more explicit as well as to draw some consequences. Thus, the overview of the topics to be examined by Team C, oriented towards "The Philosophy of Social and Cultural Sciences", starts with some remarks on what the *philosophical context* under discussion here is. This involves an explicit reflection on two different aspects: the levels of analysis and the kind of approach.

1 Some additional topics of discussion which also have a presence in Europe can be found in Harold Kincaid, "Social Sciences", in: Peter Machamer and Michael Silberstein (eds.), *The Blackwell Guide to the Philosophy of Science*, Oxford: Blackwell 2002, pp. 290-311; especially, pp. 306-307.

2 Cf. Steering Committee, *The Philosophy of Science in a European perspective* Proposal of an "à la carte Programme" to be submitted to the European Science Foundation, 24 February 2006.

F. Stadler (ed.), *The Present Situation in the Philosophy of Science*,
The Philosophy of Science in a European Perspective, DOI 10.1007/978-90-481-9115-4_18,

On the one hand, there are in fact several *levels of analysis* to be considered when dealing with the cultural and social sciences. Among them, this philosophical-methodological study has important connections with tendencies of general philosophy and methodology of science.[3] On the other hand, the *kind of approach* used to undertake this particular study – a "European perspective" – is particularly relevant and needs also some clarification, insofar as there are several ways to characterize it.

After this twofold framework of the philosophical context, there is a presentation of the main topics to be considered in each year of the program: 1) the "naturalist turn", the "social turn", and the discussion on scientific realism; 2) explanation, prediction, and confirmation: realm and limits; 3) the debate on mathematical modeling in the social sciences and consequences for experimentation; 4) the sciences that philosophy has hitherto ignored: the sciences of design; and 5) new directions in the philosophy of science. This content is accompanied by some additional remarks on the cultural and social sciences in order to make certain philosophical-methodological points explicit.

Commonly, these remarks belong to the "internal" perspective on science (language, structure, knowledge, method, etc.).[4] They assume that the foundational and methodological debate has a central role in the configuration of the cultural and social sciences. An additional study, which is beyond the limits of the present paper, could include a detailed analysis of the "external" point of view: social, political, economic, ... aspects. This contextual view requires the empirical research on the institutions that work in Europe on these topics.[5] Nonetheless, some pieces

3 On the general philosophical-methodological tendencies since 1980, see Wenceslao J. Gonzalez, "Novelty and Continuity in Philosophy and Methodology of Science", in: Wenceslao J. Gonzalez and Jesus Alcolea (eds.), *Contemporary Perspectives in Philosophy and Methodology of Science*, A Coruña: Netbiblo 2006, pp. 1-28. It also includes some remarks on the status of "methodology of science" and its present relations with "philosophy of science".

4 On the philosophical approach: from "internal" to "external", see Wenceslao J. Gonzalez, "The Philosophical Approach to Science, Technology and Society", in: Wenceslao J. Gonzalez (ed.), *Science, Technology and Society: A Philosophical Perspective*, A Coruña: Netbiblo 2005, pp. 13-20.

5 The proposal itself submitted to the European Science Foundation listed in the Introduction a number of research centers in Europe focused on philosophy and methodology of science: "Institute Vienna Circle, Vienna, Austria; Centre for Logic and Philosophy of Science, Ghent University, Belgium; Max Planck Institute for Human Development, Berlin, Germany; Zentrum für Philosophie und Wissenschaftstheorie, Konstanz University, Germany; ZIF, Zentrum für interdisziplinäre Forschung, University of Bielefeld, Germany; European Cultural Centre of Delphi, Delphi, Greece; CIRESS, Interdisciplinary Research Centre for Epistemology and History of Science, University of Bologna, Italy; Institute for History and Foundations of Science, Utrecht University, Netherlands; SCASS, Swedish Collegium for Advanced Studies in the Social Science, Uppsala, Sweden; CPNSS, Centre for Philosophy of the Natural and Social Science, LSE, London, UK". Many of them deal in one way or another with

of information of the contextual view are also contained here in connection with the "internal" perspective, which is the mainstream of the present paper.

1. The Levels of Analysis Dealing with the Cultural and Social Sciences

There are at least three different levels of analysis related to these sciences. First, the general scientific status of the cultural and social sciences, which requires us to consider such diverse elements of science (mainly language, structure, knowledge, method, activity, aims and values). Second, the scientific status of the cultural and social sciences as compared to that of the natural sciences. Among other problems, this concerns methodological controversies such as *Erklären-Verstehen*.[6] Third, the specific issues on the scientific status of each cultural and social discipline (economics, psychology, sociology, archaeology, anthropology, law, etc.). These specific aspects can be seen in every science of this realm (e.g., those related to prediction in economics). This analysis includes those sciences (some new) that have been hitherto ignored in the philosophical literature (such as the sciences of the artificial understood as sciences of design).[7]

Within this third level of analysis, there are two philosophical-methodological options: the broad approach and the restrictive position. In the case of the broad approach, there is interest in connecting the philosophical reflection (i.e., the consideration of the semantical, logical, epistemological, methodological, ontological, axiological, ethical, etc., aspects) on the particular discipline (economics, psychology, sociology, etc.) with topics on science in general. In the case of the restrictive position, the focus of attention is on the problems (mainly methodological) of the discipline analyzed without any real interest in science in general.

Depending on the degree of generality in the analysis, there is a scale of progressive specialization of the research made in philosophy and methodology of

topics related to the social sciences. Obviously, the list is not exhaustive, and requires many University Departments in History and Philosophy of Science that have been established across Europe to be taken into account.

6 Cf. Wenceslao J. Gonzalez, "From *Erklären-Verstehen* to *Prediction-Understanding*: The Methodological Framework in Economics", in: Matti Sintonen, Petri Ylikoski and Karl Miller (eds.), *Realism in Action*: *Essays in the Philosophy of Social Sciences*, Dordrecht: Kluwer 2003, pp. 33-50. In the case of the *Erklären-Verstehen*, the main names involved in the seven conceptions of this methodological controversy are Europeans: Wilhelm Dilthey, Max Weber, Carl Gustav Hempel, Peter Winch, Hans Georg Gadamer, Georg Henrik von Wright, and Karl Otto Apel.

7 Commonly, the companions and the books of readings made by European publishing houses are focused on the second and third level. See for example Stephen P. Turner/Paul A. Roth (eds.), *The Blackwell Guide to the Philosophy of the Social Sciences*, Oxford: Blackwell 2003. (The interesting thing is that, in this case, only three out of thirteen of the authors work at a European university.)

the cultural and social sciences. At the first level, there is a direct relation with philosophers of science who worked in general conceptions of science (such as Karl Popper). Frequently, these general approaches on science (Popperian, etc.) have a direct incidence on the following levels of analysis, even though some of those views (Kuhnian, Lakatosian, etc.) were not initially thought for the cultural and social sciences.

Within the second level, the interest can come from scientists and philosophers interested mostly in the cultural and social sciences (Carl Menger, Max Weber, Friedrich von Hayek, Hans Georg Gadamer, Karl Otto Apel, etc.). Thus, it is very common that they discuss topics related to the similarities and differences between the natural sciences and the social sciences. These themes include the methodological differences between "explanation" and "understanding", which also have repercussions for central issues of the social sciences such as prediction (possibility, reliability, etc.).

At the third level, the philosophical point of view is focused on each science (economics, psychology, etc.). So they can follow a broad approach on those sciences, when they are open to general issues of science (such as in the case of philosophers of economics: Mark Blaug, Uskali Mäki, etc.); or they can adopt a restrictive position (e.g., when the analysis is specifically methodological), and when they are scientists that do not have a real interest for philosophy of science (as frequently happens with experts in statistical economics and econometrics).

2. The Kind of Approach: "A European Perspective"

These three levels of analysis – general, comparative, and specific – can be considered within "a European perspective". But it is a complex issue to establish the precise sense and reference of that expression. In this regard, the initial problem is the question of the existence of a specific European approach in dealing with philosophical questions of science: is there a *tout court* "European perspective" rather than a mere viewpoint held by some representative Europeans? This ontological question is undoubtedly linked to the problem of an adequate characterization of "a European perspective" in philosophy of science. This categorization should go beyond semantical considerations to deal with epistemological and methodological aspects.

Obviously, the present program of the ESF assumes *as a matter of fact* that there is "a European perspective" in philosophy of science. In addition, the proposal sent to the European Science Foundation depicts some features of that perspective. Thus, it appears to be the combination of the amount of philosophical contributions on science made in different countries of this continent, both historically and thematically. In addition, the "European perspective" seems to be something developed with specific traits, somehow different to viewpoints of other places in the world.

Certainly we can see "a European perspective" from different angles, among them is the historical approach and the thematic view. The first one can be connected with the reality of past, whereas the second one pays more attention to the present and to the future. This means that we can consider what "a European perspective" *is* as well as how it *ought to be*. In this regard, the philosophical outlook needs to be analytical (i.e., the consideration of the past and the present) as well as prescriptive (i.e., the meta-reflection on the future).

1) Historically, we can understand that there is *de facto* a long tradition of doing philosophy in Europe.[8] Thus, among other philosophical branches, "philosophy of science" – as we conceive it now – has been developed in one way or another in many European countries. But this could also be seen as mere recognition of a *factum* rather than a form of describing something particular or characteristic of this continent. 2) We can think of this issue in a more thematic view: there is a way of doing philosophy of science that it is somehow distinctive of Europe and, therefore, different from the philosophy of science made in America or in other continents. Thus, in the latter – the contents and style of thinking – the issue of "identity" becomes more relevant than in the former (i.e., in the mentioned historical approach).

This route of searching for *a European perspective* in terms of "identity" can lead to several positions, among them are the "integrative" position, the specific view based on the historical background, and a more rigorous conception upon thematic terms. All three options are open, in principle, to the three levels of philosophical-methodological analysis already pointed out: general, comparative, and specific.

a) An "integrative" position is adopted when the focus of attention is on the search for a *common ground*, both in historical terms and in thematic ones. So the "European perspective" is the result of the contributions made by institutions located in European countries as well as the work of philosophers that either are of any European nationality (insofar as there is no actual "European nationality") or are residents in any nation of this continent.

b) A more specific view than the previous one comes from emphasizing that the "European perspective" is based on a *historical-methodological* approach. According to this standpoint, the "European perspective" is only valid for those

8 If we focus on contemporary philosophy, there are some philosophical conceptions that started on European soil and have had influence on the philosophy of the cultural and social sciences. These views include the following "styles of thought": a) the analytical philosophy of language developed by G. Frege, B. Russell, L. Wittgenstein, etc.; b) the logical empiricist views based on R. Carnap, H. Reichenbach, etc.; c) the hermeneutical tradition of thinkers interested in historical aspects, such as H. G. Gadamer; and d) the postmodern approaches of structuralist roots that have had particular relevance in French authors.

philosophical movements that originally started in this continent or have been developed mostly in Europe, even though the initial outsets were established abroad. In this case has a special weight the recognition of the research tradition.

c) If there is a particular stress on the thematic elements, then a more rigorous conception of the "European perspective" can be adopted that assumes the existence of *strict boundaries*. This means taking into account only those philosophical movements rooted in Europe and developed in this continent, otherwise it is considered that they do not have the specific features to be European or that they belong to the so-called "international philosophy".

These three options requires to take into account the Proposal of an "à la carte Programme" that was submitted to the European Science Foundation, entitled *The Philosophy of Science in a European perspective*. The document gives a guide on how the issue is to be understood in this project.[9] Moreover, the guideline directly affects the present topic of trends and problems in philosophy of social and cultural sciences because it gives a "road map" to embrace the European perspective at stake. This can be summarized in four elements:

I) The analysis of the proposal is linked to relevant scientists that were born and lived in European countries (such as Carl Menger, Ludwig von Mises, Otto Neurath, etc.). II) A special emphasis is put on the institutional contribution to this topic, both in historical terms (the Vienna Circle, the Berlin School, etc.) and in our times, pointing out the existence an increasing number of centers devoted to philosophy of science in Europe (some of them explicitly focus on the social sciences).[10] III) The general approach can be seen as an "integrative" position: the interest is mainly in what is being done across Europe without more specifications.[11] IV) There is a clear recognition of the problems regarding "identity", in spite of the amount of work done in this continent on this subject-matter: "although there (...) is a solid growth in interest and numbers of people working in the subject, there is a lack of coherence in the European research effort".[12]

Therefore, in this paper the overview of the trends and problems from a European perspective is aware of geographical and institutional components. It emphasizes an "integrative" position in the philosophy of social and cultural sciences

9 Cf. Steering Committee, *The Philosophy of Science in a European perspective* Proposal of an "à la carte Programme" to be submitted to the European Science Foundation, Introduction.

10 Among others are the Swedish Collegium for Advanced Studies in the Social Science (SCASS), Uppsala (Sweden), and the Centre for Philosophy of the Natural and Social Science (CPNSS) at the London School of Economics (LSE), London, UK.

11 "Even though Europe is no longer alone in setting the parameters for discourse in and about science, during the last few decades a renewed and increasing interest in philosophical issues has been shown again by scholars all over Europe", Steering Committee, *The Philosophy of Science in a European perspective* Proposal of an "à la carte Programme", Introduction.

12 Steering Committee, *The Philosophy of Science in a European perspective* Proposal of an "à la carte Programme", Introduction.

open to an identity factor, due to the existence of a wide variety of orientations. Consequently, the remarks will take into account the three levels of analysis pointed out on the scientific status of the social sciences – general, comparative, and specific – as well as the guidance of the proposal approved by the European Science Foundation.

Put differently, the remarks are made here on the basis of the existence of four main elements: (i) the relevant scholars related to Europe, because their nationality or place of residence; (ii) the institutions that develop research on philosophy of social and cultural sciences; (iii) the "integrative view"; and (iv) the lack of coherence in order to have a neat picture of the work done in the continent.

Because of the space available for these remarks and the explicit character of *overview* of this paper, which makes clear that it is not an exhaustive study (of names, tendencies, institutions, etc.), the focus will be mainly in the central problems of the philosophy of social and cultural sciences rather than in other aspects. The sequence of topics will be presented according the schedule for the five years of the program. Obviously, the attention will be paid in the main aspects and those thinkers with more influence so far in the topics to be discussed.

3. The "Naturalist Turn", the "Social Turn", and the Discussion on Scientific Realism

Following the three levels of analysis pointed out – general, comparative, and specific –, some key tendencies – the "naturalist turn", the "social turn", and the varieties of scientific realism – are approaches that can be seen in the first and second levels. These philosophical-methodological viewpoints are directly connected to the debates on the scientific status of the social sciences as well as to the comparison between the natural and social sciences. These trends that belong to a "post-historical turn", and they have been very influential for more than two decades.

All of them – the views that conform "naturalistic turn", "social turn", and the realist conceptualizations – have an over-arching viewpoint on science. They accept a general frame that is common for empirical sciences. From that frame can be reached any cultural or social science, even though each one of those viewpoints involves a set of different conceptions (i.e., there are different types of naturalism, a diversity of realisms, a variety of social orientations …).

Moreover, some of them can be combined, because it is possible that a thinker might be naturalist in some issues (e.g., epistemological and methodological) and realist in other points (e.g., semantical and ontological). In any case, the three general viewpoints mentioned – naturalist, social, and realist – involve some kind of "foundational" approach on social sciences – an epistemological grounding – as well as a methodological characterization of the research on social sciences.

1. Naturalism in the social sciences – the second level of analysis: the comparative position – presupposes naturalism in science in general (the first level of analysis). In this regard, we can find several kinds of naturalisms in science that have repercussions on the naturalisms in the social sciences, among those general philosophical-methodological views on science are these:

(i) Semantic naturalism, where there is an acceptance of meaning as linguistic use, because meaning is based on a practice that can be described rather than prescribed. (ii) Epistemological naturalism, which accepts that human knowledge is well oriented and assumes a continuity between science and philosophy (and, then, that a metaphysical foundation of any of them is not needed). (iii) Methodological naturalism, where the progress of science (including the case of the social sciences) can be made through processes empirically tested according to criteria used in natural sciences. (iv) Ontological naturalism, which only accepts entities that in one way or another could be observable (i.e., it denies the legitimacy of unobservable entities such as "mind", "consciousness", and the like). (v) Axiological naturalism, where the scientific values are those that come from scientific practice.[13]

Undoubtedly, naturalism in the social sciences has a long tradition in Europe. This view assumes, in principle, that social sciences, both as a whole and singularly, are grounded on an epistemological and methodological basis analogous – or even identical – to that of the natural sciences. Thus, key elements – laws, models, and regularities – can be analyzed similarly to what is done in physics, biology, etc. Each discipline (economics, psychology, sociology, archaeology, anthropology) offers examples of the naturalistic perspective, such as the defense of economic laws, reductionist models in psychology and in sociology, or causal explanations in archaeology. A very influential version of naturalism comes from the use of an evolutionary framework for explaining and predicting in the social sciences.

When an *alternative* view to naturalism in the social sciences is adopted, there could be different options, either strong (i.e., anti-naturalism) or moderate (i.e., an interpretative perspective). These possibilities lead to stress human action –individual and social – as meaningful and ruled by intentionality. Therefore, the internal factors – intentions, beliefs, goals, etc. – and the external aspects – social, cultural, political, etc. – of human action can modify the basic methodological tools (laws, models and regularities). As a consequence, a dualist methodology is assumed, either in strict terms (anti-naturalism rejects a convergence with methods of the natural sciences) or in a more flexible way (in interpretative versions).

Each discipline features examples of alternative approaches to naturalism: rhetoric in economics, interpretative sociology, narrative conception of history, symbolic social anthropology, etc. In this regard, the legacy of Wittgenstein's

13 Cf. Wenceslao J. Gonzalez, "Novelty and Continuity in Philosophy and Methodology of Science", *Op. cit.*, p. 5.

views on the language of action (either directly[14] or through some authors such as Georg Henrik von Wright[15]) or the impact of hermeneutical elements (either in a Gadamerian flavor[16] or based on any of the proposals of the Frankfurt School[17]) have worked out in favor of an alternative to naturalism in the social sciences. Intentionality and interpretation are used to distinguish cultural and social sciences from natural sciences. These views can be seen in tune with some form of *Verstehen* instead of supporting an *Erklären*.

2. Meanwhile the "social turn" in the philosophy of the social sciences is frequently connected to Kuhnian schemes,[18] even though Kuhn himself disagrees with some of these interpretations[19]. After the publication of *The Structure of Scientific Revolutions*, there was in Europe a "sociological switch" based on an expansion of Kuhnian ideas. They have followed different views, which can be summarized in the following ways:

a) The finalization thesis (the conception of the *Finalizierung der Wissenschaft*) developed by the group that worked at the Max Planck Institut at Starnberg;[20] b) the *strong programme* in sociology of science of the Edinburgh School

14 An example is Peter Winch, *The Idea of a Social Science*, London: Routledge and K. Paul 1958 (2nd edition, 1990).

15 Cf. Georg Henrik von Wright, *Explanation and Understanding*, Ithaca: Cornell University Press 1971; Georg Henrik von Wright, "Replies", in: Juha Manninen/Raimo Tuomela (eds.), *Essays on Explanation and Understanding. Studies in the Foundations of Humanities and Social Sciences*, Dordrecht: Reidel 1976, pp. 371-413; and Georg Henrik von Wright, "Probleme des Erklärens und Verstehens von Handlungen", in: *Conceptus*, 19, 1985, pp. 3-19.

16 Cf. Hans Georg Gadamer, *Wahrheit und Methode*, Tubingen: J. C. B. Mohr (P. Siebeck) 5th ed., 1986 (1st ed., 1960).

17 Cf. Karl Otto Apel, "Causal Explanation, Motivational Explanation, and Hermeneutical Understanding", in: Gilbert Ryle (ed.), *Contemporary Aspects of Philosophy*, Stockfield: Oriel Press 1976, pp. 161-176; and Karl Otto Apel, *Die Erklären-Verstehen Kontroverse in Transzendental-Pragmatischer Sicht*, Frankfurt: Suhrkamp 1979. Translated into English by Georgina Warnke: *Understanding and Explanation. A Transcendental–Pragmatic Perspective*, Cambridge, MA: The MIT Press 1984.

18 Wenceslao J. Gonzalez, "The Philosophical Approach to Science, Technology and Society", *Op. cit.*, pp. 3-49.

19 "I am among those who have found the claims of the strong program absurd: an example of deconstruction gone mad", Thomas S. Kuhn, "The Trouble with the Historical Philosophy of Science", lecture at the University of Harvard on 19 November 1991. Paper reprinted in Thomas S. Kuhn, *The Road Since* Structure: *Philosophical Essays, 1970-1993, with an Autobiographical Interview*, edited by James Conant and John Haugeland, Chicago: The University of Chicago Press 2000, p. 110.

20 Cf. Wolf Schäfer (ed.), *Finalization in Science. The Social Orientation of Scientific Progress*, Dordrecht: Reidel 1983.

led by Barry Barnes[21] (now at the University of Exeter) and David Bloor;[22] c) the *Empirical Programme of Relativism* (EPOR) endorsed by academics of the University of Bath such as Harry Collins;[23] and d) the ethnomethodology of a constructivist orientation, connected to the works of Bruno Latour[24] and Steve Woolgar,[25] who have a purely sociological vision of scientific knowledge (objects, facts and discoveries).[26]

Many of these authors are sociologists rather than philosophers, but they have had influence on philosophy, mainly in the discussions on science, technology and society. On the one hand, they have stressed the analysis of the "external perspective" on science, emphasizing social values (e.g., on aims, processes and results); and on the other hand, they have called attention to sociology and social anthropology in order to understand laboratory experimentation and science as a human practice within a social milieu. The mainstream tendency in those cases is a social constructivism, a position commonly criticized by many realists.

---

21 Cf. Barry Barnes, *Interests and the Growth of Knowledge*, London: Routledge and K. Paul 1977; Barry Barnes, *T. S. Kuhn and Social Science,* London: Macmillan 1982 (N. York: Columbia University Press 1982); and Barry Barnes/David Bloor/John Henry, *Scientific Knowledge. A Sociological Analysis*, Chicago: The University of Chicago Press 1996.

22 Cf. David Bloor, "Wittgenstein and Mannheim on the Sociology of Mathematics", in: *Studies in History and Philosophy of Science*, 4, 1973, pp. 173-191; David Bloor, "Popper's Mystification of Objective Knowledge", in: *Science Studies*, 4, 1974, pp. 65-76; David Bloor, *Knowledge and Social Imagery*, London: Routledge and K. Paul 1976 (2nd ed., Chicago: The University of Chicago Press 1991); and David Bloor, *Wittgenstein: A Social Theory of Knowledge*, London: Macmillan 1983.

23 Cf. Harry M. Collins, "An Empirical Relativist Programme in the Sociology of Scientific Knowledge", in: Karin D. Knorr-Cetina/Michael Mulkay (eds.), *Science Observed: Perspectives in the Social Study of Science*, Sage, London, 1983, pp. 85-100; and Harry M. Collins/Trevor Pinch, *The Golem: What Everyone Should Know About Science*, Cambridge University Press, Cambridge, 1993.

24 Cf. Bruno Latour, *Science in Action*, Milton Keynes: Open University Press 1987. Bruno Latour, *Les Microbes: guerre et paix, suivi de Irréductions*, Paris: A.-M. Métaillé 1984; revised and expanded English version, translated by A. Sheridan and J. Law: *The Pasteurisation of France*, Cambridge, MA: Harvard University Press 1988. Bruno Latour, *Nous n'avons jamais été modernes – Essai d'anthropologie symétrique*, Paris: La Découverte 1991; revised and augmented edition, translated into English by Catherine Porter: *We have Never been Modern,* Brighton: Harvester 1993.

25 Cf. Steven Woolgar, "Critique and Criticism: Two Readings of Ethnomethodology", in: *Social Studies of Science*, 11, 4, 1981, pp. 504-514; Steven Woolgar, *Science: The Very Idea,* London: Tavistock 1988; Steven Woolgar (ed.), *Knowledge and Reflexivity: New Frontiers in the Sociology of Knowledge*, London: Sage 1988; and Michael Lynch and Steve Woolgar (eds.), *Representation in Scientific Practice*, Cambridge, MA: The MIT Press 1990.

26 Cf. Wenceslao J. Gonzalez, "Las revoluciones científicas y la evolución de Thomas S. Kuhn", in: Wenceslao J. Gonzalez (ed.), *Análisis de Thomas Kuhn: Las revoluciones científicas*, Madrid: Trotta 2004, pp. 15-103; especially, pp. 36-43.

3. A parallel tendency is scientific realism in social sciences, which is within the second level of analysis (the comparative position) and presupposes the existence of realism on science in general (the first level of analysis). In this regard, following the list of characteristic elements of a science, we have *de facto* a large number of possibilities for scientific realism: semantical, logical, epistemological, methodological, ontological, axiological, and ethical.

Commonly, the discussion on realism pays more attention to some of these possibilities, mainly those versions related to language, knowledge, method, and reality. These factors involve a vision of the world as well as values (in the different realms: cognitive, ethical, social, economic, etc.) that influence scientific practice. In addition, every possibility of approach to scientific realism (e.g., semantical, epistemological, methodological or ontological) can receive different interpretations within a realist framework. Altogether – the list of levels and the number of interpretations in each – leads to a plethora of versions of "scientific realism".[27]

Among the contemporary diversity of scientific realisms, there are some that have roots in Europe or have a characteristic version developed by a European. John Worrall's "structural realism"[28] and Ilkka Niiniluoto's "scientific critical realism" are two examples of general conceptions of science proposed by Europeans that can have repercussions on the social sciences. In the case of the structural realism, the initial focus on the natural sciences has been enlarged *de facto*. Thus, there are some new views related to social affairs, even though the relation is rather indirect: the social concern appears through the analysis of medical problems.[29] Meanwhile, the conception of the scientific critical realism already has a general characterization of the social sciences. In addition to epistemological and methodological claims, this position is connected with an ontology of the three worlds.[30]

27 Cf. Wenceslao J. Gonzalez, "Novelty and Continuity in Philosophy and Methodology of Science", *Op. cit.*, pp. 12-13.

28 Worrall's ideas have been the starting point for new reflections on the social sciences, cf. Harold Kincaid, "Structural Realism and the Social Sciences", in: *Philosophy of Science*, 75, 5, 2008, pp. 720-731.

29 Cf. John Worrall, "Why Randomize? Evidence and Ethics in Clinical Trials", in: Wenceslao J. Gonzalez and Jesus Alcolea (eds.), *Contemporary Perspectives in Philosophy and Methodology of Science*, pp. 65-82; and John Worrall, "Why There's No Cause to Randomize", in: *The British Journal for the Philosophy of Science*, 58, 3, 2007, pp. 451-488.

30 Cf. Ilkka Niiniluoto, "Realism, Wordlmaking, and the Social Sciences", in: Ilkka Niiniluoto, *Is Science Progressive?*, Dordrecht: Reidel 1984, pp. 211-225. (Symposium on "Scientific Progress and the Social Sciences", University of Tampere, April, 1980); Ilkka Niiniluoto, "Finalization, Applied Science, and Science Policy", in: Ilkka Niiniluoto, *Is Science Progressive?*, pp. 226-243; Ilkka Niiniluoto, *Critical Scientific Realism*, Oxford: Clarendon Press 1999; and Ilkka Niiniluoto, "World 3: A Critical Defence", in: Ian Jarvie, Karl Milford and David Miller (eds.), *Karl Popper: A Centenary*

Both approaches on realism – structural and critical – have relevant European roots – Henri Poincaré and Imre Lakatos –, in the case of Worrall, and Karl Popper's third world view, in Niiniluoto's proposal. Their views go beyond the original sources insofar as they are new areas of philosophical research, such as social consequences of clinical trials (including ethical problems)[31] and sustainable development.[32] Their realist conceptions can be used to analyze a characteristic problem of the social sciences, such as the discussion on objectivity and, thereafter, the debate on truth. The philosophical discussion on realism in the social sciences depends on the possibility of objectivity (mainly in language, knowledge, method, and values) as a crucial starting point to face the problem of truth (historical, economic, sociological, psychological, anthropological, etc.)

4. Explanation, Prediction, and Confirmation: Realm and Limits

Subsequent to the reflection on the naturalist, social, and realist views, the issues for the research in the second year are those in this section. In this regard, focusing on basic science, there are nowadays two central topics in the philosophy of the cultural and social sciences related to explanation, prediction, and confirmation. On the one hand, there is a methodological controversy on explanation and description in the cultural sciences, which affects above all social anthropology and historical sciences; and, on the other hand, the problems of the realm and limits of explanation and prediction in the social sciences have a clear relevance, especially in economics, but also with repercussions on other social sciences, such as sociology or political science.

"Explanation" versus "description" is a methodological dispute that can be seen commonly in the research of cultural sciences, such as social anthropology and historical sciences (e.g., in university departments). *De facto*, many authors assume that "description" is good enough in order to have "science" or even "mature science" in cultural subject matters. At the same time, those social scientists that have been trained in general methodology of science reject that "description" could be good enough in order to have "science", because they consider that scientific undertakings require "explanation", and that science should therefore give grounded replies to "why?" questions.[33]

Assessment, vol. II, Aldershot: Ashgate 2006, pp. 59-69.

31 Cf. John Worrall, "Why Randomize? Evidence and Ethics in Clinical Trials", *Op. cit.*, pp. 76-80.

32 Cf. Ilkka Niiniluoto, "Nature, Man, and Technology –Remarks on Sustainable Development", in: Lassi Heininen (ed.), *The Changing Circumpolar North: Opportunities for Academic Development*, Arctic Centre Publications 6, Rovaniemi, 1994, pp. 73-87.

33 Cf. Wenceslao J. Gonzalez, "Caracterización de la 'explicación científica' y tipos de explicaciones científicas", in: Wenceslao J. Gonzalez (ed.), *Diversidad de la expli-*

The realm and limits of explanation and prediction in the social sciences raised another methodological questions, some of them connected with the previous discussion. On the one hand, there is a twofold problem, concerning the characteristics of "scientific explanations" in the social sciences, and what kind of scientific explanations is to be found in this sphere (causal, non causal, etc.). On the other hand, social scientists – including Nobel Prize laureates – deeply disagree about the possibility of elaborating scientific predictions in the realm of human affairs.[34] In addition, from the point of view of confirmation, there is a debate on whether explanation or prediction has a greater epistemological and methodological impact.

Regarding scientific explanation in the social sciences, even when a causal approach is being adopted, the notion of "cause" is not always understood in the same way. In economics there are clear differences between Clive G. Granger's views and other perspectives.[35] In psychology there are also differences on "causation" in connection with the links between reasons, actions and causes. In archaeology, we can find attempts to combine two different kinds of causes (connected with means and ends).

Prediction is a key issue in the epistemology and methodology of social sciences. It can be used as a test of basic science (e.g., of economic theory) and as a tool for social policy in applied science (e.g., of applied economics). The realm and limits of predictions are crucial topics for *applied sciences*,[36] such as econom-

---

cación científica, Barcelona: Ariel 2002, pp. 13-49. In this regard, see Amparo Gomez, *Filosofía y Metodología de las Ciencias Sociales*, Madrid: Alianza Editorial 2003, and Robert C. Bishop, *The Philosophy of Social Sciences*, London: Continuum 2007, chapter 15, especially, pp. 326-331.

34 A paramount example is Sir John Hicks, Nobel Prize in economics in 1972. Cf. John Hicks, "A Discipline not a Science", in: John Hicks, *Classics and Moderns. Collected Essays on Economic Theory, v. III*, Cambridge, MA: Harvard University Press 1983, pp. 364-375; and John Hicks, "Is Economics a Science?", in: Mauro Baranzini/Roberto Scazzieri (eds.), *Foundations of Economics. Structures of Inquiry and Economic Theory*, Oxford: B. Blackwell 1986, pp. 91-101.

On this issue, cf. Wenceslao J. Gonzalez, "Prediction as Scientific Test of Economics", in: Wenceslao J. Gonzalez and Jesus Alcolea (eds.), *Contemporary Perspectives in Philosophy and Methodology of Science*, pp. 83-112; especially, pp. 88-90 and 98-100. See also Dieter Helm, "Predictions and Causes: A Comparison of Friedman and Hicks on Method", *Oxford Economic Papers*, new series, 36, 1984, pp. 118-134.

35 Cf. Clive W. J. Granger, "Testing for Causality, a Personal Viewpoint", in: *Journal of Economic Dynamics and Control*, 2, 1980, pp. 329-352; Clive W. J. Granger, "Where are the Controversies in Econometric Methodology?", in: Clive W. J. Granger (ed.), *Modelling Economics Series: Readings in Econometric Methodology*, Oxford: Clarendon Press 1990, pp. 1-23; and Clive W. J. Granger, "Time Series Analysis, Cointegration, and Applications", in: Tore Frängsmyr (ed.), *From Les Prix Nobel. The Nobel Prizes 2003*, Nobel Foundation, Stockholm: Nobel Foundation 2004, pp. 360-366.

36 Cf. Ilkka Niiniluoto, "The Aim and Structure of Applied Research", in: *Erkenntnis*, 38, 1993, pp. 1-21.

ics, sociology or social psychology. The importance of prediction is even greater insofar as it is connected to prescription (i.e., what ought to be done),[37] as is the case in economics (as the ongoing international financial crises that break out in 2008 has emphasized).

But the need for prediction in social sciences is usually counterbalanced by its unreliable nature, as Nobel Prizes in economics (Herbert Simon, Amartya Sen) have repeatedly pointed out,[38] even though some social events (e.g., public transport) can have very reliable predictions. In addition, there is always the possibility that a social prediction can affect the social phenomenon itself, a problem that affects sociology as well as political science.

All these issues related to explanation, prediction, and confirmation, their realm and limits, should be discussed from a European perspective. The approach should take into account an "integrative" viewpoint. This involves the contributions made by institutions located in European countries and the work of philosophers of any of the European nationalities. In addition, the three levels of the philosophical-methodological analysis – general, comparative, and specific – are at stake here.

5. The Debate on Mathematical Modeling in the Social Sciences and Consequences for Experimentation

A traditional controversial area of discussion for the cultural sciences is the role of probability and statistics in the study of human events in the cultural milieu, because the predominant view is in favor of qualitative aspects (e.g., trends, symbols, etc.) and against the use of experiments on cultural factors. Meanwhile, in the social sciences there is commonly a different attitude on mathematical modeling and on the role of experiments. The third year research of this program will deal precisely with the debate on mathematical modeling in the social sciences and its contribution to experiments on social issues.

This methodological debate on mathematical modeling is to be analyzed taking into account at least three kinds of problems: a) the very status of mathematical modeling (i.e., mathematics as a "language" used to establish knowledge, which includes the procedures of proof, or mathematics as a "heuristic tool" connected

37 See Herbert A. Simon, "Prediction and Prescription in Systems Modeling", in: *Operations Research*, 38, 1990, pp. 7-14; reprinted in Herbert A. Simon, *Models of Bounded Rationality. Vol. 3: Empirically Grounded Economic Reason*, Cambridge, MA: The MIT Press 1997, pp. 115-128.

38 On this issue, cf. Wenceslao J. Gonzalez, "Prediction and Prescription in Economics: A Philosophical and Methodological Approach", in: *Theoria*, 13, 32, 1998, pp. 321-345. See also Herbert A. Simon, "Forecasting the Future or Shaping it?", in: *Industrial and Corporate Change*, 11, 3, 2002, pp. 601-605.

to discovery through abstract representation);[39] b) the issue of symmetry or asymmetry between mathematical modeling in the natural sciences as compared to the social sciences; and c) the methodological controversy between qualitative and quantitative models within the social sciences.

Usually, experimental approaches in the social sciences need to deal with these kinds of problems on modeling (status; symmetry or asymmetry; and qualitative versus quantitative). Certainly it is the case in experimental economics, where Nobel Prizes (such as Reinhard Selten or Vernon Smith) are experts in game theory.[40] It is also the case in experimental psychology, where statistical techniques (Classical or Bayesian) have a crucial role. In addition, there are interdisciplinary studies based upon those criteria (e.g., in the combination of psychology and economics).[41]

Even though mathematical modeling is used in most social sciences (economics, psychology, history, sociology, political science), there is also an ongoing controversy about its realm and limits. The difference in the responses on the role of mathematical modeling can be seen in many ways, such as the diversity of methodological approaches in history (e.g., American "New History" versus "narrative history") or in psychology (e.g., in psychobiology and psychology of personality).

---

39 Herbert Simon used to distinguish two uses of mathematics. On the one hand, mathematics as a language of proof, where rigor is essential, a guarantee that conclusions are correct (Tjalling Koopmans, Gerard Debreu, Kenneth Arrow, etc.). And, on the other hand, mathematics as a language of discovery, as a tool to arrive at new ideas, where solutions reached with its help should be checked for correctness. The second one is Simon's preference, and he considers that it is physicists' mathematics and engineers' mathematics instead of mathematicians' mathematics. Cf. Herbert A. Simon, *Models of my Life*, N. York, NY: Basic Books 1991, pp. 106-107.

40 Cf. Wenceslao J. Gonzalez, "The Role of Experiments in the Social Sciences: The Case of Economics", in Theo Kuipers (ed.), *General Philosophy of Science: Focal Issues*, Amsterdam: Elsevier 2007, pp. 275-301. Robert Aumman is another Nobel Prize in economics expert in game theory and interested in epistemological and methodological matters, cf. Robert J. Aumann, "Rationality and Bounded Rationality", in: *Games and Economic Behavior*, 21, 1-2, 1997, pp. 2-14.

41 Cf. Daniel Kahneman/Jack Knetsch/Richard Thaler, "Experimental Tests of the Endowment Effect and the Coase Theorem", in: *Journal of Political Economy*, 98, 1990, pp. 1325-1348; and Daniel Kahneman, "Maps of Bounded Rationality: Psychology for Behavioral Economics", in: *American Economic Review*, 93, 5, 2003, pp. 1449-1475.

6. The Sciences that Philosophy has hitherto Ignored: The Sciences of Design

Among the fields that have been hitherto ignored by philosophy are the sciences of the artificial, understood as "sciences of design".[42] It happens that an important part of what is usually called "social sciences" belongs, *de facto*, to the realm of the "sciences of the artificial" (e.g., library science, communication, or economics). In addition, there are other sciences within the territory of the "sciences of design", such as pharmacology, that are also sciences of the artificial – in the sense proposed by Herbert Simon – and have a clear link to social sciences. Commonly, these sciences have been hitherto ignored by philosophy (at least, in the mainstream philosophy of science), yet they have relevant links with the cultural and social sciences that should be analyzed.

These sciences of the artificial, which are usually design sciences, come from what Ilkka Niiniluoto calls "scientification", i.e., a process to change a professional practice into a scientific discipline.[43] Thus, they come from a social practice that requires scientific support in the form of a design. Habitually, they are "applied sciences", and this feature is relevant to study them, because traditionally philosophy of science has focused more on basic science than on applied science.

Following the differences and similarities between the sciences of the artificial and the cultural and social sciences, it is possible – in the fourth year of research of this program – to shed light on some sciences that were not commonly studied in this area. These are library science, communication, pharmacology, economics, etc., which can have epistemological and methodological problems insofar as they are applied sciences of design. These problems are not purely cognitive, because they involve other values (social, cultural, economic, ecological, etc.).

7. New Directions in the Philosophy of Science

A European perspective should be open to the future. The research of Team C includes an explicit reflection of new directions in the philosophy of science. *Prima facie*, this may be understood in different ways, according to which it might be considered as "novelty", which includes at least three possibilities. (i) There is a novelty in ontological terms, when something has happen after the original moment; in this case, new conceptions after the approval of the program by the European Science Foundation. (ii) There is a novelty in epistemological terms, when something is going on de facto but we are not aware of the innovative character of

42 Cf. Herbert A. Simon, *The Sciences of the Artificial*, 3rd ed., Cambridge, MA: The MIT Press 1996 (1st ed., 1969; 2nd ed., 1981).

43 Cf. Ilkka Niiniluoto, "The Aim and Structure of Applied Research", *Loc. cit.*, pp. 8-11.

its contents. (iii) There is also a heuristic novelty, when there are new patterns or original paths to address old problems.

Looking at five years from now, it seems rather obvious that there will be some new directions in the philosophy of science in any of these three possible options of "novelty". In addition to the study of new branches of the social sciences (such as rising subdisciplines), these emerging philosophies of science can be on new approaches in the social concern on science. They may be connected to the social constructivism and the realism on the cultural and social sciences. Thus, in the final year of the program, Team C will investigate the novel views on the influence of social constructivism and realism on the social sciences.

From a methodological viewpoint, the opposition between social constructivism and realism affects the social sciences as a whole as well as each one of them (anthropology, archaeology, economics, etc.). It is a debate that has not only theoretical consequences, but also practical effects. One example is how to understand medicine, due to new approaches to medicine as a cultural and social science: the alternative medicine, which is a relevant aspect of the social concern on medicine as a science.[44] This issue should be studied with an eye to the European perspective understood in the "integrative" terms pointed out.

On the one hand, social constructivism is a widespread tendency in the social sciences, as Ian Hacking has emphasized.[45] Understood as an epistemological and methodological approach – and often also as an ontological conception –, social constructivism is adopted nowadays by virtually every social discipline. Social constructivism could be open to methodological individualism (e.g., in cognitive approaches in psychology) or might adopt a clear holistic methodology (e.g., in social psychology or in social policy). On the other hand, realism offers a variety of alternatives to social constructivism. The revival of "scientific realism" in recent years emphasizes the possibility of objectivity regarding social phenomena.

8. Final Remarks

My previous lines suggest that any attempt to analyze the trends and problems in philosophy of social and cultural sciences from a European perspective should take into account several aspects. First, the level of analysis in the philosophical-methodological approach used (general, comparative, or specific). Second, what might be reasonably considered from a philosophical point of view as "a European perspective", which can be understood in "integrative" terms.[46] Third, the

44 Cf. Donald A. Gillies, "El problema de la demarcación y la Medicina alternativa", in: Wenceslao J. Gonzalez (ed), *Karl Popper: Revisión de su legado*, Madrid: Unión Editorial 2004, pp. 197-219.

45 Cf. Ian Hacking, *The Social Construction of* What?, Cambridge, MA: Harvard University Press 1999.

46 If we focus our attention on journals that deal with topics related to philosophy of

current stage of the philosophical considerations after the decline of the "historical turn". In this regard, there is now an active competition among the naturalist conceptions, the followers of the "social turn", and the supporters of the scientific realism. These general philosophical-methodological approaches have direct repercussions on the views of the cultural and social sciences.

They appear as topics of discussion for the first year. Thereafter, some other relevant topics, which have been pointed out, are scheduled for the following years. My proposals have been predominantly within levels two and three of the philosophical analysis. And they have been focused on what should be the *context* of discussion in order to move along the lines of the suggestions of submitted to the European Science Foundation. In this regard, I have suggested that "European" can be understood from an "integrative" viewpoint, where a wide range of conceptions can be included (insofar as there is a search for a *common ground*, both in historical terms and in thematic ones). Moreover, the contributions made by institutions located in European countries and the work of philosophers of any of the European nationalities can be used to build up an "identity" for the philosophy of the cultural and social sciences in Europe.[47]

Bibliography

Karl Otto Apel, "Causal Explanation, Motivational Explanation, and Hermeneutical Understanding", in: Gilbert Ryle (ed.), *Contemporary Aspects of Philosophy*, Stockfield: Oriel Press 1976, pp. 161-176.

Karl Otto Apel, *Die Erklären-Verstehen Kontroverse in transzendental-pragmatischer Sicht*, Frankfurt: Suhrkamp 1979. Translated into English by Georgina Warnke: *Understanding and Explanation. A Transcendental-Pragmatic Perspective*, Cambridge, MA: The MIT Press 1984.

Robert J. Aumann, "Rationality and Bounded Rationality", in: *Games and Economic Behavior*, 21, 1-2, 1997, pp. 2-14.

Barry Barnes, *Interests and the Growth of Knowledge*, London: Routledge and K. Paul 1977.

economics, there is a relevant list of journals directed by European editors or run by European publishing houses. The list includes *Economics and Philosophy*, *Journal of Economic Methodology*, and *Journal for the History of Economic Thought*. In addition, there are other publications with a clear presence of Europeans in the editorial board, such as *Politics, Philosophy and Economics*. In an "integrative sense", these journals might be considered as "Europeans". But, if we follow restricted criteria, it is hard to say that they are *tout court* "Europeans".

47 This issue is not disconnected from the philosophical discussion on the European unity. In this regard, there are many contributions. Among them are those initiated in the first half of the twentieth century, cf. Harold C. Raley, *José Ortega y Gasset: Philosopher of the European Unity*, Alabama: University of Alabama Press 1971.

Barry Barnes, *T. S. Kuhn and Social Science,* London: Macmillan 1982 (N. York: Columbia University Press 1982).

Barry Barnes/David Bloor/John Henry, *Scientific Knowledge. A Sociological Analysis*, Chicago: The University of Chicago Press 1996.

Robert C. Bishop, *The Philosophy of Social Sciences*, London: Continuum 2007.

David Bloor, "Wittgenstein and Mannheim on the Sociology of Mathematics", in: *Studies in History and Philosophy of Science*, 4, 1973, pp. 173-191.

David Bloor, "Popper's Mystification of Objective Knowledge", in: *Science Studies*, 4, 1974, pp. 65-76.

David Bloor, *Knowledge and Social Imagery*, London: Routledge and K. Paul 1976 (2nd ed., Chicago: The University of Chicago Press 1991).

David Bloor, *Wittgenstein: A Social Theory of Knowledge*, London: Macmillan 1983.

Harry M. Collins, "An Empirical Relativist Programme in the Sociology of Scientific Knowledge", in: Karin D. Knorr-Cetina/Michael Mulkay (eds.), *Science Observed: Perspectives in the Social Study of Science*, London: Sage 1983, pp. 85-100.

Harry M. Collins/Trevor Pinch, *The Golem: What Everyone Should Know About Science*, Cambridge: Cambridge University Press 1993.

Hans Georg Gadamer, *Wahrheit und Methode*, Tubingen: J. C. B. Mohr (P. Siebeck) 5th ed. 1986 (1st ed. 1960).

Donald A. Gillies, "El problema de la demarcación y la Medicina alternativa", in: Wenceslao J. Gonzalez (ed.), *Karl Popper: Revisión de su legado*, Madrid: Unión Editorial 2004, pp. 197-219.

Amparo Gomez, *Filosofía y Metodología de las Ciencias Sociales*, Madrid: Alianza Editorial 2003.

Wenceslao J. Gonzalez, "Prediction and Prescription in Economics: A Philosophical and Methodological Approach", in: *Theoria*, 13, 32, 1998, pp. 321-345.

Wenceslao J. Gonzalez, "Caracterización de la 'explicación científica' y tipos de explicaciones científicas", in: Wenceslao J. Gonzalez (ed.), *Diversidad de la explicación científica*, Barcelona: Ariel 2002, pp. 13-49.

Wenceslao J. Gonzalez, "From *Erklären-Verstehen* to *Prediction-Understanding*: The Methodological Framework in Economics", in: Matti Sintonen, Petri Ylikoski and Karl Miller (eds.), *Realism in Action: Essays in the Philosophy of Social Sciences*, Dordrecht: Kluwer 2003, pp. 33-50.

Wenceslao J. Gonzalez, "Las revoluciones científicas y la evolución de Thomas S. Kuhn", in: Wenceslao J. Gonzalez (ed.), *Análisis de Thomas Kuhn: Las revoluciones científicas*, Madrid: Trotta 2004, pp. 15-103.

Wenceslao J. Gonzalez, "The Philosophical Approach to Science, Technology and Society", in: Wenceslao J. Gonzalez (ed.), *Science, Technology and Society: A Philosophical Perspective*, A Coruña: Netbiblo 2005, pp. 3-49.

Wenceslao J. Gonzalez, "Novelty and Continuity in Philosophy and Methodology of Science", in: Wenceslao J. Gonzalez and Jesus Alcolea (eds.), *Contem-*

porary Perspectives in Philosophy and Methodology of Science, A Coruña: Netbiblo 2006, pp. 1-28.

Wenceslao J. Gonzalez, "Prediction as Scientific Test of Economics", in: Wenceslao J. Gonzalez and Jesus Alcolea (eds.), *Contemporary Perspectives in Philosophy and Methodology of Science*, A Coruña: Netbiblo 2006, pp. 83-112.

Wenceslao J. Gonzalez, "The Role of Experiments in the Social Sciences: The Case of Economics", in: Theo Kuipers (ed.), *General Philosophy of Science: Focal Issues*, Amsterdam: Elsevier 2007, pp. 275-301.

Clive W. J. Granger, "Testing for Causality, a Personal Viewpoint", in: *Journal of Economic Dynamics and Control*, 2, 1980, pp. 329-352.

Clive W. J. Granger, "Where are the Controversies in Econometric Methodology?", in: Clive W. J. Granger (ed.), *Modelling Economics Series: Readings in Econometric Methodology*, Oxford: Clarendon Press 1990, pp. 1-23.

Clive W. J. Granger, "Time Series Analysis, Cointegration, and Applications", in: Tore Frängsmyr (ed.), *From Les Prix Nobel. The Nobel Prizes 2003*, Nobel Foundation, Stockholm: Nobel Foundation 2004, pp. 360-366.

Ian Hacking, *The Social Construction of* What?, Cambridge, MA: Harvard University Press 1999.

Dieter Helm, "Predictions and Causes: A Comparison of Friedman and Hicks on Method", *Oxford Economic Papers*, new series, 36, 1984, pp. 118-134.

John Hicks, "A Discipline not a Science", in: John Hicks, *Classics and Moderns. Collected Essays on Economic Theory, v. III*, Cambridge, MA: Harvard University Press 1983, pp. 364-375.

John Hicks, "Is Economics a Science?", in: Mauro Baranzini/Roberto Scazzieri (eds.), *Foundations of Economics. Structures of Inquiry and Economic Theory*, Oxford: B. Blackwell 1986, pp. 91-101.

Daniel Kahneman/Jack Knetsch/Richard Thaler, "Experimental Tests of the Endowment Effect and the Coase Theorem", in: *Journal of Political Economy*, 98, 1990, pp. 1325-1348.

Daniel Kahneman, "Maps of Bounded Rationality: Psychology for Behavioral Economics", in: *American Economic Review*, 93, 5, 2003, pp. 1449-1475.

Harold Kincaid, "Social Sciences", in: Peter Machamer and Michael Silberstein (eds.), *The Blackwell Guide to the Philosophy of Science*, Oxford: Blackwell 2002, pp. 290-311.

Harold Kincaid, "Structural Realism and the Social Sciences", in: *Philosophy of Science*, 75, 5, 2008, pp. 720-731.

Thomas S. Kuhn, "The Trouble with the Historical Philosophy of Science", lecture at the University of Harvard on 19 November 1991. Paper reprinted in Thomas S. Kuhn, *The Road Since* Structure: *Philosophical Essays, 1970–1993, with an Autobiographical Interview*, edited by James Conant and John Haugeland, Chicago: The University of Chicago Press 2000, pp. 105-120.

Bruno Latour, *Science in Action*, Milton Keynes: Open University Press 1987.

Bruno Latour, *Les Microbes: guerre et paix, suivi de Irréductions*, Paris: A.-M. Métaillé 1984. Revised and expanded English version, translated by A. Sheridan and J. Law: *The Pasteurisation of France*, Cambridge, MA: Harvard University Press 1988.

Bruno Latour, *Nous n'avons jamais été modernes – Essai d'anthropologie symétrique*, Paris: La Découverte 1991. Revised and augmented edition, translated into English by Catherine Porter: *We have Never been Modern,* Brighton: Harvester 1993.

Michael Lynch and Steve Woolgar (eds.), *Representation in Scientific Practice*, Cambridge, MA: The MIT Press 1990.

James W. McAllister, "Editorial: Contours of a European Philosophy of Science", in: *International Studies in the Philosophy of Science*, 22, 1, 2008, pp. 1-3.

Ilkka Niiniluoto, "The Aim and Structure of Applied Research", in: *Erkenntnis*, 38, 1993, pp. 1-21.

Ilkka Niiniluoto, "Realism, Wordlmaking, and the Social Sciences," in: Ilkka Niiniluoto, *Is Science Progressive?*, Dordrecht: Reidel 1984, pp. 211-225. (Symposium on "Scientific Progress and the Social Sciences", University of Tampere, April, 1980.)

Ilkka Niiniluoto, "Finalization, Applied Science, and Science Policy," in: Ilkka Niiniluoto, *Is Science Progressive?*, Dordrecht: Reidel 1984, pp. 226-243.

Ilkka Niiniluoto, "Nature, Man, and Technology –Remarks on Sustainable Development," in: Lassi Heininen (ed.), *The Changing Circumpolar North: Opportunities for Academic Development*, Arctic Centre Publications 6, Rovaniemi, 1994, pp. 73-87.

Ilkka Niiniluoto, *Critical Scientific Realism*, Oxford: Clarendon Press 1999.

Ilkka Niiniluoto, "World 3: A Critical Defence", in: Ian Jarvie, Karl Milford and David Miller (eds.), *Karl Popper: A Centenary Assessment*, vol. II, Aldershot: Ashgate 2006, pp. 59-69.

Harold C. Raley, *José Ortega y Gasset: Philosopher of the European Unity*, Alabama: University of Alabama Press 1971.

Wolf Schäfer (ed.), *Finalization in Science. The Social Orientation of Scientific Progress*, Dordrecht: Reidel 1983.

Herbert A. Simon, "Prediction and Prescription in Systems Modeling," in: *Operations Research*, 38, 1990, pp. 7-14. Reprinted in Herbert A. Simon, *Models of Bounded Rationality. Vol. 3: Empirically Grounded Economic Reason*, Cambridge, MA: The MIT Press 1997, pp. 115-128.

Herbert A. Simon, *Models of my Life*, N. York, NY: Basic Books 1991.

Herbert A. Simon, *The Sciences of the Artificial*, 3rd ed., Cambridge, MA: The MIT Press 1996 (1st ed., 1969; 2nd ed., 1981).

Herbert A. Simon, "Forecasting the Future or Shaping it?", in: *Industrial and Corporate Change*, 11, 3, 2002, pp. 601-605.

Steering Committee, *The Philosophy of Science in a European perspective* Proposal of an "à la carte Programme" to be submitted to the European Science Foundation, 24 February 2006.

Stephen P. Turner / Paul A. Roth (eds.), *The Blackwell Guide to the Philosophy of the Social Sciences*, Oxford: Blackwell 2003.

Peter Winch, *The Idea of a Social Science*, London: Routledge and K. Paul 1958 (2nd edition, 1990).

Georg Henrik von Wright, *Explanation and Understanding*, Ithaca: Cornell University Press 1971.

Georg Henrik von Wright, "Replies," in: Juha Manninen / Raimo Tuomela (eds.), *Essays on Explanation and Understanding. Studies in the Foundations of Humanities and Social Sciences*, Dordrecht: Reidel 1976, pp. 371-413.

Georg Henrik von Wright, "Probleme des Erklären und Verstehens von Handlungen," in: *Conceptus*, 19, 1985, pp. 3-19.

Steven Woolgar, "Critique and Criticism: Two Readings of Ethnomethodology", in: *Social Studies of Science*, 11, 4, 1981, pp. 504-514.

Steven Woolgar, *Science: The Very Idea*, London: Tavistock 1988.

Steven Woolgar (ed.), *Knowledge and Reflexitivity: New Frontiers in the Sociology of Knowledge*, London: Sage 1988.

John Worrall, "Why Randomize? Evidence and Ethics in Clinical Trials", in: Wenceslao J. Gonzalez and Jesus Alcolea (eds.), *Contemporary Perspectives in Philosophy and Methodology of Science*, A Coruña: Netbiblo 2006, pp. 65-82.

John Worrall, "Why There's No Cause to Randomize", in: *The British Journal for the Philosophy of Science*, 58, 3, 2007, pp. 451-488.

University of A Coruña
Faculty of Humanities
Dr. Vazquez Cabrera Street, w/n
15403 Ferrol
Spain
wencglez@udc.es

Arto Siitonen

State of the Art

A Commentary on Wenceslao J. Gonzalez' Contribution, "Trends and Problems in Philosophy of Social and Cultural Sciences: A European Perspective"

Introduction

Scientific research proceeds from problems to their supposed solutions, which in turn raise further problems. A typical research process goes through the stages of invention and testing. Its results are presented in theories; these justify the results and make them understandable. Theories often have applications, for instance in technology, commerce, social organizations and in teaching. Research is a cultural activity pursued in international co-operation among individuals and groups. It both follows traditions and renews them, and takes its direction according to both its internal logic and also external factors. The assumed direction at a certain time makes up the trend that is characteristic for science in general, or for a given discipline. Trends are typical ways of approaching problems and designing theories.

Cultural and social sciences in particular are reflective, in that they are part of the processes they investigate, and in that these sciences' cultivation contributes to the development of culture and society. Each of these groups of sciences has in its background certain branches of philosophy – i.e., philosophy of cultural sciences and philosophy of culture; philosophy of social sciences and social philosophy. Moreover, the idea of humanism can be seen behind the cultural sciences, and behind philosophy of culture. One may wonder whether the doctrine of socialism has a similar role in the philosophy of social sciences and in social philosophy. There is an important difference, however. Humanism involves the cultivation of literature as well as dedication to learning and teaching, whereas the word 'socialism' has a strongly political connotation. Cultural sciences are called "humanistic disciplines", but social sciences cannot be characterized as "socialistic disciplines".

Sometimes the cultural and social sciences are classified together as human or moral sciences, since the research object of both areas concerns human action and its consequences. In this case, the term 'human sciences' differs from the terms 'humanistic sciences' and 'humanities', the latter being tantamount to 'cultural sciences'. Moreover, the humanities are also called 'arts' or 'art subjects' (cf. the titles 'Bachelor of Arts', 'Master of Arts'). The boundaries between cultural sci-

F. Stadler (ed.), *The Present Situation in the Philosophy of Science*,
The Philosophy of Science in a European Perspective, DOI 10.1007/978-90-481-9115-4_19,

ences and social sciences are not sharp. For instance, in some cases history is considered a social rather than a cultural science.[1]

Culture and language differences introduce a further complication, because the German tradition of the Geisteswissenschaften differs from the Anglo-Saxon conception of sciences. In the latter, the terms 'science' and 'natural science' are implicitly understood as synonyms. Thus, the German word 'Wissenschaft' has a much broader scope than the English word 'science'. In respect to this awkward situation, one may suggest that there is a distinction between humanities and cultural sciences. The distinctive mark of the latter is that they follow accepted methods of science – construct hypotheses and test them, using deduction, induction and abduction.

Besides identifying various branches of science, we may also ask where and how these have been cultivated. These are questions pertaining to traditions, styles and geography. Terms such as 'Mesopotamian', 'Egyptian', 'Greek', 'American', etc. to modify the word 'science' are needed in order to answer such questions.

The following comments concern Wenceslao J. Gonzalez analysis of the fields of cultural and social sciences and the problems, trends, turns and directions in their philosophy, as seen from a European point of view.

1. Tasks ahead

Gonzalez thoroughly and insightfully reviews and maps the social and cultural landscape in addition to the sciences dedicated to clarifying it. He completes his review with a plan of tasks to be accomplished, year by year. The relevant disciplines – social and cultural sciences and their philosophy – are to be examined in respect to their basic concepts and methods. He emphasizes that there are several levels of analysis and sets himself the task of clarifying what can be considered as "a European perspective".

Social sciences make up a spectrum that stretches from psychology, social psychology and sociology to the study of international relations. Social sciences are dedicated to society, politics and economy. On the other hand, cultural sciences comprise linguistics and all philologies (there are in principle as many of these as there are natural languages), archaeology and social or cultural anthropology. Literary and folkloric studies, history, musicology and the study of fine arts belong to humanistic disciplines. Cultural sciences and humanistic disciplines are dedicated to sounds, signs and symbols and the spoken and written word. The basic methods of cultural sciences and humanities are documentation, interpretation and explanation.

1 "History, or some aspects of history, are sometimes classified as a social science." Bernard Williams, *Philosophy as a Humanistic Discipline*. Princeton and Oxford: Princeton University Press, 2006, p. 180.

As to mathematics and natural sciences, these belong mostly to the background than to the content of the planned study. Mathematics and logic provide tools for analysing formal structures. Statistics and probabilistic reasoning are central to the methodology of economics, sociology and other social sciences. The research areas of natural sciences and of social and cultural sciences sometimes overlap, whereas the sciences' approaches differ. These common factors will receive a due consideration on the first two levels of study. Thus, level i) of Gonzalez's analysis concerns "the general scientific status of the cultural and social sciences", whereas in level ii) he focuses on clarifying of „the scientific status of the cultural and social sciences as compared to that of the natural sciences" (cf. Gonzalez′ text).

Such a status-determination and comparison are highly demanding tasks by themselves. Nevertheless, level iii) of the analysis virtually exceeds the general and the comparative levels in terms of details to be considered: it concerns "the specific issues on the scientific status of each cultural and social discipline". This can be seen in both "the broad approach" and in "the restrictive position". The broad approach involves philosophical reflection on a given particular discipline and its relations to science in general, whereas the restrictive approach focuses specifically on "the problems of the discipline analyzed".

Thus, Gonzalez's main distinctions are: i) general, ii) comparative and iii) specific. Here the specific can be the broad approach and the restricted position. One may draw a figure of this by first distinguishing general – specific, and second making a subdivision of the specific-category into broad – restrictive. Moreover, an extra line can be drawn between the general and the specific to mark the comparison of the group of cultural and social sciences to that of the natural sciences.

In any given discipline, the various problems' identification, analysis and possible resolution, as well as the clarification of methods and theories with their ramifications, are huge tasks in themselves (cf. the restrictive approach). Society, economy, politics, jurisdiction and culture continuously generate new problems, and so do the social and cultural sciences themselves that are dedicated to those problems' analysis and resolution. Some relief is given by Gonzalez's qualification that the focus will mainly be on methodological problems. The broad approach will have to connect the analyses of problems to an overall philosophical study of the universal features of scientific research – and of the specific character of a given social or cultural science. All of this is to be related to an analysis of the general status of cultural and social sciences, and to be compared with natural sciences. Thus, the requirements ahead are demanding but by no means impossible to meet.

2. On Europe and 'European'

Historians of science or philosophy typically raise five kinds of question: where?, when?, who?, what?, and how? The desiderata of these questions are place, time, persons, content and style. Concerning the desideratum of the ‚where' question, it should be noted that localization of such entities as 'Europe', 'America', 'Asia', etc. first requires a clarifying of their boundaries, then a specifying of an area within these boundaries – for instance: where is Europe?, where in Europe? These are geographic issues. Chronological determination in turn varies in its scope, depending on whether the research object is, say, a stream of thought, an institute, a university, a research group, or certain periods in their development or in the lives of their members. The who-question concerns individual scientists and philosophers, their life and works. The question 'what?' is addressed to the achievements of these persons. The how-question, finally, concerns methods and styles of thought.

According to Gonzalez, analyses on all three above-mentioned levels – i) general, ii) comparative and iii) specific – should be done within, or from the integrative point of view, of a "European perspective" that is assumed de facto in the programme of the European Science Foundation. This raises complex questions as to "the precise sense and reference of that expression." One may think that the reference of the word 'European' is the concrete geographic entity with its past, present and future, while the sense is constituted by the features that make the philosophy cultivated on that continent a specifically European one. Thus, Gonzalez speaks of an historical and a thematic approach that complement each other. The former is addressed to the tradition of philosophy on the European continent, whereas the latter is concerned with the style of thinking that has created, and creates, identity, and is thus peculiar to Europe.

Logically speaking, several positions can be assumed in respect to the explication of a European perspective. Gonzalez distinguishes three possible standpoints. And a fourth one could be added; but if it were assumed then it would be incompatible with any consolidated picture of European philosophy of science. The widest alternative he calls "the integrative position". It outlines a general approach to actual philosophical work done in Europe with the goal of finding common ground. The second view is more specific, since it requires that one may call 'European' only those streams that originated or mainly developed in Europe. The third view is even more rigorous, in substituting the 'or' by an 'and', so that the boundaries are especially strict. Finally, according to the fourth view, there is not enough coherence among the various philosophical occupations in Europe to sustain a unified picture.

One may note that choosing among these four alternatives is not only an issue of contents or merely a matter of facts of culture, but is also based on voluntary decision and commitment. Accordingly, why should we not adopt the integrative position and design our arguments so that we can defend it? This appears to be the choice made by Gonzalez and encouraged by the ESF steering committee.

Let us also mention a geographic-historical consideration of Europe, and two earlier analyses of the concept of European philosophy. 'Europe' is a geographic concept. Europe consists of a certain part of the Eurasian continent, of North Sea isles (Iceland, Great Britain, etc.) and of Mediterranean isles. It is a matter of convention where the eastern boundaries of Europe are drawn; customarily these are thought to run across the Uralian mountains and Bosporus. Present-day Europe, as a politically, economically and culturally conceived entity, may be characterized in terms of the European Union agenda (cf. the area of the present EU and its expansion programme).

The very concept of European philosophy, in its historical meaning, seems to include such cities as Miletus, Ephesos and Alexandria, and thus parts of Middle East. European philosophy forms a part of Western philosophy. This tradition was inaugurated in the city states of Ancient Greece, continued in medieval cities and sites (alongside the Eastern, Arabic tradition), and modernized since the Renaissance. Linguistically, it was rooted in Classical Greek, and continued in Latin and in various national languages (Romanic, Germanic, Anglo-Saxon, Slavonic, etc.) The geographic and chronological characterization of 'Europe' is completed by 'Europe' as experienced psychologically and socio-culturally, and as expressed symbolically.

A specifically European philosophy as well as perspective can be gleaned from these elements. European philosophy and its perspective are clearly distinguishable from the Indian and Chinese philosophical traditions, albeit far less definitely from the American and Australian philosophies. However, although the latter can be traced back to the European tradition, they are cultivated elsewhere than in Europe. In this sense – but only in this sense – does the Austrian philosophy, for instance, belong to European philosophy, whereas the Australian philosophy does not.

We are thus allowed to fix some boundaries by tracing a line from Plato's Academy through Medieval cloisters and universities to modern and contemporary European universities, research institutes, groups and projects, as well as by naming representative individuals and works. All of these are identifiable as the defining factors of European philosophy and European science.

Among explicit presentations of the European philosophical and scientific heritage, one may mention Innocent M. Bochenski's work, *Contemporary European Philosophy*,[2] and Edmund Husserl's cultural criticism in his work, *Die Krisis der europäischen Wissenschaften und die transzendentale Phänomenologie*.[3] Husserl's work is based on his lecture in Vienna on May 7th, 1935. That year saw

2 Berkeley and Los Angeles: University of California Press 1957. (Originally: *Europäische Philosophie der Gegenwart*. 1947).

3 Haag: Martinus Nijhoff 1962 (Husserliana Band VI).

migration of many intellectuals, and German science and philosophy were losing their leading position in the world.

Bochenski divides the streams to be considered into six main classes: philosophy of matter, mind, idea, life, essence, existence and being. He includes the pragmatism of William James and John Dewey in the chapter on philosophy of life, justifying this by admitting that they "are Americans, but their influence upon European thought has been so important that they at least deserve touching upon".[4] If one adds that these philosophers not only influenced European thought but were strongly influenced by it, the integrative position (cf. above), widely applied, may be seen to accommodate them.

Husserl is stressing the freedom of scientific and philosophical thinking that was gained in Renaissance as a rediscovered gift of the Classical culture. The actual (i.e., until 1935) crisis of European philosophy and science has resulted from blind formalistic thinking and one-sided service to technology. The way out becomes available, when formalization is complemented by transcendental reflection, and external analysis by the study of intentionality.[5]

3. On "Turns"; on Naturalism and Realism

Turns in general may be said to be decisive phases in cultural development; they introduce trends. Gonzalez studies the naturalist turn and the social turn alongside with scientific realism. These turns are said to be part of – or perhaps even to constitute – a "post-historical turn", by which he presumably means the developments since T. S. Kuhn's influential theory of the formation of scientific paradigms and revolutions.

Gonzalez distinguishes five types of naturalism in science, and seven types of realism. In both cases, an implication holds: naturalism in the social and cultural sciences presupposes naturalism in science in general – and realism in the cultural and social sciences presupposes realism in science in general. The types that naturalism shares with realism concern semantics, epistemology, methodology, ontology and axiology. The additional types of realism concern logic and ethics. Naturalists stress description, empirical testing, observation and practice, and their stance is anti-metaphysical (in the substantive sense). 'Reducibility' is the key word here: of meaning to use, of value to fact, of prescription to description. Realists, on the other hand, focus on possible relations between language and reality, and between science and facts. This leads them to question assumptions of truth and objectivity in natural, cultural and social sciences.

There are two kinds of alternative to naturalism: anti-naturalism and a moderate view, the latter being tantamount to "an interpretative perspective". In this

4 Bochenski, Loc. cit., p. 114.

5 Cf. Husserl, Ibid., esp. sections 3, 10-14, 34, 61 and 68.

respect, it seems that yet another "turn" can be posited: the interpretive one. This claim may be sustained by mentioning the achievements of Hans-Georg Gadamer[6] (historical understanding), Peter Winch[7] (an inward approach to social sciences), and Georg Henrik von Wright[8] (clarification of the relations between explanation and understanding). Cf. also the work entitled *The Interpretive Turn*.[9]

As a predecessor to these developments, one may mention Heinrich Gomperz's article "Interpretation"[10], in which he distinguishes the historical sciences from sociology and analyzes interpretation as „one particular method of historical research".[11] This would be addressed to conduct, language and texts. Of these, he analyzes conduct and characterizes the methods of its interpretation as either subjective or objective. It would be fruitful to continue Gomperz's programme, analysing language and texts according to his guidelines.

A much earlier attempt to clarify the concept of interpretation is Bernard Bolzano's analysis in §§ 385-387 of his main work. Here, he discusses the following procedures: inquiry into judgments by other persons, discovery of intentions behind given actions, and interpretation of given signs.[12] He uses the term 'hermeneutics' in his remarks in the text.[13] This is an objectivistic analysis to which attention is due. One may also mention Franz Brentano's study of intentionality in his 1924 work on empirical psychology.

Gonzalez applies the concept of interpretation to 'naturalism' and 'realism' themselves, because the various types of naturalism and realism can be seen as so many interpretations. As he says, there is "a plethora of versions of scientific realism". He mentions two of these versions: the structural realism of John Worrall and the scientific critical realism of Ilkka Niiniluoto. One may mention another variant: the ontological realism by Hans Reichenbach that was based on probabilistic considerations. According to him, these indicate "the world as it is, objectively speaking."[14]

6 *Wahrheit und Methode*. Tübingen: J.C.B. Mohr (Paul Siebeck) 1965 (originally 1960).

7 *The Idea of a Social Science and its Relation to Philosophy*. London: Routledge and Kegan Paul, 1958.

8 *Explanation and Understanding*. Ithaca–New York: Cornell University Press, 1971.

9 David R. Hiley / James F. Bohman / Richard Shusterman (Eds.), *The Interpretive Turn*, Ithaca–London: Cornell University Press, 1991.

10 *Erkenntnis*, 7, 1937/38, pp. 225-232.

11 Gomperz, „Interpretation", p. 226.

12 *Wissenschaftslehre*. Band 3. Aalen: Scientia Verlag, 1981 (originally 1837).

13 Ibid., p. 553.

14 *Experience and Prediction*. Chicago: The University of Chicago Press, 1938, p. 220.

4. Describing, Explaining, Predicting, and Confirming

Empirical sciences concern facts, events and processes. The point of view from which these are approached determines the differences between natural, social and cultural sciences, and between the many disciplines they include. Each discipline applies its concepts to the observations made in its own sphere, and raises questions in ways that have been established in its own tradition – which in turn is reinvigorated by new inquiries and challenged by creative researchers. What is important in the events to be clarified, are minute details on the one hand, and patterns, regularities and trends on the other. It is the task of a science of any kind to account for these – in other words, to make sense of them.

How is the account to be given? Each discipline does so in its own way; but the question also concerns typical features that characterize science as science. Every investigation must describe facts and put forward suppositions in response to the questions raised. Descriptions answer the 'what' and 'which' questions, hypotheses the 'how' and 'why' questions. Insofar as the hypotheses are confirmed, they yield an explanation. A successful prediction helps to confirm a hypothesis.

A task of philosophy of science is to study the capacities of description, explanation, prediction and confirmation. There are two kinds of question about these procedures within cultural disciplines and social sciences: (1) How far do cultural sciences proceed by describing phenomena, and how far by explaining them? (2) What are the limits of explanation and prediction within the social sciences? Concerning (1), one may also ask about differences among the cultural sciences. (According to Gonzalez, the description/explanation controversy affects especially to the cultural sciences, among them to social anthropology and historical sciences). Moreover, in such cases as grammar, it appears that not only description but also prescription is relevant. Grammar can be seen as a part of applied linguistics, not as a science. As applied linguistics, it requires prescription, and as applied science, it requires resolution of concrete problems. As for (2), one may wonder if prediction plays any role in cultural sciences. Certainly we have expectations not only about natural events, but also about society and culture. Sometimes human facts can be predicted even better than events in nature. Thus, we rely more on the constancy of our friends' character than on next week's weather forecast. With respect to both nature and culture, certain observed regularities sustain our anticipations.

In social sciences, description presumably has a minor role compared with explanation and prediction. The following questions arise: how far do these proceed by explaining phenomena, how far by predicting these – and how far by prescribing them? Are there differences in this respect among the disciplines of social science? How are the results of social scientific research confirmed? How are various causal hypotheses tested? What interpretations can be given to the notion of causality in social affairs? How can the gamut of actions, causes and reasons in psychology and social psychology be thoroughly clarified? Further complications are self-fulfilling prophesies. How to account for prescriptions that are presumably based on predictions? (Examples abound in sociology and political science).

One has good reason to expect both social and cultural sciences to raise, and to give well-founded answers to, not only ‚what' and ‚which' questions, but also questions that require explication and explanation – i.e. 'how' and 'why'.

5. On Mathematics and Experimentation

Further problems are raised by the roles of mathematics and experimentation in social sciences, and in humanities. These two issues are closely connected, because one may suppose that devising and organizing experiments is enhanced when performed within a mathematical framework. Mathematics is concerned with orders, structures, relations, classes and numbers; or, from a pragmatic point of view, with ordering, classifying, deducing and counting. Reality can be represented by mathematical models.

Humanists and social scientists seem to have different attitudes towards using mathematics and experiments in research. According to Gonzalez, "the predominant view" among humanists favours the qualitative, such as symbols and trends, over mathematics. Those who adopt this view also do not believe that experimentation might be useful in studying cultural factors, whereas social scientists usually have a more positive attitude to mathematical models and experiments in the clarification of social issues.

In the research of symbols and cultures, as for instance in social anthropology or in art history, mathematics is not used but the approach focuses on qualitative aspects. On the contrary, in contemporary history or in sociology, quantitative factors such as economic issues are in key role, with constant use of statistics.

Relevant problems include the following (cf. philosophy of mathematics): a) The very status of mathematical modelling. Is mathematics to be understood as a "language" used to establish knowledge by procedures of proof; or as a "heuristic tool" connected to discovery? b) Do natural sciences differ from social and cultural sciences in respect to mathematical modelling? c) What is the relation between qualitative and quantitative models within the social sciences?

Concerning the alternatives given in a), the question is: Is this 'or' exclusive or inclusive? One may suggest, "both and". Mathematicians fruitfully interpret mathematics as a language of proof, whereas physicists and engineers consider it as a tool for developing new ideas to be tested. One may presume that various heuristic procedures are useful also elsewhere in scientific research.

Problem b) raises a new difficulty, because now the terms of comparison are natural vs. social/cultural sciences. This problem can be tackled in respect to history of science, or within a systematic dimension. Mathematical models were earlier developed in natural sciences and technology, rather than in the study of societies.

However, statistics has received a key role in modern social sciences. Indeed, there is no empirical area that could be shown incapable of being treated mathematically.[15]

Problem c) concerns the relation between quantitative and qualitative models in the social sciences. It is reasonable to adopt a "both and" approach. According to the problem to be tackled, statistics and mathematical models may be needed, or an explication of qualitative differences, or both. Game theory is a fruitful mathematical basis for analysing social issues. In information theory and cybernetics, the qualitative and the quantitative aspects are both relevant. The same is true of economics.

As to humanities, for instance history can be enriched by statistics of human, animal and plant populations in the area and era with which the research is concerned. In linguistics as well as literary research, the study of word frequencies is useful for some purposes. The most convincing example is musicology. Music and mathematics were closely connected already by the Pythagoreans, because in both of them, the notions of measure, number and harmony are central.

The question of experimentation is closely connected to the issue of mathematical modelling. Gonzalez mentions as examples experimental economics and experimental psychology. One may add that in human sciences, tests and experiments are used, for instance, in linguistics and phonetics. Furthermore, there are no obstacles to their use in musicology or in the study of literature or art, in archaeology, etc.

6. Ignored Branches

Basic science is often distinguished from applied science; as, for example, pure mathematics is from applied mathematics. Gonzalez maintains that philosophy of science traditionally focused on basic science, whereas applied sciences in general have received less attention than basic science. "Among the fields that have been hitherto ignored by philosophy are the sciences of the artificial understood as 'sciences of design'." One may mention communication research, library science, computer science, economics, medicine, psychiatry, pharmacology, and all branches of engineering sciences, including genetic engineering. A distinction can be drawn between the concepts 'natural' and 'artificial'; and thus natural sciences may be distinguished from such applied disciplines as "sciences of the artificial" and "sciences of design". The latter deserve, with good reason, to become new objects of a philosophical analysis whose task would be to clarify their connections to cultural and social sciences.

---

15 Cf. Gerhard Frey, *Die Mathematisierung unserer Welt*. Stuttgart: Kohlhammer, 1967, p. 120.

Computer technology has contributed to the emergence of the virtual world, which has an artificial character. There is a corresponding "science of the virtual" cultivated in so-called "multi-media" laboratories. A relevant philosophical question is: what is the difference between the virtual and the fictive? Both can be contrasted to the real world. Logicians may also wonder how possible worlds are related to the virtual world. As to pharmacology, one may note that its early philosophical analysis was accomplished by the controversial physician Paracelsus in the 16th century. He criticized the traditional "iatrocentric" (physician-centred) conception of medicaments and thus helped to make pharmacology a science in its own right.

All in all, in the sciences of design and of the artificial, practice and theory proceed in a fruitful, mutual relation.

7. Novelty and the Future

Lastly, Gonzalez addresses the question of emerging and future directions in philosophy of science; these, he says, "may be new approaches in the social concern on science". A remarkable propelling force behind this projected development is the tension between realism and social constructivism. Another strain in methodology concerns individualism and holism.

These issues are relevant to all branches of scientific research and their interdisciplinary connections, while a special focus will be given to cultural and social sciences. In this respect, it is remarkable that medicine can be understood, because of its new approaches, as a cultural and social science.

At the end of Section 7, Gonzalez suggests that the revival of scientific realism is connected with the possibility of objectivity in social and cultural sciences. One may wonder where its opposite force, constructivism, is supposed to lead. To subjectivism and relativism, perhaps? However, there may be some objective constraints that would enable us to identify viable constructions, and to proceed towards an objective way of regarding social and cultural phenomena.

8. Review

The final remarks by Gonzalez concern the background factors of the philosophy of social and cultural sciences. These include the development of society and culture and, correspondingly, social philosophy and the philosophy of culture. Other branches of science are also relevant here; notably, mathematics and natural sciences. This is because mathematics plays a role in all disciplines, and because social and cultural sciences are comparable, in issues such as explanation and prediction, to natural sciences.

History enjoys a special role among the background factors of sciences, including 'history' as one of the cultural disciplines. In a timeless frame, it is possible to speak of 'things' and 'facts', but in using such concepts as 'events', 'processes' or 'trends', we are presupposing a time frame and with it, the basis for history. That facts become scientific facts and develop through the process of inquiry, was the basic insight of Ludwik Fleck.[16] This idea was further developed by Thomas S. Kuhn into the theory of paradigms and scientific revolutions.[17] Kuhn says that he encountered by "random exploration" the "almost unknown monograph" by Fleck, and saw that it "anticipated many of my own ideas"[18]. Kuhn's work, in turn, was continued by Imre Lakatos's theory of research programmes.

This development introduces two more turns. Gonzalez refers to the present stage of philosophy of science "after the decline of the 'historical turn'". Currently there is "an active competition" among (1) naturalists, (2) those who subscribe to the "social turn", and (3) realists.

Of course, this does not mean that historical research (conceived as 'history') cannot be further cultivated and its substantial foundation (conceived as history) considered. Gonzalez mentions history among social sciences and distinguishes between the methodologies of "New History" and "narrative history" in Sec. 5 of his contribution. The former is an area in which mathematical models are used, whereas in narration, such models presumably are not employed. Thus, the narrative history is predominantly qualitative and focuses on individual or social agents. The New History is "impersonal", focusing on quantitative factors such as production and transportation.

9. The programme ahead

The topics of discussion and tasks of research ahead, on a year-by-year basis, are the following:

(1) Clarify the cultural and social sciences and the philosophical-methodological approaches to them.

(2) Analyze the methodological controversy on description, explanation and prediction and their limits in cultural and social sciences.

(3) Study of the debate on mathematical modelling in social and cultural sciences and its contribution to social issues.

(4) Analyze the differences between the cultural and social sciences, on the one hand, and the sciences of the artificial (e.g. library science, communication research, pharmacology, economics) on the other.

---

16 Cf. his work *Entstehung und Entwicklung einer wissenschaftlichen Tatsache*. Frankfurt/Main: Suhrkamp, 1980 (originally 1935).

17 *The Structure of Scientific Revolutions*. Chicago: The University of Chicago Press, 1970 (originally 1962).

18 Ibid., pp. vi-vii.

(5) Investigate the novel views about the influence of social constructivism and realism on social sciences.

Bibliography

Innocent M. Bochenski, *Contemporary European Philosophy*. Berkeley and Los Angeles: University of California Press 1957.

Bernard Bolzano, *Wissenschaftslehre*. Band 3, Aalen: Scientia Verlag 1981 (first published 1837).

Ludwik Fleck, *Entstehung und Entwicklung einer wissenschaftlichen Tatsache*. Frankfurt am Main: Suhrkamp 1980 (first published 1935).

Gerhard Frey, *Die Mathematisierung unserer Welt*. Stuttgart: Kohlhammer 1967.

Hans Georg Gadamer, *Wahrheit und Methode*. Tübingen: J.C.B. Mohr (Paul Siebeck) 1965 (first published 1960).

Heinrich Gomperz, "Interpretation", *Erkenntnis*, 7, 1937/38, pp. 225-232.

David R. Hiley / James F. Bohman / Richard Shusterman, *The Interpretive Turn*. Ithaca and London: Cornell University Press 1991.

Edmund Husserl, *Die Krisis der europäischen Wissenschaften und die transzendentale Phänomenologie*. Haag: Martinus Nijhoff 1962 (Husserliana Band VI).

Thomas S. Kuhn, *The Structure of Scientific Revolutions*. 2nd ed., Chicago: The University of Chicago Press 1970 (first published 1962).

Hans Reichenbach, *Experience and Prediction*. Chicago: The University of Chicago Press 1938.

Bernand Williams, *Philosophy as a Humanistic Discipline*. Princeton and Oxford: Princeton University Press 2006.

Peter Winch, *The Idea of a Social Science and its Relation to Philosophy*. London: Routledge and Kegan Paul 1958.

Georg Henrik von Wright, *Explanation and Understanding*. Ithaca–New York: Cornell University Press 1971.

University of Helsinki
Department of Philosophy
Siltavuorenpenger 20 A
SF-00014 Helsinki
Finland
arto.siitonen@helsinki.fi

Matti Sintonen

Scientific Realism, the New Mechanical Philosophers, and the Friends of Modelling

1. Of Mechanists and Modellers

A spectre is haunting philosophy of science in general and philosophy of the social sciences in particular. In highly simplified terms, there is a consensus that the syntactic or statement of scientific theories (the Received View, RV), with problems identified and defined in terms of their representations in some formal language, is beyond pale. Its most serious formal rival, the semantic view (the structuralist view is its European variant) is claimed to be insensitive to how scientific inquiry is actually conducted (see the articles in e.g. Morgan and Morrison 1999). Here NMPs have been joined in by Friends of Models of (FoMs, in brief) who maintain that the focus should be moved to models and model building instead of theories in the sense of RVs. Hopes therefore run high that the the alliances of NMPs and FoMs will change our conception of the results and the processes of scientific inquiry.

Neither NMPs nor FoMs subscribe to any single doctrine. Rather they form somewhat loose groups of philosophers who wish to redirect the efforts from that of formulating universal laws and increasingly truthlike universal theories to that of explicating mechanisms and building models of narrower scope. Such a goal is highly congenial to many social sciences, since they neither possess nor express interest in possessing universal laws and comprehensive theories. However, like the life sciences the social and cultural sciences abound with mechanisms that are used for the purposes of explaining. Social scientists also often think of themselves as engaged in building models that in turn aims at describing underlying mechanisms (for the notion of modeling mechanisms, see Glennan 2005). If there is a notion that collects the life sciences and the social sciences under one metatheoretical umbrella it seems to be this: they are in the business of building models of mechanisms that generate more or less regular or indeed sometimes lawlike connections between phenomena. Nor does this metatheory confine to these sciences. On the contrary, by drawing attention to the weaknesses of the research programme initiated by RV the NMPs and FoMs open the cases of many other sciences: many of the physical and technical sciences are in the same conceptual boat. This is also the reason why I shall start with a general examination of mechanisms since mechanisms was the pattern of intelligibility for the Old Mechanical Philosophers (see section 3).

F. Stadler (ed.), *The Present Situation in the Philosophy of Science*,
The Philosophy of Science in a European Perspective, DOI 10.1007/978-90-481-9115-4_20,

Adopting NMP, and embracing the point of view of model building, has consequences for one of the most persistent issues in philosophy of science in general and in the social sciences in particular. The alliance of NMPs and FoMs calls for a more realist view of what scientists actually do. At the same time it conducts a mutiny against realism as it came to be defined in RV and its aftermath. Realists traditionally maintain that the world is "out there", independently of the inquirer. The direction of fit is from word and thought to world, not the other way around, for the inquirer´s epistemic (or conative) attitudes towards the world do not have direct (causal) impact on what things, properties and relations there are. This part of realism causes no alarm amongst NMPs and FoMs, but the statement or theory variant of realism which maintains that science aims at theories that capture the complete literal truth in field does. The criticism is not that empirical adequacy, warranted acceptability, or some other vegetarian notion should dislodge truth as the central concept that marks a successful theory. Rather, the NMPs and FoMs claim that ascribing scientists the goal getting at the complete truth in the sense described, of getting closer to the truth, is misguided.

But whilst the statement or theory variant of realism might miss the mark vis-à-vis the social sciences in particular, the credentials of *causal realism* are altogether different (the above distinctions do not of course exhaust the varieties of realism). Many social scientists as well as philosophers of social science hold that these sciences only enter the company of sciences proper if they are able to refer to causal events and processes that *really* exist and, perhaps also, truly are responsible for their social explananda, singular and general. In the natural sciences the realism/antirealism debate of Johannes Kepler and Ursus marked the historical beginnings of philosophy of science, since it was the first conscious debate over the goals of science: for realists the goal is physical truth of *causae verae*, and not mere mathematical elegance in saving the phenomena (or indeed of assuming more than is needed).[1] The social sciences are latecomers on the scene, but the same strategy might work here.

1 Philosophy of science is said to have born when Johannes Kepler defended Tycho Brahe's proposal for a new world system. The debate between Kepler and Ursus involved accusations of plagiarism, hurt pride but also substantial disagreements and philosophically sophisticated arguments over the goals of science. Kepler argued that a physical interpretation and a physical explanation was a legitimate concern in astronomy, and ability to save the phenomena by mathematical means was not enough (Jardine 1991, p. 134-135). Later Isaac Newton formulated a methodological view that animated science for centuries: "we are to admit no more causes of natural things than such as are both true and sufficient to explain their appearances." (Newton 1934, 398).

The requirement of getting hold of real causes is the cornerstone of Newton's methodological legacy, and it was highly influential at least till Charles Darwin´s search for the causes of evolution. We need to distinguish between several logically speaking different kinds of claims. First, one can claim that a type of causal process is something that really exists in nature, something that is a *vera causa* and not just

There is no way that all aspects of truth, realism and mechanisms in the social science could be treated adequately within one paper. It is impossible to cover all of the social sciences (or the sciences of man, or *humaniora*) – although there is, for instance, an interesting debate in the philosophy of law over the reality of laws that are not sanctioned or enforced. I shall not dwell on the ontological and epistemological consequences of the fact that cultures and societies only exist by courtesy of human beings, and hence are, in an important sense, mind-dependent. I cannot pause to consider the methodological problem that, for ethical as well as practical reasons, carrying out controlled experiments in many of the cultural and social sciences is difficult or impossible (see e.g. Gonzalez 2007).

I shall start with the role of laws or nomic generalizations (#2) and I then move on to consider NMP (3#). The following section (#4) deals with analytic sociology which not just defines what the social sciences need to rise from (or lower to) grand general schemes to social sciences proper. Section #5 addresses, very briefly, some types of social mechanisms. Here the question is whether the social sciences indeed meet the requirements of NMP, however loose. A crucial ingredient in this is the notion of a mechanical explanation, although, following Jaakko Kuorikoski's recent proposal, it needs to take into account that mechanical explanations in the social sciences are a special variant of NMP (Kuorikoski 2009) (#6). I shall then take up action theory which, I suggest, could profit from strategic advice from analytic sociology (#7). I shall then expand this pattern of mechanical explanation to the communicative mechanisms proposed by Paul Grice and Deirdre Wilson and Dan Sperber (#8). The third example is economics, a trade where model building is the bread and mechanisms the butter (#9) – and a field where debates about realism have played a prominent role. I shall then take a look at the issue of realism from the point of view of the pragmatic turn advocated by some recent FoMs (#10). To conclude I shall transfer to the interrogative and inquisitive mode. Much as I appreciate the advantages of NMP, there is no one coherent notion of a mechanism to cover the various sciences. Nor is it obvious that NMP works as an independent metatheory for science. It may well be that NMPs and FoMs have kicked away the syntactic and semantic ladders a bit too early (#11).

the inquiring mind's wishful thinking. Secondly, one can claim that a type of cause is such that it can, in principle, have an effect of a certain sort. And third, after matters of principle have been settled, one can claim that a particular type of cause in fact is responsible for a particular type of effect. Jonathan Hodge (1977, 1987) calls these existence, adequacy, and responsibility claims respectively. Although this tripartite analysis might have much to offer in the social sciences, all I wish to point out is that the legacy insists on the causes being real as well strong enough to bring about the effects, and not something that are feigned.

2. Down With Laws!

There is little interest in laws and universal theories in the social and cultural sciences – and in acknowledging this they can find comfort from the realization that they are in good company. The life sciences claim to manage without laws. They are different but almost equal. Nor is there, for this reason mostly, interest in metatheory that builds on that view. Indeed NMP joins those who maintain that RV was an artifice of the tool-kit of logical empiricism. The logical empiricist relied on the rigor of modern symbolic logic ("logistic") in the explication of fundamental metatheoretic concepts. Hence theory structure, confirmation, and explanation were all dealt with in terms of logic, with logical inference as a unifying theme.

In RV laws or nomic generalizations – theories essentially consisted of sets of laws – were needed to systematize the claims of a field, to make prediction and to enable control, and, most importantly, to guarantee explanation. But as has been shown again and again, more is needed than nomicness. Confronted with a black bird I might ask "Why is this bird black?" and be given the following law-based explanation: "It is a raven and all ravens are black". This can be turned into a perfect deductive argument for the target explanandum. Take another example, this time from Hempel´s classic paper. I observe that something in my car leaks and turn to inspect and ask again "What?" It turns out that the cooler has cracked and I ask "Why?" I am then offered an account in terms of the laws that govern e.g. the expansion of water and contraction of metal when the temperature drops below zero centigrade – again the putative explanation can be turned into a perfect deductive argument.

The first example of an explanation is highly unsatisfactory. Merely knowing that all ravens are black does not explain why this bird, a raven, is black – at least if we have in mind scientific explanation. We would like to know some causal story, in terms of proximate causes, as to how the raven's genetic constitution and mechanisms of gene expression, yield this particular colour. Or we would like to know another type of causal story, in terms of distal causes, of what sort of selective pressure or other evolutionary forces have brought about the genetic constitution. The offered potential explanation sounds cheating because it is not clear, first of all, that it in fact is nomic or even true, and secondly, it fails to specify any intelligible mechanism for the colour. As has been pointed out, these are not the type of answers sciences are interested in – nor are such questions in fact raised in biology.

The claim that laws are not needed in all explanation is no news. William Dray (1957) challenged the orthodoxy by arguing that historians do not mention laws in their explanations because there are, in history, no laws. Dray maintained that historical explanations succeed (when they do) in bringing understanding because they rely on a peculiar view of the actions of historical agents. An action, he said, is understood not by subsuming it under one or more general laws but by seeing that it was, under the circumstances, the right or rational thing to do. The outcome was parallel to the tradition of the logic of the *Geisteswissenchaften* in that it insisted on driving a wedge between natural scientific explanation based on (causal) laws and explanation in the sciences of man that is based on the *sui generis* pattern of *Verstehen* (for an illuminating discussion on different approaches to this notion see Gonzalez 2003).

Both Carl Hempel and Karl Popper, the two main proponents of the nomothetic view of explanation, admitted that historians seldom if ever refer to laws in their explanations. Popper's reasons for this sorry state was that history in fact needs laws if it explains, but these laws are not peculiar historical laws but rather commonplace trivialities or perhaps simple psychological laws. Hempel in turn insisted that the laws required for explicit historical explanations are too complex to spell out. Both Popper and Hempel therefore subscribed to the "No laws, no explanation!" view, although for different reasons.

My reasons for making this brief detour is not the wish to drive home any deep conclusions about two types of logic of explanation. Rather, I hope that it serves as a reminder of a couple of facts. The "No laws, no explanation!" stance has enjoyed the status of a dogma for long, but not without reason. We do have the intuition that a reasonable amount of regularity and some degree of lawlikeness is required. However, it is not obvious what lawlikeness is, or what laws are.[2] The latter are sometimes glossed as exceptionless generalizations that meet a variety of conditions: – they only contain qualitative predicates, they are projectivle and support subjunctive and counterfactual conditionals, they are universal or at least very wide in scope, they enable predictions and have a crucial role in the systematization of knowledge, and of course they underlie explanatory practices (see e.g. Woodward 2000, section 6 for a discussion). But the issue is not just with lack of deterministic laws, since there are or may be probabilistic laws that carry similar or analogous nomic weight. Yet the first impression is right: deterministic or probabilistic generalization both are supposed to make the explanandum phenomenon expected, though to different degrees, and it is this general requirement that calls for trouble. Nomicness as such gives no clue to, and no intelligible account of, why a generalization holds.

The two putative nomothetic explanations are not, in fact, on the same footing. We do have the feeling that the explanation of the cracking of the cooler is illuminating, though perhaps too much so for someone´s concern. The reason it

2 Certainly lawlikeness or *Gesetzmässigkeit* was a different matter, a much wider notion, in classical German science and philosophy than it is in contemporary philosophy of science – and this is a very European thing! In section 5 below I take up von Wright's action theory which was based on the view that intentions are not Humean causes of action. They are not tied together by causal laws. Yet von Wright did not maintain that action are not lawful consequences from the intentions, and he even accepted that, as some others used the notions, actions could be caused. But for von Wright the lawlikeness was *sui generis*, neither logical nor causal.

works is that it refers to well-understood mechanical principles, and not because there are universal laws. This is what the NMPs claim. Woodward's own account, not the only one on the market, builds on the idea that explanatory relationships, unlike unexplanatory ones, can be used to manipulate and control the explanandum phenomenon. Explanatory generalizations are not exceptionless laws but rather invariances. Invariances in turn are robust and resilient under various types of interventions – and they therefore admit of degrees. Woodward's view of explanation can easily be incorporated into NMP.

3. Long Live Mechanisms!

To put the claims of NMPs in perspective it is useful to note that they are mostly concerned with scientific explanation: sciences should aim at explanation and not at mere description. Furthermore, reaching this aim is best achieved by revealing the mechanisms that generate the phenomena to explained, whether singular events and facts or generalizations and laws. To see what this amounts to, let us agree that all explanations are answers to explanation seeking questions: we ask why a certain event happened, or why a generalization obtains, or how a phenomenon was born (see Sintonen 1989, 2005). NMP holds that the crucial feature that makes for explanatoriness is that the answer to the question "Why P", viz. "P because Q" refers to a (or the) mechanism that produces P. This is to bring the "cause" back to the "Because", but with the extra requirement that it refers to a causal mechanism.

There is a sense in which we are witnessing a return to old ideas (this is one meaning of "revolution"), although in a substantially modified form. The standard of intelligibility favoured by the Old Mechanical Philosophers, OMPs, such as Descartes and Gassendi and Robert Boyle, was a mechanism.[3] OMPs were inspired by the metaphysical view that all causes and interactions of natural bodies were mechanical, and some like Thomas Hobbes embraced the extreme metaphysics that all there is is matter in motion.[4] The exemplar mechanism for OMPs was a machine or complicated contrivance such as the Strasburg clock. To explain its elaborate movements one did not need occult entities or forces but mechanical force, pulleys and gears. And of course it did not house mysteries because it was

3 Julian Reiss (2007) uses the abbreviation NMP for the "new mechanical perspective". I have wanted to make the connection, however loose, to the "old mechanical philosophy" of 17th Century.

4 Here is a quotation from Hobbes' Introduction to *Leviathan* (Introduction): "For seeing life is but a motion of limbs, the beginning whereof is in some principal part within, why may we not say that all automata (engines that move themselves by springs and wheels as doth a watch) have an artificial life?"

a man-made artifact. With its inner constitution exposed, one could see *how* it produced the actions.[5]

This ontology no longer has great following, and the strict pattern of intelligibility has been given up for more encompassing notions (see Allen, 2005, and the other papers in Craver and Darden, 2005, for the uses of "mechanism"). There is no single characterizations of what a mechanism is, and what mechanical explanation amounts to.[6] Yet there is the consensus that, to understand a phenomenon is to see how regularities about that phenomenon arise through the workings of an underlying mechanism. Such understanding requires an account of the entities that are involved, and the ways these entities are orchestrated, in time and space, to produce the regularities. Let us start with Bechtel's and Abrahamson's (2005, 423) definition: "A mechanism is a structure performing a function in virtue of its component parts, component operations, and their organization. The orchestrated functioning of the mechanism is responsible for one or more phenomena". Bechtel and Abrahamsen propose that an analysis proceeds via functional and structural decomposition, in this order: one first looks into what a system does and then what parts go into this doing. The first stage therefore consists of functional decomposition: one starts from overall functions and then proceeds to component operations. The second stage consists of structural decomposition: here the inquirer identifies the component parts that do the work, paying due respect to the operations.

A mechanisms can also be nested within another one, thus leading to a research strategy in which ever new functionally described black boxes are subjected to the treatment. Here we can appreciate the virtues of NMP over and above the advance it allows in the theory of explanation. Part of the RV was a particular view of a scientific theory: it was to consist of an uninterpreted set of formulas that

5 Thus Boyle (1688, 1968), p. 397-398: "The world is like a rare clock, such as may be that at Strasbourg, where all things are so skillfully contrived, that the engine being once set a moving, all things proceed, according to the artificer's first design, and the motions of the little statues, that at such hours perform these or those things, do not require, like those of puppets, the peculiar interposing of the artificer, or any intelligent agent employed by him, but perform their functions upon particular occasions, by virtue of the general and primitive contrivance of the whole engine". It should be noted, though, that Boyle combined teleology with mechanisms in a way that the NMPs do not, by arguing that when the contrivance view is extended to natural things such as organisms it is necessary to resort to an intelligence that had designed it. Strictly speaking the standard of intelligibility for CMN included a devine designer!

6 A useful anthology for mechanisms in biology is Craver and Darden 2005, for the brain and the neurosciences see Craver 2007, for cognitive neuroscience see e.g. Bechthel 2008 in particular. A classic plea for mechanisms in the social sciences is Jon Elster 1989, where he refers to causal mechanisms as the basic units of the social sciences, but see also Elster 1999 where he brings the notion of a mechanism to bear on emotions. Some of the increasing literature for mechanisms in the social sciences are mentioned in the sequel, but see Hedström and Swedberg 1999 in particular, as well as the Hedström 2005. For a critical account of the sufficiency of mechanisms in the social sciences, see Reiss 2007.

were then interpreted by help of observation sentences and bridge laws (as well as semantic rules) to give a true (of false) theory ("picture") of a part of the world. Progress was seen as increase in generality or scope and was effected via concept and theory reduction: concepts of a reduced theory could be defined on the basis of a reducing one, and laws of the former inferred from those of a more fundamental reducing theory. Theoretical scientific progress, then, would consist of a series of reductions moving towards the level of fundamental physics that served as the glimmering rock-bottom. Although discovery was not on the official agenda of RV it did have an unofficial one: try and find terms and bridge laws etc. that make such reduction possible.

This view of inquiry has been universally rejected, and the attraction of NMP proposes a positive alternative: scientists, and working life scientists and social scientists in particular, are not in the business of formulating law or exceptionless regularities, nor explaining by deriving explananda from these laws (and initial conditions), nor are they formulating formal derivations between terms and theories at different levels. Theories are not linguistic entities. Rather, scientists are in the business of finding and articulating mechanisms. And when NMPs are joined by FoMs the result is: scientists are in the business of designing *models of mechanims*.

NMP also provides a natural setting for incorporating pragmatics into the picture. If the standard of intelligibility is not that of derivation of singular or general facts from (more) fundamental laws but rather that of exposing mechanisms, scientists and scholarly communities are no longer bystanders who witness that there is a particular relationship between two sets of sentences, the explanandum and the exaplanans, but participants in action. Not only do they construct mechanisms but they also use these mechanisms when they explain to their audiences why and how the explanandum phenomenon takes place. What this amounts to is important for understanding of the credentials of NMP, and for the issue of realism. The standard of intelligibility might mean that the objects studied (individual people and societies) *are* mechanisms and that it is the task of scientists to uncover their nuts and bolts. Or it might mean that although the objects studied are not mechanisms (indeed, machines or at least contrivances, in the most intelligible of all possible worlds), parts of their ways of working can be explained by referring to mechanical principles. The former reading would require specifying exactly what a mechanism is – and this seems rather a hopeless task. However, it is the latter claim that a serious NMP wants or should want: resorting to mechanisms, and building models of mechanisms, need not make hefty metaphysical claims. Rather, scientists and those who utilize their results to practical and educational purposes can use mechanisms and models of mechanisms to *explain* how things work. Exactly what resources are needed to reach the required result, the epistemic state in which the explainee understands why a phenomenon arises and how it works, need not be specified once and for all for all possible cases. All manner of linguistic and non-linguistic devices, graphs, pictures, videos, live specimen, models, anything that works is allowed. For the question of realism this means: the tools used to represent a problem and the solution need not be true of the world. What counts is that claims made by help of them are true, and that the causal mechanisms that they stand for are the way they are claimed to be.

4. Analytical sociology

The Old Mechanical Philosophy, OMP, found its shape (or shapes) during the Scientific Revolution. The focus was on natural philosophy (i.e. science), but the claim was more universal and also encompassed man and society. However, up until the 19th Century this reflection took the form of speculation rather than science. There were social and political philosophies that focused on what an individual should do to carry out a good life, or what form a society should take to enable individuals and the society to flourish. Or they could build on some particular philosophical anthropology or metaphysics of human nature and society, advocating e.g. organisms or machines as exemplars. Descartes thought that brutes are just complex machines, and in so far as humans are treated from the point of view of their corporeal being (as in the study of reflex action) the same method could be followed. An extreme form was Hobbes' Body Politic in *Leviathan* where even the society or state was conceived as a superorganism. But how did the ideal find its way to the social and cultural sciences?

On one important analysis not even the classics of social science, such as the functionalists or Marxists, managed to raise social sciences to the level of theory. According to Peter Hedström (2005) they still were too general to reach the level of explanation (see also Hedström and Ylikoski, 2009). A social theory worth the title must be able to dissect the complex totalities into their component constituents and activities and to show, with precision and clarity, how the social explananda arise from the intentional actions of individuals. Hedström and Richard Swedberg (1998) point out that this goal can be achieved by help of middle-range theories, a notion introduced by Robert Merton. A middle-range theory steers between grand conceptual schemes (such as functionalism, not to speak of armchair philosophies) on one hand, and mere descriptions on the other hand. It is not enough to claim that all institutions have a function. What is needed are theories that focus on explananda that are between the society as a whole and "thick descriptions" of particular phenomena. Middle-range theories also eschew overly simplistic explanantia, such as rational choice theory, if these are supposed to provide full-blown explanations of all manner of social phenomena. Due to biases and inadequaces as well as distracting mechanisms people´s choices and actions can fail the requirements of rationality but still be amenable to explanation through tools in the kit of the analytic sociologist (see Hedström 2005, p. 61).

This is where NMP gets into the picture. There are few laws in the social sciences, but plenty of generalizations and mechanisms that are invoked to explain them, from early characterizations of market mechanisms and invisible hands to more recent game-theoretic accounts. Again there is no single account of what a social mechanism is, nor one notion of explanation in terms of mechanisms. Jon Elster, a pioneer of mechanisms in social science, writes that mechanims are, roughly speaking, frequently occurring and easily recognizable low-level "causal patterns that are triggered under generally unknown conditions or with indeterminate consequences." (Elster 1999, p. 1). In what follows I shall focus on Hedström's analytical sociology since it perhaps better than others reflects the spirit of NMP: it urges social scientists to dissect mechanisms into entities and their workings and hence to reveal their manner of working.

Social sciences frequently refer to explanations couched in terms of the variables that represent features of an individual or the environment in which the individual acts. Hedström and Swedberg admit that survey analyses and statistical techniques that are standardly used in chasing and formulating generalizations that capture the influence of social conditions on individuals and groups are highly valuable. However, the "because" of the resultant explanations, even if causal in the sense of causal modeling, do not carry the force of an intelligible connection that would, or should, satisfy a rational ignoramus who is looking for understanding. It recently turned out, in Finland, that most abused group of youths are boys rather than girls. Something is known of the statistically relevant factors but still there is the further question of how the generalizations arise from the behaviour of individuals. Jon Elster puts this constraint on intelligibility eloquently: to satisfy an explanation must, in the end, be anchored in hypotheses about individual behaviour. If we ask why consumers buy less of a good when its price goes up a social scientist needs to adopt and test a specific assumption about the reactions to individual consumer to changes in prices. (Elster 2005, p. 45).

5. Types of Social Mechanisms

What are these mechanisms, then? Do they fit the analyses of NMPs, such as Bechtel and Abrahamsen (or Stuart Glennan or James Woodward)?[7] In Hedström's view the explanatory power of sociology does not rest on deterministic (or indeterministic) laws but on social mechanisms, understood as constellations bof "entities and activities that are linked to one another in such a way that they regularly bring about a particular type of outcome" (Hedström, 2005). By these constella-

7 For reasons of space I have not been able to deal with Glennan's or Woodwards's (2000, 2003) important accounts, but here is Glennan's definition: "A mechanism for a behavior is a complex system that produces that behavior by the interaction of a number of parts, where the interactions between parts can be characterized by direct, invariant, change-relating generalizations." (Glennan 2005, p. 445)

tions Hedström refers to generative models that are precise and explicit enough to tie together entities and activities, such as agents, actions, and interaction of agents. What these models manage to do is show the plausible mechanisms (there may be more than one) that yield the outcome, the described social explanandum phenomenon.

Swedberg and Hedström provide a typology of social mechanisms that essentially draws on the type of mechanism at work. A situational mechanism explains the way a specific social situation shapes e.g. the beliefs and opportunities of an individual. Take as an example the belief forming mechanism in work when an initially false belief turns into a truth. There are actual examples of this mechanism at work in recent economic history: suppose that an individual hears a rumor, or is given a clue however weak, that her bank is in deep trouble and faces possible insolvency. In order to save her savings a customer may decide to take all her money out. Someone else may hear of this, and when others "inevitably" join in the result is a massive withdrawal of money from the bank and indeed a state of insolvency. Apart from situational mechanisms there are also action-formation mechanisms (to be discussed in the next section) and transformational mechanisms in which individuals interact with each other and in which these individual actions conspire to produce an unintended consequence. Swedberg's and Hedström's examples of the latter type of mechanism include standard prisoner´s dilemma models such as the tragedy of the commons.

Elster, Swedberg and Hedström all align themselves with the realists. In economics, for instance, abstract theories are highly valued but they are often treated with as useful instruments rather than, as in the tradition of Kepler and Newton, as representations of real causal processes or mechanisms (or realistic representations of these processes and mechanisms). But the virtues of simplicity and elegance, though important in model-building, should not beguile an inquirer into fictionalism. A theory in the social sciences that aims at understanding must therefore combine methodological individualism with causal realism: it must specify "the set of causal mechanisms that are likely to have brought about the change", and these mechanisms must be those that do the actual work, "not those that could have been at work in a fictional world invented by the theorist" (Hedström, 2005)[8]. Here the mechanism-based view differs from rational choice models since in the latter individual actors are represented by proxy and not by individuals with real life profiles (see Lehtinen and Kuorikoski 2007).

8 Here we see the vera causa strategy (see footnote 2) at work again. In the same spirit, Elster (1989) insists that the social sciences should not be happy with story telling, with how things could have happened, but should try and find out how they they actually took place. And here causal mechanisms, or causal chains, have a place of pride.

6. Social Mechanisms as Abstract Forms of Interaction

Agreed mechanisms are the entities we live by, there are a couple of questions that we must raise. First, how close to the exemplars of OMP or NMP, mechanical systems or even a contrivances, do we get? And secondly, how should we conceive the distinction between a mechanism and mechanical explanation. As Jaakko Kuorikoski (2008) has shown, the answer to the first question is: Not very close. Individuals are not mechanisms, although they might house biological or psychological mechanisms at some level. Societies are not mechanisms either, nor is there much substance to the claim that they house mechanisms that would consist of component parts and their operations, orchestrated in space and time in the required sense. What, then, becomes of the social sciences' coming of age through identifying mechanisms that really are there, generating social regularities?

Kuorikoski's diagnosis is that the notion of a mechanism suitable for the social sciences is not the same as that in some quarters of biology. The characterization of Bechtel and Abelson (2005, see section 2 above) fits two more specific and hence different types of mechanisms. In e.g. molecular biology complex systems are studied through decomposition and localization (see Bechtel and Richardson 1993). Here the study objects are, as Kuorikoski puts it, systems that are nearly decomposable into component parts, and these parts perform their tasks in accordance with the intrinsic causal powers of the parts. Kuorikoski calls these mechanisms *componential causal systems* (CCSs), and since their conditions of identity refer to the causal powers of the parts. The research strategy that comes with CSSs calls for the opening of black boxes at more fundamental level (which is why study in molecular biologists easily leads to cooperation between biologists, biochemists and biophysicists). The mechanisms therefore are really there in the sense that they contain orchestrated and localizable parts that produce the regularities (not universal laws, since the working of a mechanism is contingent on internal and external matters).

In Kuorikoski's view social mechanisms are not CCSs but *abstract forms of interaction*, AFIs. An AFI mechanism, such as a mechanisms that regulate price formation on the market, are abstract in that they only take into account some causally relevant factors. AFIs are also decomposable into parts and their operations that contribute to the system's overall performance, but this they do *not* do in accordance with their intrinsic causal powers. The component operations of social mechanisms cannot be paired with clearly localizable parts as in CCSs, but AFIs are as real as pieces of the furniture of the world as CCSs.

The spectre of NMP is haunting the world, and the call for all NMPs (and FoMs) of the academic world to unite makes sense. Mechanistic explanation is to rule the world, reduction in the sense of identifying more fundamental mechanisms with their own component parts and their orchestrated operations is the strategy to follow. But although the call to unite is not hollow it comes with a disclaimer: the underlying ontologies and methodologies are different.

7. Explanation and Understanding of Action

Understanding individual actions, we saw, is a crucial constraint on social theory, and Hedström's and Swedberg's action formation mechanisms are designed to help to see how specific actions arise. Situational mechanisms in turn focus on a specific social situation affects the agent. Here the focus is on specifically social action, but one can ask if the notion of a mechanism could throw light on individual intention formation and hence on how an agent's actions (not just social ones) are determined.

G. H. von Wright set out to do in his action theory. He challenged, in an original way, both the "No law, no explanation" view and the idea that actions are causally explainable to begin with. A leading interpretativist and intentionalist he emphasized that action explanations are *sui generis* and cannot be reduced to causal explanation by help of nomic laws. Here is how he described his view:

> To explain an action is, broadly speaking, to give a truthful answer to the question of why the action was done (performed, undertaken) ... As a common name for all the factors which explain action I shall use the technical term determinants of action. (von Wright 1980, p. 27)

Von Wright distinguished between actions that are internally determined and those that are externally determined. Suppose we have the question "Why did agent *A* do *p*?" One type of potential answers to such questions are of the form " *A* did *p* because she intended to obtain *q*, and took *p* to be necessary for (or sufficient for, or at least helpful towards) the obtaining of q". The intentionalist model of action explanation discussed in detail in von Wright's classic *Explanation and Understanding* was geared to such internally determined actions (von Wright 1971). Here putative explanations refer to *A*´s volitional attitudes (willing, wanting) and to her cognitive or epistemic attitudes (believing) which concern the required means (here *p*) of obtaining the state of affairs wanted (q). Together these two types of mental attitudes form a ground or reason for doing *p*.

Von Wright's (1971) was criticized for not bringing in the actions of others, since his practical syllogism focused on the relationship between the intention of an agent *A* to bring about *q*, the action *p* a that *A* thinks is needed to bring about *q*, and *A*´s setting out to *p*. An internal determinant is so-called because the reason here, as a or the determinant of action, "is a combination of two mental attitudes *with* an agent" (von Wright 1980, p. 28). The notion of an externally determined action is a partial response to criticisms of excessive internalism. External determinants are characteristically responses symbolic challenges. Such actions are often, but not always, verbal, and they result from participation in institutional practices. Learning to respond to such challenges is learning these practices through socialization.

Symbolic challenges can be personal (like answers to a question) or anonymous (like reacting to the "challenge" posed by a traffic-light's turning red). Posing such a challenge is an external determinant of action. It is typical for such actions that the agent does not form a prior intention: she simply reacts to the challenge. When asked for the reason she could respond: "You asked me!" or "The light turned red!" An explanation of a person´s action to brake because the traffic light turned red can therefore be a complete one: the external challenge is a compelling ground for performing the action, although we would not usually say that it forces one to perform it. Another form of participation in an institutional practice is behaviour which follows a rule. Rules can be laws, moral rules or rules governing manners. Some of these are constitutive while others are prescriptive. The former define institutional practices and enter into action explanations only indirectly. The latter are, in von Wright's view, also determinants of action.

It is easy to see why (though perhaps not how) an internal determinant leads to action, since on von Wright's view there is a sort of conceptual connection between the determinant and the action. But this does not extend to externally determined actions: that a person understands a rule (or an institutional practice), and recognizes that the situation calls for compliance, does not suffice for action. There must be further means of securing this. And here von Wright follows a lead by Jürgen Habermas. To secure public interests the society tries to make its members participate in patterns of communicative action. And here von Wright makes explicit reference to a social mechanism: "To this end a special "motivational mechanism" has to be invented. Its efficacy is what I here call the normative pressure in a society." (Von Wright 1980, p. 45). If A does not respond in an appropriate manner, she will be subject to unpleasant consequences, anything from legal sanctions to raised eyebrows. But people do not in the main respond to challenges in order to avoid punishment. The members of the community *internalize* institutional practices and simply respond to the challenges without feeling them as restraints on their freedom to act.

It is clear, by the lights of analytic sociology, that we here have causal social mechanisms at work. Not only do we have the determinants of actions, but there are mechanisms that explain the formation of intentions, the entities that on causal accounts cause actions directly. Why do individuals form the particular intentions they do? Ultimately, von Wright thought, they arise from two sources, viz., from wants and duties. Wants may have natural objects (such as health and happiness) but also contingent ones, as when someone is treated with something and acquires a taste. But people frequently perform actions that they do not particularly fancy doing or want to perform, such as reading piles of term papers. Why do they to this? What explains intentions to do things that one does not want? The reason is that people also have duties that are an important kind of determinant in intention generation. But whence does this come? The explanation, again, is the same motivational mechanisms that makes people participate in communicative action: the aura of normative pressure. People feel the pressure but eventually, if things run smoothly, internalize the voice of duty and form the intentions that society calls for in particular circumstances.

But given that we could draw a flow-chart of determinants of action, with internal and external determinants in place, and given that these determinants could be represented by boxes that might still be opened to reveal yet deeper belief- and intention-forming mechanisms, why did von Wright wish to steer away from causal accounts and, presumably, explanations in terms of causal mechanisms? To see why, note how quotation in the beginning of this section continues:

> In colloquial language these factors [determinants] are sometimes referred to as causes, sometimes as grounds or reasons, sometimes as motives. ... The term "cause" itself is used with a multitude of meanings. For reasons of expediency, I shall reserve the term for what is often called "Humean" or nomic causes. They are, roughly speaking, causes related to their effects by a law which is an inductive generalization (von Wright 1980, loc. cit).

Actions for von Wright are not in the causal realm, and wants do not serve as causal springs of action because they both presuppose free will, freedom to act in accordance with the intention, and freedom to make choices in accordance with one´s wants. This is why also the determinants of action have an instrumentalist flavour. They are not part of the furniture of the causal world. Intentions and their determinants are not *Humean* causes of actions; intentions and their determinants are not in the head, which is why they cannot be detected independently of and prior to action. In a sense intention explanation are like *ex post actu* rationalizations of action: we see someone do something and we look for a motivational background that might make the action intelligible or understandable. But there is no way we could, equipped by the psychologists', cognitive neuroscientists' or social scientists' instruments, capture the entities. However, the patterns of explanation that pertains to the *outward* aspect of intentional action, the bodily part, was according to von Wright amenable to Humean laws. Von Wright's view therefore was a dualistic with respect to the explananda: on one had there were the bodily movements as well as the brain events, on the other hand there were the actions.

8. Analytic Action Theory

The suggestion that causation and mechanisms are incompatible with intentionalist accounts of action builds on a metaphysically loaded notion of a cause. If this tie is severed, action theory might well be incorporated into the causal-mechanical picture canvassed by analytic sociology. Consider, to start with, von Wright's experimentalist notion of causation:

> [T]o think of a relation between events as causal is to think of it under the aspect of (possible) action. It is therefore true, but at the same time a little misleading to say that if p is a (sufficient) cause of q, then if I could produce p I could bring about q. For that p is the cause of q, I have endeavored to say here, means that I could bring about q, if I could do (so that). (von Wright 1971, p. 74)

This view of causation is essentially the one which e.g. James Woodward (2003) started with in his elaboration of the manipulationist notion of causation: causes are something that can be used to bring about some state of affairs. Knowing the cause of a phenomenon enables us to *manipulate* the world.

Now ignoring, for a while, von Wright's distinction between the inner and the outer (bodily) aspect of action as well as the so-called logical connection argument, von Wright's view fits NMP reasonably well. Given that the ability to bring about events or states of affairs is the touchstone of a cause – causal reality shall we say, what reason could there be for the exclusion of this notion from action theory? Surely there is no objection as such to maintaining that wants (or desires) and beliefs bring about intentions, and intention bring about, in appropriate circumstances, actions. Furthermore, there is no need to think, pace Davidson, that there are strict physical laws between wants, beliefs and actions, when properly construed.

The proposal I suggest though cannot argue for here, is that action theory, and speech act theory for that matter, could benefit from the distinction between CCSs and *abstract forms of interaction*, AFIs. Recall that the former, but not the latter, are identified through the intrinsic causal powers (or dispositions) of the parts that perform the operations. They are therefore stuff-dependent and their parts could be localized in space and time. Now if there is, in a CCS system, an event and that event causes another one, they are separately identifiable token occurrences, and hence Humean (I do not wish to take a strong stand as to exactly what Hume maintained, though). Now the mechanisms that generate intentions, through the components of belief- and want-generation, are not CCSs but AFIs which is why someone equipped with a thick notion of cause might find it objectionable to call these components causally efficatious. The regularities generated need not be lawlike. And of course here we have the extra difficulty of not being able to intervene with the parts to see how the system operates, although the new imaging techniques make us think differently in the future.

The crucial question is: could one nevertheless bring about determinants of action, that is, beliefs, wants and intentions in another person? Certainly. I can induce in you a belief that it is raining, when you are not able to check the weather for yourself, by uttering "It is raining". I can therefore, and thereby, bring about intentions and actions in you, in a rather reliable way. For instance I could bring about, in you, the intention to take along the umbrella when you go out. Inducing wants and desires in other people may be a more difficult matter, but not hopeless. By displaying a chocolate box on the kitchen table I can pretty reliably bring it about that my children come to have the desire to have chocolate, and often the intention and its execution follow. The desire, and the belief, therefore produce intention and action. Ditto for symbolic challenges. When you are about to walk under a moving car, I yell "Watch out!" and save you from injury. When I am asked, in dinner table, to pass the salt via the verbal request "Could you please pass the salt!" I comply. In fact just about all performances of speech acts fall within a causal theory of action suitably construed: an utterer can perform a locutionary, illocutionary or perlocutionary act by way of uttering a token utterance, and these are successful to the extent the speaker manages to utter a token expression, or manages to get the hearer recognize her communicative intention, or manages to get the hearer respond in the way she wishes.

There is no reason to deny that we can bring about mental states in other people, and indeed we do so regularly in human verbal and nonverbal communication. It may of course be objected that manipulation of other people's actions by producing in them certain mental states is to be morally condemned. Indoctrination is a case in point, and, arguably, using social science in marketing to get people buy products which they don't need, or did not even know before, might be found objectionable. But manipulation as such is morally neutral and as the innocent examples above show we employ manipulation techniques all the time to bring about morally good or neutral results.

9. The Science of Communication

The so-called Gricean theory of communication in fact is an account of the *mechanism* that enables us to explain thy way face-to-face communication works, whether linguistic or non-linguistic (Grice 1989; see also for Bach and Harnish 1979 for an early account of the details; again, new imaging techniques might well bring about facts about the neuro-cognitive basis of these mechanisms – and hence help to decide between alternative ones). There had been rival proposals for the mechanism before, and there have been improvements and novel proposals since. Moreover, the explanatory strategy fits NMP also in that we can distinguish submechanisms in the overall mechanisms.

Very briefly, the standard mechanism for communication that reigned prior to Grice can be captured with the code model. According to this model a speaker is equipped with a device that consists of several components such as the semantic, syntactic and the phonological component. This device codes she speaker's thought into a form that can be transmitted *via* vibrations of air. The listener has an identical contrivance that works in the opposite order. The vibrations reach the ear and then the brain where the message is decoded, starting with the operation of the phonological component, and eventually reaching the semantic one.[9]

9 How much of the cognitive neuroscientific mechanisms are to be included in the model

Paul Grice challenged this model by arguing that communication takes place through intentions and the recognition of these intentions. In Gricean communication the speaker and the listener have some mutual context-dependent interpretation strategies or maxims; similarly, they have some mutual and other beliefs concerning reality, including what other people want and believe. In an attempt to communicate the speaker modifies the listener's acoustic (or visual) environment by offering direct or indirect evidence about her communicative intentions. The listener reasons, on the basis of verbal evidence and non-verbal clues, relying on mutual beliefs, what the speaker's communicative intentions at any one time are.

The Gricean model differs substantially from the code model of communication, since the mechanisms it postulates is based on reasoning rather than explicit coding. To the extent the reasoning and the principles on which this reasoning follows can be made explicit – and this has been the target of speech act theory since Grices's (and Austin's) seminal papers – there is reason to think that also speech act theory fits NMP. There have also been later proposals that either improve or replace the Gricean mechanism or its submechanisms by different ones. Dan Sperber and Deirdre Wilson (1976) propose, in their Relevance Theory, that the Gricean maxims can be replaced by a simpler Cognitive Principle of Relevance (CPR): Human cognition tends to be geared to the maximisation of relevance . When someone utters something the hearer, guided by CPR, tries to find the interpretation that is most relevant in the context. This is not the proper place to go into the details, suffice it to point out that there is indeed promise that it leads to empirically controlled results – and that it indeed fits NMP. The Relevance theory claims considerable predictive and explanatory power in a wide range of applications. And Wilson and Sperber (2004) write:

> The universal cognitive tendency to maximize relevance makes it possible, at least to some extent, to predict and manipulate the mental states of others. Knowing of your tendency to pick out the most relevant stimuli in your environment and process them so as to maximize their relevance, I may be able to produce a stimulus which is likely to attract your attention, to prompt the retrieval of certain contextual assumptions and to point you towards an intended conclusion. For example, I may leave my empty glass in your line of vision, intending you to notice and conclude that I might like another drink.

---

of communication is a matter of choice. Here is how Peretz and Zatorre (2005, 90) sum up the path, much the same for speech and music: "A sound reaching the eardrum sets into motion a complex cascade of mechanical, chemical, and neural events in the cochlea, brain stem, midbrain nuclei, and cortex that eventually – but rapidly – results in a percept. The task of auditory cognitive neuroscience is to figure out how this happens. Musical sounds and all other sounds share most of the processing stages throughout the auditory neuraxis."

10. Realism and the Pragmatics of Model Building

Some voices in the alliance of NMPs and FoMs claim – so it seems at least – that the philosophy of science of yester year put too much emphasis on claims and theories as units of analysis. The complaint was that neither RV nor the semantic view gives real-life scientists or scholarly communities, with their aims and possibly differing explanatory ideals and interests, their due. And not only do they leave out the pragmatic dimension which describes cognitive processes and social interaction within a community, the theory-centered and proposition-obsessed views do not pay enough attention to the material aspect of inquiry.

Given the focus on models (and mechanims), the goal of getting (closer to) the Truth – highly important for critical scientific realism for instance – is about to lose its pride of place in the agenda of philosophy of science (for this, see Niiniluoto 1999).[10] Thus Ronald Giere (1988) poposed a model-based account of science in which scientists build idealized systems or "theoretical models" (or "models", for short) without detours via their linguistic representations. To the extent there is, for instance, a suitable relationship between, say, equations and their corresponding models one can, Giere writes, speak of truth. But truth here has no epistemological bite: the equations do give a true account of the corresponding model, but only because they are so defined as to satisfy the equations (Giere 1988, p. 79). As to the relationship between the real system (later often called the target system) and the model, the relationship is not one of truth (since neither is a linguistic entity) or isomorphism but rather one of similarity. Giere later summed up his long goodbye to RV by maintaining that truth is a "carryover from an older picture of science." (1999).

Again this is highly congenial to philosophers of social science, and the pressure to jettison the notion of truth has been visible in economic theory in particular. Philosophy of economics therefore contains one of the best-focused realism debates in the entire field of social sciences. Econometric models are often said not to represent reality as it is, and hence to give a finger to instrumentalism. Some have thought that this amounts to a betrayal of the cause, whereas others hold that it is a good thing and only shows that economics is like any science: it aims at distilling the truth, and doing this requires removing irrelevant detail. Econometric models need not be true of economic reality but they rather simplify, idealize, etc by omitting a lot of detail.

10 I should perhaps make it clear at the outset that I do not subscribe to all these criticisms of the old philosophy. This should become obvious towards the end of the paper. But very briefly, from the logical point of view the alleged divide between the syntactic and the semantic views is not all that great, since they can the two types of formalizing theories are in the same logical boat. A theory given through its models can be given a linguistic interpretation and vice versa. Secondly, one Niiniluoto's points is precisely to develop a theory of truthlikeness in which closeness to truth or verisimilitude admits of degrees.

I can here only pick up a couple of developments that seem pertinent to the theme. In a forthcoming paper Uskali Mäki confesses to being a realist in his philosophical outlook but points that the overall picture is more complicated than is usually admitted. Economic models, he writes, are no different from physical theories in that they refer to idealizations: where physicists talk of frictionless planes and perfectly elastic molecules, economists feel free to make highly idealized or straightforwardly false assumptions concerning e.g. complete information and zero transaction costs. But instead of just praising or criticizing such an apparently reckless bent of mind Mäki points out that false idealizations are a means to a noble (indeed, more truthlike) end, viz., that of "theoretically isolating causally significant fragments of the complex reality" (Mäki, fortgcoming; see also Mäki 1994, 1996 as well as Lehtinen and Kuorikoski 2007). One therefore need not deny ontological realism – that the economic world has a certain structure that is independent of the economic theorizer, nor indeed that theories and models are true to the extent they manage to capture those structures. What matters to Mäki are not the features that scientific realists normally wish to celebrate, such as approximate truth of actual theories or predictive and referential success, but rather that things in the world are in one way rather than another, and that a science worth the title should try and find out what that way is.

It is this noble goal, then, that justifies using falsehoods. Unrealistic assumptions abound in economic modelling but this should not cause alarm since these idealizing assumptions are exploited for the purposes of theoretical isolation. Mäki's pet example is J. H. von Thünen's model of agricultural land use that assumes cities without any resptrictions on dimensions, or neighbouring towns, as well as uniform fertility and climate etc. (Mäki, forthcoming, p. 79). There are no such cities and evearyone knows that. But can anything more be said about the noble goal of truth? In Mäki's case it is realism about the *mechanisms* that are claimed to operate not just in the imagined model but in the real or target system. Needless to say, the target system was not any particular town (theoretical sciences seldom are concerned about particular systems, only particular systems that are used to make more general points). This being the case, the falsities adopted are just a means to building a model that is realistic in the sense that it captures the mechanisms in operation in the target systems.

There is feature that FoMs are fond of emphasizing, as of recently, viz. the pragmatic aspects of model building. I shall conclude with it since it highlights a more general feature. I have assumed, above, that scientific explanations can be usefully conceived as answers to why- and how-questions. This indeed was Hempel's gambit in his classic nomothetic account. But Hempel also made it clear that he wanted to capture the logical and not the psychological or anthropomorphical ascpecs of understanding, and hence of explanation. Now Hempel's logic of explanation was intended as an explication of the logical aspect of explanation, and it was carried out by construing the relationship of the explanandum and the explanans as a two-placed relationship between (sets of) sentences. Now critics were eager to point out that what counts as an explanation depends on who the explainee is and what she already knows (and wishes to know, perhaps). Later, this criticism lead to pragmatic models of explanation that made the explanation relationship a 3-placed one ("S explains to H why Q"), or incorporated even more ingredients into the structure. Matti Sintonen (1989) argued that there can be no two-placed formal relations to capture explanatoriness, but claimed that bringing in subjects and their knowledge as well as their interests and the context does not lead to pernicious subjectivism or relativism.

These pragmatic features are now being brought into model-building. Ronald Giere's cognitive approach to scientific theorizing already left this door open. The notion of similarity is crucial to that account but, as Giere notes, this raises the problem of respects and degrees of similarity. Indeed, he thought that "it is respects of similarity, not degrees, that primarily separate realists from anti-realists" (Giere 1988, p. 93). But assessing respects inevitably brings in the scientists and their communities, for they are the agents that do the assessment work here. Giere later makes this pragmatic commitment explicit by allowing purposes to enter the picture in his attempt to understand representational practices in science. He writes that "we are looking at a relationship with roughly the following form: *S* uses *X* to represent *W* for purposes *P*". As already suggested, and as Giere notes, this proposal draws an exact parallel in the theory of explanation. Giere indeed allows *S* stand for "an individual scientist, a scientific group, or a larger scientific community".

To get a handle on admissible or fruitful ways of falsifying reality (my term) we must note that model building is not just a two-place relations between the target object or system and the model – whether a physical model or, more often, system of equations of whatnot. Rather, models are designed by people with particular purposes and aspects in mind. Similarly, there is a particular audience for which the model, or the results of the modeling, are addressed, and perhaps also further variables. I have argued, elsewhere, that taking into account pragmatic aspects of explanation does not lead to subjectivism or antirealism. Similarly, bringing pragmatics into the business of building models (of mechanisms, often) is not a denial of realism or of the requirement of reality. As Mäki puts it in the case of economic modelling, it is but a concession to the fact that modeling as well as simulation are conducted with particular purposes in mind. Here Mäki wants to go deeper on the pragmatic path and bring in further variables to the effect that his account of representation becomes -placed: "Agent *A* uses object *M* as a representative of some target system *R* for purpose *P*, addressing audience *E*, (potentially) prompting issues of resemblance to arise."

11. In the Interrogative or Inquisitive Mood

We started this brief tour into realism in the social (and cultural) sciences from the claim of the NMPs and FoMs to provide a philosophy of science in general, and in the social sciences in particular, that could be more realistic than RV in giving an account of what scientists actually do. The cases examined only constituted a possibly one-sided sample: social theory or sociology, action theory and the theory of communication, and economics. It remains to ask if the NMPs and FoMs have been successful in their claims. The claim of the NMPs stands strong at least in this sense: the strategy of finding and refining (causal) mechanisms certainly is superior to previous accounts of what social scientists conceive themselves as engaged in. As to the claim of the FoMs goes, the question whether models should be the unit of appraisal is moot. Clearly also their claim of being more faithful to actual practice is well justified.

Yet there are some conceptual clouds in the NMP (cum FoM) horizon, and it concerns the ultimate conceptual autonomy or sufficiency of mechanisms and modelling vis-à-vis some specifically philosophical tasks. RV had a specific philosophical mission: a canonical way of specifying the content of a theory. This canonical tool extended to the notion of an explanations (what is an explanation), to progress through reduction, or to inductive support provided by observation statements to a theory. These accounts were given a linguistic and a formal guise, and one could then ask about the truth or truth-likeness of a theory, or compare the contents of rival theories and debate their merits with respect to each other. The major motivation for using RV notions in a description was that these notions could provide an objective handle on what counts as a scientific claim (as against loose talk), or an explanation (as against subjective sense of comprehension) or confirmation (which would be independent of pragmatic factors). Especially now that pragmatic aspects, people with their knowledge and interests, are explicitly welcomed into metatheory one could raise some nagging questions: first, does the alliance of NMPs and FoMs give us a normative handle, so important for philosophy of science, to discuss the merits of alternative proposals. If it does, does it do so without resorting to RV or, shall we say, formal notions in an essential way?

References

Garland Allen, "Mechanism, vitalism and organicism in late nineteenth and twentieth-century biology: the importance of historical context", Stud. Hist. Phil. Biol. & Biomed. Sci. 36, pp. 261-283.

Kent Bach and Robert M. Harnish, *Linguistic Communication and Speech Acts*. Cambridge, Massachusetts and London, England: The MIT Press 1979.

William Bechtel, *Mental Mechanisms: Philosophical Perspectives on Cognitive Neuroscience*, Routledge, 2008.

William Bechtel and Adele Abrahamsen, "Explanation: a mechanist alternative", Stud. Hist. Phil. Biol. & Biomed. Sci. 36, pp. 421-441, 2005.

Carl F. Craver and Lindley Darden, "Introduction" in: *Studies in History and Philosophy of Biological and Biomedical Sciences*. 36, 2005, pp. 233-244.

Carl F. Craver, *Explaining the Brain: Mechanisms and the Mosaic Unity of Neuroscience*. Oxford: Clarendon Press. 2007.

Newton C. A da Costa and Steven French, "The Model-Theoretic Approach in the Philosophy of Science", *Philosophy of Science* 57, 1990, pp.248-265.

William Dray, *Laws and Explanation in History*. Oxford: Oxford University Press. 1957.

Jon Elster, *Nuts and Bolts for the Social Sciences*. Cambridge: Cambridge University Press 1989b.

Jon Elster, *Alchemies of the Mind: Rationality and the Emotions*. Cambridge: Cambridge University Press 1999.

Steven French and James Ladyman, "Reinflating the Semantic Approach" in: *International Studies in the Philosophy of Science* 13, 2, pp.103-121. 1999.

Stuart Glennan, "Modeling Mechanisms" in: *Studies in History and Philosophy of Biological and Biomedical Sciences* 36, 2005, pp. 443-464.

Ronald N. Giere, *Explaining Science: A Cognitive Approach*. Chicago and London: The University of Chicago Press. 1988.

Ronald N. Giere (ed.), *Cognitive Models of Science*. Minnesota Studies in the Philosophy of Science, Vol. XV. Minneapolis: University of Minnesota Press. 1992.

Ronald N. Giere, *Science without Laws*. Chicago and London: The University of Chicago Press. 1999.

Ronald N. Giere, "How Models Are Used to Represent Reality" in: *Philosophy of Science* 71 , pp.742-752. 2004.

Wenceslao J. Gonzalez, "From Erklären-Verstehen to Prediction-Understanding: The Methodological Framework in Economics", in Sintonen, M., Ylikoski, P. and Miller, K. (eds), *Realism in Action: Essays in the Philosophy of Social Sciences*, Kluwer, Dordrecht, 2003, pp. 33-50.

Wenceslao J. Gonzalez, "The Role of Experiments in the Social Sciences: The Case of Economics", in Kuipers, T. (ed), *General Philosophy of Science: Focal Issues*, Elsevier, Amsterdam, 2007, pp. 275-301.

Paul Grice, *Studies in the Way of Words*. Cambridge, Mass., Harvard University Press 1989.

Rom Harré, "Material Objects in Social Worlds" in: *Theory, Culture & Society* 19 (5/6): 2002, pp. 23-33.

Peter Hedström, *Dissecting the Social. On the Principles of Analytical Sociology*. Cambridge University Press. 2005.

Peter Hedström and Richard Swedberg, "Social Mechanisms: An introductory essay", in Peter Hedström and Richard Swedberg (Eds.) *Social Mechanisms. An Analytical Approach to Social Theory*. Cambridge: Cambridge University Press. 1998.

Peter Hedström and Lars Udéhn, "Analytical sociology and theories of the middle range", in Peter Bearman and Peter Hedström (eds.) *The Oxford Handbook of Analytical Sociology*. (Forthcoming).

Peter Hedström and Petri Ylikoski, "Analytical Sociology" in: *SAGE Handbook of Philosophy of Science* (edited by I. Jarvie and J. Zamora-Bonilla), SAGE, New York, forthcoming, 2009.

Carl Hempel, *Aspects of Scienftific Explanation and Other Essays in the Philosophy of Science*. New York: The Free Press 1965.

M. J. S. Hodge, "The Structure and Stretegy of Darwin's Argument". *British Journal for the History of Science* 10. Pp. 237-246. 1977.

M. J. S. Hodge, "Natural Selection as a Causal, Empirical and Probabilistic Theory". In L. Krüger (ed.), *The Probabilistic Revolution*, vol. 2. Cambridge, Massachusetts: MIT Press. Pp. 233-270. 1987.

Nicholas Jardine, Jardine *The Scenes of Inquiry: On the Reality of Questions in the Sciences*, Clarendon Press, Oxford. 1991.

Jaakko Kuorikoski, "Two Concepts of Mechanims: Componential Causal System and Abtract Form of Interaction", *International Studies in the Philosophy of Science*. (forthcoming).

Aki Lehtinen, and Jaakko Kuorikoski, "Unrealistic Assumptions in Rational Choice Theory", *Philosophy of the Social Sciences*, vol. 37, no. 2, pp. 115-138. 2007.

Uskali Mäki, "Isolation, idealization and truth in economics", *Idealization in Economics*, edited by Bert Hamminga and Neil de Marchi, special issue of *Poznan Studies in the Philosophy of the Sciences and the Humanities*, 38, pp.147-168. 1994.

Uskali Mäki, "Scientific Realism and Some Pecularities of Economics", in R.S. Cohen, R. Hilpinen and Qiu Renzong (eds.), *Realism and Anti-Realism in the Philosophy of Science*. Dordrecht: Kluwer, pp. 427-447. 1996.

Uskali Mäki, "My philosophy of economics: Realistic realism about unrealistic models", in *Handbook of the Philosophy of Economics*, ed. Harold Kincaid and Don Ross. Oxford University Press. Forthcoming.

Mary S. Morgan and Margaret Morrison (eds.), *Models as Mediators. Perspectives on Natural and Social Science*. Cambridge: Cambridge University Press. 1999.

Ilkka Niiniluoto, *Critical Scientific Realism*. Oxford: Clarendon. 1999.

Isabel Peretz and Rober J. Zatorre, "Brain Organization For Music Processing", *Annual Review of Psychology*. 56: 89–114. 2005.

Julian Reiss, "Do We Need Mechanisms in the Social Sciences?" *Philosophy of the Social Sciences*, Volume 37, Number 2, pp. 163-184. 2007.

Matti Sintonen, "Explanation: In Search of the Rationale", in: Philip Kitcher and Wesley C. Salmon (eds.), *Scientific Explanation. Minnesota Studies in the Philosophy of Science*, Vol. 13. Minneapolis: University of Minnesota Press. pp. 253-282. 1989.

Matti Sintonen, "From the Logic of Questions to the Logic of Inquiry", in: Randall E. Auxier and Lewis E. Hahn (eds.), *The Philosophy of Jaakko Hintikka*. Library of Living Philosophers. Illinois: Carbondale. 2005.

Dan Sperber, and Deirdre Wilson, *Relevance*. Cambridge, Mass: Harvard University Press. 1986.

Mauricio Suárez, "Scientific Representation: Against Similarity and Isomorphism", *International Studies in the Philosophy of Science* 17, pp.225-244. 2003.

Frederick Suppe, *The Semantic Conception of Theories and Scientific Realism*. Urbana: University of Illinois Press 1989.

Deirdre Wilson and Dan Sperber, "Relevance Theory", in G. Ward and L. Horn (eds.) *Handbook Of Pragmatics*. Oxford: Blackwell, 607-632, 2004.

James Woodward, "Explanation and Invariance in the Special Sciences" in: *British Journal for the Philosophy of Science* 51, 2000, pp.197-254.

James Woodward, *Making Things Happen: A Theory of Causal Explanation*, Oxford: Oxford University Press. 2003.

G. H. von Wright, *Explanation and Understanding*, London: Routledge & Kegan Paul. 1971.

G. H. von Wright, *Freedom and Determination*, Acta Philosophica Fennica, Vol. XXXI. 1980.

Petri Ylikoski, "Social Mechanisms and Explanatory Relevance", in *Social Mechanisms and Analytical Sociology* (edited by P. Demeulenaere), Cambridge University Press. Forthcoming.

Petri Ylikoski, "Illusions in Scientific Understanding", in: De Regt, Leonelli & Eigner (eds.) *Scientific Understanding: Philosophical Perspectives*, Pittsburgh University Press 2009. Forthcoming.

Professor of Philosophy
University of Helsinki
P.O. Box 24
00014 University of Helsinki
Finland
matti.sintonen@helsinki.fi

Daniel Andler

Is Naturalism the Unsurpassable Philosophy for the Sciences of Man in the 21st Century?

Jean-Paul Sartre famously wrote, nearly 50 years ago, that Marxism "remains the philosophy of our time. We cannot go beyond it." In his critic Raymond Aron's words, Marxism was for Sartre the "insurpassable [or, in other translations: unsuperable] philosophy of our time.[1]" Taken in context, Sartre's pronouncement was at once descriptive and prescriptive: it was, according to him, neither objectively possible for the philosopher to leave the confines of Marxism, nor ethically permissible to attempt to do so.

This 'thick' or hybrid modality was characteristic of dialectical materialism: the eventual overthrow of capitalism, the dictatorship of the proletariat, the subsequent disappearance of the state, these were stages which were at once inevitable, and the proper aims of political action at successive moments of the historical process. Dialectical materialism was at once a theory of the historical and social process, an overarching perspective, a methodology for arriving at the truth regarding these matters, and finally a practical (ethical and political) norm. The underlying necessity was material, not metaphysical: ideas were thought to be the necessary byproduct of objective economic conditions, in particular of the production relations and the accompanying class struggle. One mechanism which was supposed to underlay or implement this necessitation was a principle of ideological reflection:[2] dialectical materialism, as a theory or perspective, a train of thought, was thought to be secreted by the economic set-up, becoming the ideology of the proletariat, thus motivating its members to undertake the revolutionary activities which would eventually lead to the overthrow of capitalism, etc. Indeed, Sartre's quote above is truncated: "[Marxism] remains [...] the philosophy of our time *because we have not gone beyond the circumstances which engendered it.*"[3]

1 The French word is "indépassable".

2 The label is one which I am coining for present purposes. I am not a Marxian scholar and as will be immediately obvious my goal in this paper has nothing to do with political philosophy or history. In particular, although I am aware of the distance between Marx himself and later forms of Marxism such as dialectical materialism, and of differences between various forms, 'vulgar' and otherwise, of Marx-inspired thought, I have no use here for such distinctions. Interested non-specialist readers might like to consult http://marxmyths.org/.

3 Jean-Paul Sartre, *Critique de la raison dialectique*, vol. 1: *Questions de méthode*, Paris: Gallimard 1960, p. 29. English translation *Search for a Method.* New York: Alfred Knopf 1963, p. 30. In French: "Il [le marxisme] reste donc la philosophie de notre temps: il est indépassable parce que les circonstances qui l'ont engendré ne sont pas

F. Stadler (ed.), *The Present Situation in the Philosophy of Science*,
The Philosophy of Science in a European Perspective, DOI 10.1007/978-90-481-9115-4_21,

Thus Sartre, faithful to materialism, recognized at once the inevitability of a certain train of thought at a given moment of human history, and its contingent character: the process of which it was a part would eventually lead to a new situation, in which a different train of thought would become available (and would in fact *inevitably* be taken up, thus presumably becoming the insurpassable philosophy of the new epoch).

This remembrance of things past motivates this paper's title and its general direction, as I will try to make evident presently. But first I need to make perfectly clear that I do not intend to base a value judgment on naturalism on the parallel I am drawing with Sartre's version of Marxism. It is perfectly obvious that there are continuities between Marxism and contemporary naturalism, but I do not intend to draw them out in this paper. Marxism, especially of the Sartrian sort, is held in low esteem in many quarters nowadays, in particular among a majority of committed naturalists. I am emphatically not suggesting that what (at least until the recent economic events) appeared to most people as history's negative judgment on Marxism has any bearing on contemporary naturalism. In fact, I will be defending a position which falls in the ballpark of 'liberalized' naturalism. One of the differences between my position and stronger or stricter forms of naturalism concerns the modal status of the naturalistic stance, and this is where the parallel with Marxism comes in, merely as a heuristic or expository device.

1. Naturalism: Descriptive and normative

What is variously known as scientific or philosophical naturalism in the context of contemporary analytic philosophy appears, at least to our eyes which do not yet have the benefit of hindsight, as one of those bicentennial groundswells which sweep the entire philosophical scene. As many authors have stressed (it has indeed become an *idée reçue*, a ready-made morsel of philosophical conversation), nearly everyone (in the English-speaking world) is a naturalist *of sorts*. It turns out that there are different kinds of naturalism. Some say that different philosophers *mean* different things by naturalism,[4] but I prefer to think that philosophers have different views about the *nature, structure and scope* of naturalism, conceived as a very general stance towards human knowledge and the role played by the natural sciences. In the most general sense, I see naturalism as the recommendation that

encore dépassées." I am indebted to Thomas Flynn for locating the exact passages in the French original and the English translation. Italics in the text are mine.

4 "It is a commonplace that 'Naturalism means many different things to many different people'." Mario De Caro/David Macarthur (Eds.), *Naturalism in question*. Cambridge, MA: Harvard University Press 2004, Editors' introduction, p. 3. The embedded quotation is from Lawrence Sklar, "Naturalism and the interpretation of theories", in: *Proceedings and Addresses of the American Philosophical Association* 75, 2, 2001, pp. 43-58.

natural science be taken with utmost seriousness. From this as starting point, the routes to more evolved, fleshed-out philosophical positions are many. The two main branches are ontological or metaphysical, whose central tenet is that nothing truly exists but what the natural sciences purport to provide knowledge about; and epistemic or methodological, which take the natural sciences as a paradigm of all knowledge-seeking activity. These branches fork in turn, again and again, and the resulting paths can meet again or remain disjoint. Describing the resulting and ever-changing landscape is not a task which I will undertake here, although I will have to engage in a while in a bit of simple taxonomy.[5] The point here, besides fixing some terminology, is to stress the philosophical nature of the various strains of naturalism thus construed: they are conjectures regarding how certain (abstract) things actually are, research programs aiming at showing them to be this or that way, arguments purporting to provide evidence that they are, etc. In other words, naturalism is seen as an theoretical enterprise.

But seen in another light, as when naturalists are engaging their philosophical opponents, it sometimes looks as if naturalism regards itself as a cause, whose opponents, whether outright antinaturalists or mere skeptics or agnostics, are wrong, not just theoretically wrong, but ethically wrong. Thinkers on both side can sound as if they believed that it is their duty to defend, as the case may be, naturalism or antinaturalism: not merely an intellectual duty, but a moral and a political one as well. The other side is often seen as committing the sin of scientism, or, to the contrary, of antiscientific obscurantism. There is a symmetrical disagreement about which side is truly defending the humanity in mankind: Is it the naturalist who is denying humans their basic dignity by putting them on par with (nonhuman) animals or (biological, naturally evolved) machines? Or is it the antinaturalist who, whether she means it or not, "objectively" (as the Marxist would say) sides with the opponents of progress and the defenders of the established, superstitious or violent order, by preventing the lights of science from shining over the ills of humankind and bring about some cures?

The militant naturalist is also a crusader, one who believes that it will necessarily turn out, in the long run, that the world falls entirely within the scope of the natural sciences, and that those not in his camp, ranging from declared opponents to the uncommitted and weak-kneed, are standing in the way thus retarding the eventual triumph at the cost of unnecessary intellectual toil and human suffering. The militant anti-naturalist in turn sees the naturalist as divesting philosophical energy in a direction which will turn out to be eventually fruitless, and in the meantime as slowing down her attempts to bring out the theoretical, and even the practical conditions of a more humane world.

It might be thought that this militant streak is nothing over and above the normal passion which enlivens the truly important philosophical debates. Granted,

5 See. *e.g.*., Philip Kitcher, *The Advancement of Science*. New York & Oxford: Oxford University Press 1993; David Papineau, *Philosophical Naturalism*. Oxford : Blackwell 1993; De Caro/Macarthur, *ibid.*

the debate around naturalism is not a unique occurrence of militant excitement in the history of philosophy. But nor is it always the case that a disagreement, however deep, in philosophical orientation is accompanied by such militancy. An example in point, in the philosophy of science, is the issue of realism: the divide is deep, positions are entrenched, yet nobody in either camp believes that, should the opponent prevail, something terrible would happen to science, to philosophy, or to humanity. While, as Steven Horst writes in a recent book, "some naturalists [...] claim, in essence, that the mind *must* be naturalized, or else something unseemly follows".[6] This is the normative form of naturalism which sometimes (not always of course) shadows the epistemic, philosophically legitimate form of naturalism. Normative naturalism (in this very peculiar sense of the phrase, quite distinct from the various doctrines put forth and illustrated under that label by such authors as Larry Laudan or Joseph Rouse[7]) has a prescriptive and an eschatological dimension, just like dialectical materialism: it will *necessarily* come about, you *must* help it along, or else ...

We are now in a position to understand why the problem of naturalism can be posed on two levels. First, we can ask whether naturalism (in one or another of the available varieties) is true. Second, we can ask whether the question may reasonably be asked, or whether, as some would have us believe, it is the 'insurpassable philosophy of our time'. This paper does not propose direct answers to either question. Rather, it defends a stance which permits to pursue the first question while taking the second into account. I will plead neither for or against naturalism in its strong form. In particular, as I don't think that even in its strong form, naturalism is committed to reductionism, I do not feel the need to raise the anti-reductionist flag. Nor do I believe that naturalism is necessarily 'normative', although it is an important part of the current dialectical situation that it sometimes is, and thus I certainly don't think it should be combated as such. It should be discussed on its merits as a philosophical thesis, and as a fruitful metaphysical research program, in Popper's sense.[8]

There is however a tension which the reader cannot have failed to notice between the claim that nearly everyone today is a naturalist and the description of the epic struggle which naturalists wage against hardened opponents and meek skep-

6 Steven Horst, *Beyond Reduction: Philosophy of Mind and Post-Reductionist Philosophy of Science*, Oxford: Oxford University Press, 2007, p. 15; quoted by D. Gene Witmer, in *Notre Dame Philosophical Reviews*, 2008: http://ndpr.nd.edu/review.cfm?id=128. Witmer disputes the aptness of Horst's characterization, arguing that the norm in question is nothing over and above the perfectly reasonable commitment of the honest scholar to bring her and others' beliefs in line with what appears as the best available evidence provided by science.

7 Larry Laudan, "Normative Naturalism", in: *Philosophy of Science* 57, 1990, pp. 44-59; Joseph Rouse, *How Scientific Practices Matter*. Chicago: University of Chicago Press 2002.

8 Karl R. Popper, *Realism and the Aim of Science*. London: Hutchinson, 1983, §23, pp. 89-193.

tics. The tension can be partly relieved by relativizing the near-universal quantification to a subpopulation of philosophers, and concomitantly noticing that the remaining population is far from empty, comprising at least analytic philosophers of a Wittgensteinian bent, and a vast majority of Continental philosophers and social scientists. There *is* a struggle, but at this point in time the naturalists seem to have the upper hand; why this is the case is a question worth asking, and we will get back to it presently. Some considerable tension remains however, and it is borne by an increasingly visible group of 'liberalized' naturalists,[9] philosophers who subscribe to the most fundamental inspiration of naturalism, yet see major objections to some of the tenets of the pure sort of scientific naturalism which seems prima facie, and often claims to be, the only naturalistic game in town. There is indeed a gnawing suspicion that liberalized naturalism is a more or less subtly disguised form of antinaturalism. The suspicion does not arise only in the critical mind of the 'card-carrying' naturalist: it sits at the heart of the liberalized naturalist herself.[10] In what follows, I will present my own form of liberalized naturalism, and try to show that despite its combining insights from both sides, it is a principled position. But I don't expect to convince the reader that it is a perfectly stable position, not being fully convinced myself.

2. Caught in the middle

Here in a nutshell is the predicament. The fact is that right before our eyes naturalism inspires, and receives reciprocal support from fruitful scientific work and interesting philosophical explorations; moreover, there is synergy between the science and the philosophy, which rather comforts the metatheoretical aspect of naturalism, the idea that philosophy and science are, as the expression goes, 'continuous'. This seems to require the responsible philosopher to be naturalistic *enough.* On the other hand, general arguments in favor of naturalism, and specific proposals, whether broadly philosophical or broadly scientific, are far from entirely successful; some general antinaturalist arguments do seem to cut some ice; and there is at present no sign at all that ongoing naturalistic programs can reach much beyond the area for which they are tailored: the assumption that they will eventually generalize seems at the present stage quite implausible. Our responsible philosopher must therefore not be *too* naturalistic. Finally, he should not merely steer a middle course, unpalatable to both sides: he should make the antinaturalist side of his position acceptable or at least intelligible to his naturalist friends, and vice-versa, the naturalist side comprehensible to his antinaturalist friends.

---

9 Such is the label which De Caro and Macarthur claim for the diverse group of authors in their collection.

10 See the editors' attempt to justify their attachment to naturalism while reaffirming a rejection of "scientific" or "scientistic" naturalism: *op. cit.*, pp. 13-14.

The fair-minded philosopher must keep firmly in mind the two main features of the dialectical situation. One is that, granting that the assessment just proposed is correct, there is an uneasy stand-off between naturalists and anti-naturalists, with strong arguments on both sides. The second is that even the non-naturalist can sense a dissymetry in the debate: the initiative is on the side of the naturalists, and to simply discard the evidence and the ongoing scientific and philosophical work on which they base their case would be irresponsible and, yes, somehow wrong. The antinaturalists can't claim that ignoring their evidence and the work in progress on their side would be unreasonable or intellectually unethical *in the same way,* because all or most of it is of a negative nature. Not only are naturalists doing most of the moving on the field, but when they lose control of the ball (as for example when they are subjected to devastating criticism, sometimes from their own ranks) they promptly pick it up: there is a certain self-sustaining robustness in their game which makes their way of playing the right way to play. This is perhaps the sense in which naturalism is indeed the 'unsurpassable philosophy of our time'.

The fair-minded philosopher's task is thus fairly straightforward: to give due credit to the ongoing research programs which are inspired by, or are grist to the mill of scientific naturalism, and yet redress the game so as to block the unsurpassibility thesis. It might be objected that the responsible philosopher has a more pressing duty: to arbitrate as best he can between the two sides, and come up with his own considered judgment, issuing in a verdict. But this would be precisely forcing his choice in a way reminiscent of a familiar militant technique. There is a legitimate third way between throwing one's arms in the air and choosing sides, which is to examine the arguments, find them inconclusive and ask not, once again, Who is right, all things considered?, but, What should my stance be, all things considered? The 'minimal naturalism' which I will proceed to defend is my answer to the second question.

Naturalism, I suggested above, in the most general sense, takes the form of a recommendation: *Take natural science with utmost seriousness*. In R. B. Perry's terms, it is "the philosophical generalization of the sciences". But what does this entail? No interesting form of naturalism stops at such general declarations. What makes an avatar of naturalism interesting is the problem situation it proposes[11]. It

11 As Huw Price has recently been arguing (see his "Naturalism without representationalism", in: De Caro/Macarthur, *op. cit.,* pp. 71-88), the very first and most basic choice may well consist in deciding between two possible *targets* of the generalization: should it be the *subject* whose position in the overall scheme of things is to be characterized as that of an inhabitant of nature as science reveals it; or the *objects* which we talk about and represent linguistically which must, as he puts it, *placed* in the natural realm? In Price's own words, subject naturalism asserts that "We humans are natural creatures; human knowledge is itself a natural phenomenon", while object naturalism insists that "whatever exists exists in the natural realms". Although Price makes no mention of it, it seems to me that McDowell's earlier proposal that we conceive of "thinking as the exercise of powers possessed [...] unmysteriously by a thinking being itself, an animal

starts with a certain contrast between a class of prima facie "natural" entities, and a class of prima facie non-"natural" entities. The first class fixes the reference of "nature" at the start of the game, the second specifies the problem at hand. So for example, the first class might be the set of entities postulated by our best current physical theories, and the second might contain mental states and processes. The corresponding naturalistic stance is expressed, on the ontological plane, by some kind of physicalism about the mental, and on the epistemic plane, by the demand that psychology, or cognitive science, be included in the natural sciences. Or the first class might consist in the ontology of the natural sciences, and the second might be the set of social processes: naturalism in this situation might amount ontologically to the rejection of an autonomous sphere of social entities, and epistemically to the rejection of a bifurcation between the natural and the social sciences. And so forth: the structure of the issue consists in the specification of Class I and Class II, and the claim that Class II is in fact included in Class I. A final condition for a form of naturalism to be worth investigating is that this inclusion relation be non-trivial: it must be moot and require serious scientific and/or philosophical work. Dogmatic assertions of the form "Everything is (or: is at bottom) natural" trivialize the problem and deserve no consideration, no more than such counterparts as "Everything is (or: is in the last analysis) socially constructed."

I now come to a distinction which is not made explicit, as far as I know, in the literature, perhaps because it is too obvious, perhaps because most philosophers focus on just one side of the distinction, and/or take the other side for granted.

Some forms of naturalism, regardless of their choice of Classes I and II, include a proposed strategy for establishing the inclusion of Class II in Class I. Due to the non-triviality condition, it is not one bit *obvious* that the strategy will work, but the naturalist means to argue that in the fullness of time, every member of class II will be shown, by some clever application of the proposed strategy, to belong to Class I. Programs of this sort are instances of what I call *anchored* naturalism. *Free-floating* naturalism, by contrast, consists in arguments of a general nature purporting to establish, in one fell swoop, the inclusion relation, so that, once the argument is accepted, there is no work left to be done. Although of course there is nothing to prevent a philosopher, or a scientist for that matter, from proposing both specific reductive strategies and general arguments, sometimes in different

that lives its life in cognitive and practical relations to the world" is a form of subject naturalism. (Quote taken from Hilary Putnam, *Words and Life*. Cambridge, MA: Harvard University Press 1994, pp. 307-8; the reference provided is to McDowell's Auguste Comte lecture to the LSE of 2 February 1993). See also John McDowell, *Mind and World*. Cambridge, MA: Harvard University Press, 1994, in which McDowell distances himself with what he calls 'bald naturalism'.

Price's naturalism is clearly of the 'liberalized' sort, and is developed so as to explicitly reject (object, mainstream, contemporary scientific or bald) naturalism. What I discuss in the sequel is object naturalism in Price's sense; my Class I/Class II problem is what he calls a 'placement problem'. My own version of liberalized naturalism, to be sketched presently, is (I believe) compatible with Price's.

writings or at different moments of her itinerary, in fact philosophers tend to fall on one or the other side of the fence. Thus Quine, and those working in the naturalized epistemology tradition, tend to be free-floaters, while Putnam, whether in his former, optimistic mood regarding naturalism, or in his present, pessimistic mood, is interested in anchored naturalism. And he becomes impatient with free-floating naturalism, which he finds "very puzzling", or a form of outright "I know not what". Poincaré's verdict on a related matter is just as final. Poincaré is concerned with the unity of nature: "The question we must ask is not whether nature is one, *but how it is one*."[12]. In other words, the only question worth asking, if, for example, we are worried about the 'imponderable fluids' such as caloric, phlogiston etc., is not, *Can* we reduce or eliminate them?, but rather, *How* do we do it? In other words, what is called for is hard scientific work, genuine scientific imagination, possibly prodded and supported by philosophy, rather than overarching, armchair considerations to the effect that necessarily, some research strategy *or other* will succeed in accomplishing the called-for naturalization.

3. Methodological naturalism

The following 2-by-2 logical table may be thought to provide a simple, natural way of locating various forms of naturalism:

Epistemology ➔ 🡻 Ontology	Yes	No
Yes	**1-1** (full-blooded naturalism)	**1-0** (caution, type B)
No	**0-1** (caution, type A)	**0-0** (full-blooded anti-naturalism)

Table 1: A logical space of attitudes towards naturalism based on ontology and epistemology

Position 1-1 is occupied by full-blooded naturalists, such as a vast majority of contemporary philosophers of mind, who believe both that everything belongs to the natural order and that natural science is the sole means to acquire a true picture of the world. Position 0-0 is occupied by full-blooded anti-naturalists, who believe neither. Position 1-0 in particular by those who are doubly cautious, being willing neither to posit non-natural entities nor to rule out epistemic resources outside natural science (if, for example, they doubt that natural science can provide a full and faithful picture of the world, and they believe or suspect that we do have a grasp, however imperfect, of certain aspects of it, crystallized in common wisdom, practical know-how, non-natural science, literature, etc.). Finally, position 0-1 is defended by another type of cautious philosopher, who sees no reason to deny

12 Henri Poincaré, *La science et l'hypothèse,* Paris: Flammarion 1902, chap. IX, p. 161; my italics.

the possibility of non-natural entities, but considers natural science to be the only valid way to acquire genuine knowledge.

One clear limitation of this table is that it does not distinguish between two ways of ruling out the 'No' answers: positive denial or agnosticism. To decline to be committed to the non-existence of non-natural things, or of non-natural-scientific sources of knowledge, is one thing; to be committed to the existence of non-natural entities, or non-natural-scientific sources of knowledge, quite another. A 3-by-3 table would remedy this shortcoming. But another limitation is that it leaves no room for positions which distinguish between commitment to science and commitment to *natural* science; a 3-by-4 table would be needed to take care of that problem.

The table I propose instead to use in order to locate intermediate positions, including the one I want to defend, is a bit simpler. It is based on a pair of dimensions which are orthogonal to the ones used in Table I. On the horizontal axis figures the contrast class, viz. what 'natural' is contrasted with. The basic possibilities are: non-natural equals non-physical or non-material; not accessible by scientific means; and not accessible by the means of the natural sciences. Note that the contrast can be construed either ontologically or epistemically or both. On the vertical axis one finds the two basic stances, or values of commitment towards the naturalist thesis: rejection and acceptance; and in between, a half-way, cautious stance which I label 'methodological'.

Contrast class: natural *vs* ➔ ➔ Commitment	Non-physical (non-material)	Not scientifically accessible	Not natural-scientifically accessible
Rejection	Supernaturalism (theological dualism, non-religious spiritualism)	Atheistic pluralism	Scientific dualism (bifurcationism)
Methodological stance	Methodological naturalism, standard sense	Methodological naturalism, my sense (MENA)	Methodological naturalism, Chomsky's sense, or anti-bifurcationism
Acceptance	Atheological naturalism (anti-supernaturalism)	Broad scientific naturalism	Strict scientific naturalism (physicalism or pluralistic naturalism)

Table 2: A logical space of attitudes towards naturalism based on commitment and contrast class

Let us look at the first line. The anti-naturalist may affirm or refuse to rule out either non-material entities (column 1); or entities undetected or unregimented by science *tout court* (column 2); or, more restrictively, by natural science (column

3). The first cell is typically defended by religious believers and by non-religious dualists, who believe in the existence of a spiritual realm separate from the material realm; the second, by philosophers (and lay folks) who believe there are provinces of reality which are out of the reach of science altogether; the third, by philosophers and social scientists who believe there are provinces of reality which are out of the reach of natural science but not inaccessible to all forms of science. Occupants of cell 0-0 in Table 1 straddle the first and second cells of this line, while the occupants of cell 1-0 straddle the second and third.

The third line's three cells, by contrast, need not be occupied by theorists with different beliefs, but by thinkers with different opponents in mind. In the first position, the naturalist insists on rejecting the notion of a non-material realm (either by discrediting the very idea, or by arguing that the purported non-material realm is part of nature after all). In the second position, the target is the pluralist who believes that science leaves out entities or phenomena which belong to the real world. In the third position, the main target is the view of a realm consisting of meaning, norms, values, history, culture which is separated from the natural realm yet is the proper object of study of the sciences of man. This third position is defended by the typical scientific naturalist (the occupant of cell 1-1 in Table 1).

The second line is the one I wish to draw attention to. I have dubbed it "methodological" to indicate its intermediate position on the commitment scale: any doctrine which stops short of committing to naturalism or to anti-naturalism, conceived as clear-cut views based on positive (rather than merely skeptical) arguments, belongs, on my taxonomy, in the methodological group. However, the phrase "methodological naturalism" is used in the literature in at least two different, more restricted senses. What I call the standard sense is the one used in the debate concerning the practice of natural science by Christian believers. It is the stance recommended to someone who is both a believer and a scientist: qua believer, he is a committed supernaturalist, but qua scientist he suspends his belief in the supernatural and conducts his scientific business *as if* only the material world existed. As Michael Martin puts it, "in the context of [scientific] inquiry only natural processes and events exist".[13]

Quite another meaning is given by Chomsky in his recent discussions of naturalism:[14] for him, methodological naturalism is the rejection of 'methodological dualism', the view that human cognitive and (in particular) linguistic processes are subject to a dual description, one provided by the natural sciences, the other by other rational sources. To Chomsky, theoretical understanding of whatever order of phenomena has but one source and takes but one form, whether the topic be the formation of waterfalls, the collision of electrons or the human mind and

13 "Justifying methodological naturalism" (2002) (available online at www.infidels.org).

14 Chomsky, *New horizons in the study of langage and mind.* Cambridge: Cambridge University Press, 2002 and subsequent writings discussed inter alia by Pierre Jacob, "Chomsky's naturalism: its scope and limits", forthcoming (French version in: *Chomsky,* special issue of *Cahiers de l'Herne,* Paris, 2007, pp. 202-214).

its various components. Chomsky's doctrine here is at base nothing particularly new – it is in fact the position of the typical occupants of column 1 in Table 1: it constitutes a detailed and updated version of the monist view of the sciences of man, defended and attacked for at least 150 years[15]. But while Chomsky has felt the need, throughout his career, to reassert his stand in the face of what remains (at least on the Continent) a majority view in the human and social sciences, his main concern in his recent writings has been to refute the ontological naturalism which is espoused by the overwhelming majority of his fellow cognitive scientists: unlike them, he defends cell 0-1 in Table 1.[16] The Christian and Chomskyan construals of methodological naturalism occupy the first and third cell respectively of line 2 in Table 2. What will serve as my starting point is the position which occupies the middle cell.

Methodological naturalism, in my sense (which I will abbreviate as MENA), is distinct from the other two varieties inasmuch as it withholds both a negative and a positive final judgment on ontological naturalism. It is radically non-committal in that sense, and also because it takes no stand on the bifurcation thesis: the main opponent here is not the defender of a niche for non-naturalistic human science, but one who denies the legitimacy of a scientific approach to all sorts of (prima facie non-natural) things. No great leap of philosophical imagination, MENA in one or another form is constantly re-discovered by philosophers who are either uncomfortable about the majority ontological doctrine in philosophy of mind, or uneasy about the underlying philosophy of science, or unnerved by the unending 'war of methods' in the sciences of man, or (like myself) about all three. I should say right off that I will not end up defending MENA as I am about to present it, but I take it as a first approximation of the position which I will recommend.

MENA is expressed in the following maxim:

> *Engage in whatever inquiry, at any given stage of the scientific problem situation, is recommended by scientific naturalism with the aim of securing a positive result, but refrain from any commitment, explicit or implicit, regarding the outcome of the inquiry.*

15 Chomsky is emphatically *not* claiming that natural science is the only source of knowledge concerning mankind: "Someone committed to it [methodological naturalism] can consistently believe (I do) that we learn much more of human interest about how people think and feel and act by reading novels or studying history than from all of naturalistic psychology, and perhaps always will." Noam Chomsky, "Chomsky, Noam", in: Samuel Guttenplan (Ed.), *A Companion to the philosophy of mind.* Oxford: Blackwell, 1994, p. 153. But natural science does have an exclusive responsibility, according to him, in developing a *theoretical understanding* of mind and langage. There is no *third way*.

16 My interpretation of Chomsky's ontology might be questioned, but nothing central to the present paper hinges on it; the point here is terminological: methdological naturalism is used by Christians on the one hand, by Chomsky on the other, in different though related senses, and the notion I propose to defend is yet something else, albeit in the same ballpark.

By a 'positive result', I mean a demonstration of the fact that some member of the class of prima-facie non-natural entities actually belongs to the other class. Non-commitment means: no assumption made that the naturalistic inquiry about one particular entity or process is bound to succeed, let alone that there exists one strategy, which we are bound to discover eventually, which will work on each member of the non-natural class. Indeed, MENA rejects any commitment to a thesis of the form: everything is really, at bottom, natural, or to any other manifestations of free-floating naturalism. Indeed, in a secundary sense, 'methodological' also expresses an implicit demand for a method or family of methods.

Two objections might be raised right away. The first is the well-known precariousness of 'as if' positions generally. While, as the example of instrumentalism in the philosophy of science would tend to show, there is no logical inconsistency in defending MENA while rejecting ontological naturalism, there is a threat of pragmatic incoherence in (i) adopting a maxim or strategy or heuristic whose success depends on the existence of certain entities or processes while (ii) invoking ontological abstinence with respect to those entities and processes. My answer is this: MENA encourages piecemeal attempts, on hopeful candidates from Class II, to bring them into Class I, but it does so without postulating universal in-principle success. It even supports a strategy of considering not-hopeful candidates in order to zero-in on what gets in the way. The history of logic provides an example which shows that this can be a good strategy: proving that certain number-theoretic functions are computable (or feasibly computable) is a goal which one can rationally pursue without believing that all functions are computable (or feasibly computable). And failing to show that a function is computable (or feasibly computable) is not necessarily failure tout court; it helps one get the knack, grasp a pattern, and guess whether the case at hand fits the pattern.

The second objection is that despite its apparent opposition to free-floating naturalism, MENA actually comes perilously close to it, due to its non-commitment to any particular naturalization strategy. In combination with its as-if character, this lack of constructive content makes it too bland to be of much theoretical or practical help. MENA needs to be strengthened. I will try and show that an intermediate position with a little more bite can be reached, which is the *minimal naturalism* which I think we should countenance. But first I need to say a few words about the obstacles which, as I see it, block both a brutal upgrading or a brutal downgrading of MENA.

3. On reforming rather than overthrowing MENA

In this section I will limit myself to the naturalization of the mind. There are, as I mentioned at the outset, other entities which one might want to naturalize, and which in fact are being subjected to naturalistic approaches. Most of them however are connected to, or dependent on the naturalization of the mind and the concepts deployed to that end.

I will begin with two sets of considerations blocking full ontological naturalism, the first from philosophy of science, the second from philosophy of mind. I will move on to arguments against outright rejection of naturalism. The upshot will be that MENA should be be reformed, rather than rejected in favor of one or the other position occupying the logical space outlined above. The discussion will perforce be highly condensed.

a. Limits from above, 1: The record of cognitive science

It is sometimes thought that cognitive science is the eating proof of the naturalistic pie. It contributes in no small measure to making naturalism 'unsurpassable'. It is said to be 'naturalizing' the mind right before our eyes. Thanks to the long-awaited conceptual and instrumental tools which, unlike its predecessors in pre-scientific and scientific psychology, cognitive science finally has secured, it shows every sign of being successful. As we shall see shortly, there is some definite merit to this appraisal, but for now the focus is on its more dubious parts.

One familiar but deep philosophical question bears on how naturalistic cognitive science really is. To what extent has it freed itself, and can it free itself, of its hybrid vocabulary, part intentional psychology plus information processing, part neuroscience plus mathematical-physical modeling? To some philosophers, the lack of a clear-cut answer to this question pretty much closes the case: cognitive science is and is bound to remain a non-natural science. But perhaps they are laboring under too narrow a notion of the natural, just as Locke, before he got "convinced by the judicious Mr Newton's incomparable book",[17] was laboring under too narrow a notion of the physical. Rather than pursue this difficult matter, I choose a more pedestrian route.

First, the bare empirical fact is that cognitive science in its present state of development presents a characteristically 'gappy' structure. It is most clearly successful for 'input systems' (Fodor's 1983 expression[18]), and at least until recently progress on 'higher' or 'central' processes was widely regarded as less than impressive. Fodor's particular way of drawing the contours of the gap is outdated, and there has been, partly as a response to Fodor's grim assessment, a wealth of interesting work in areas (such as social and moral cognition, emotions, consciousness ...) formerly all but closed to cognitive science. Yet the very abundance of new concepts, paradigms, results, originating in cognitive neuroscience but also at the interface of developmental, cognitive and social psychology, evolutionary biology, anthropology, retroactively proves how deeply ignorant we were all along

17 Locke, *Reply to Stillingfleet*, 1699, quoted in Roberto Torretti, *The Philosophy of physics*. Cambridge: Cambridge University Press, 1999 p. 75 (full title of Locke's opuscule in Torretti, p. 478).

18 Jerry L. Fodor, *The Modularity of Mind*, Cambridge, MA: MIT Press 1983.

about countless functions of the brain/mind. What we are acquiring right now is not only new knowledge in cognitive and brain science, but reasons to suspect that we are still today, as it turns out we were yesterday, more ignorant than knowledgeable.

Second, the general strategy followed by cognitive science since its inception is the one at work in biology in general. It consists in combining a top-down and a bottom-up approach: the first identifies a set of functions and their interrelations; the second identifies the corresponding set of 'forms' (material structures or systems) and their causal interconnections. Bottom-up naturalization of the mind is the goal of neuroscience. Top-down naturalization is offered by two distinct research programs: functionalism (the information-processing paradigm, also known as computationalism, classical or otherwise), and evolutionary psychology. Full vindication of ontological naturalism regarding the mind requires a triple success: completion of the top-down analysis; completion of the bottom-up analysis; and, importantly, articulation of the two approaches. In the pioneering stage of cognitive science, this articulation was thought to be provided by an existence proof. The computer was seen as an information-processing, mechanically realized cognitive organ: however defective in its details, the model did show what sense could be made of an articulation of the top-down and bottom-up analyses. That was the whole idea, the grand idea of functionalist neo- or Turing mechanism, articulated by such founders of the field as Newell and Simon, Marr and Fodor.[19]

The hope was that suitable complexification of this paradigm would yield a satisfactory notion of a natural mind. Now while success has been notable, there are conceptual problems, and reasons to doubt that cognitive science is on its way to solving them. It would be futile to try and review here the ever-expanding critique of the classical approach in cognitive science, but it is worth stressing that the key idea of the articulation between the top-down and bottom-up approaches is under attack.[20] Neuro-imagery, evolutionary theorizing, psychology, mathematical/informational simulation, no longer seem to fit together in the way proposed by functionalism, nor is there at present any clear alternative framework. Further,

19 Newell, A. & Simon, H.A. (1976), "Computer science as empirical enquiry: Symbols and search", in: *Comm. Am. Ass. Computing Machinery,* 19, pp. 113-126; repr. in John Haugeland (Ed.), *Mind Design.* Cambridge, MA: MIT Press 1981. David Marr, *Vision: A Computational Investigation into the Human Representation and Processing of Visual Information,* San Francisco: Henry Holt & Company 1982. Jerry L. Fodor, *The Language of Thought.* New York: Thos. Crowell 1975; repr. Cambridge, Mass: Harvard U.P.; Jerry A. Fodor, *Representations: Philosophical Essays on the Foundations of Cognitive Science.* Cambridge, Mass.: MIT Press, 1981.

20 See *e.g.* Jaegwon Kim, *Mind in a Physical World: An Essay on the Mind-Body Problem and Mental Causation.* Cambridge, MA: MIT Press, 1998 ; Lawrence A. Shapiro, "Reductionism, Embodiment, and the Generality of Psychology" in: H. Looren de Jong & M. Schouten (Eds.), *The Matter of Mind.* Malden, MA: Blackwell Publishing 2006, pp. 101-120; Denis Mareschal *et al., Neuroconstructivism: How the Brain Constructs Cognition.* New York: Oxford University Press, 2007.

the 'deep' faculties (consciousness, intentionality, spontaneity) seem to resist both bottom-up and top-down naturalization, despite the numerous attempts made to this day, and ongoing.

The moral to draw, it would seem, is that the knowledge we have acquired about the mind, considerable as it is, has not reached the level where we can confidently predict the vindication of ontological naturalism about the mind. And following, *inter alia,* Chomsky's recommendation, the naturalistic spirit itself recommends heeding this consideration.

b. Limits from above, 2: The argument from context

Running on roughly parallel tracks with several philosophers of note, such as Charles Travis,[21] but with a different starting point, I have over the years sketched a contextualist, or situated, view of cognition in general, and of rational inquiry in particular. In itself, contextualism appears to raise no insuperable obstacle against naturalism: naturalistic models of context-sensitivity are in fact quite an active area of investigation. However, the way I propose to think about contextuality is as an irreducibly normative dimension of thought, and this, it would seem, does get in the way of all but the most hybrid naturalization programs, programs whose claim to naturalness are extremely dubious.[22]

If there is any merit to my arguments, they seem to imply severe in-principle limitations on the very idea of *prescriptive* natural models of higher cognitive processes, as opposed to *permissive* models. This would by no means spell the end of ongoing efforts to specify such permissive models (which determine the envelope of feasible cognitive acts), but it would very much dull the ontological and ethical teeth of naturalism: who ever doubted that there are natural constraints on what we can think? Of course, as we gain empirical knowledge and conceptual sharpness on these natural constraints, our picture of thought processes can undergo profound revisions. But they will continue, one might well reckon, to contain a non-natural dimension – second nature, culture, spontaneity, history, …

c. Limits from below: the record of cognitive science, revisited, and the liveliness of philosophical psychology

Despite its limited success, and despite its shaky foundations, cognitive science is thriving. It is leading philosophy of mind and epistemology (briefly, philosophical psychology) in corners which they hadn't visited before, despite centuries of hard work. As a result, our views about the mind and its natural underpinnings are undergoing profound changes.

21 Charles Travis, *Occasion-Sensitivity: Selected Essays.* New York: Oxford University Press, 2008.

22 Daniel Andler, "The normativity of context", in: *Philosophical Studies* 100, 2000, pp. 273-303; Daniel Andler, "Context: the case for a principled epistemic particularism", in: *Journal of Pragmatics,* 35, 3, 2003, pp. 349-371.

Arguing in favor of such a view is certainly not impossible. But first it is hard to do in a few sentences, and second, it is even more a matter of judgment than of argument. Familiarity with the field, not just its results but its inner processes, its ongoing discussions, its speculative energy, induces a strong impression of a thriving research program. Of course, the same impression was certainly conveyed to optimistic, not to say gullible, witnesses or participants in scientific programs which have since reputedly gone bankrupt, such as Gestalt psychology, behaviorism or classical artificial intelligence. Indeed, as we will see in a minute, this consideration plays a crucial role in the stiffening of my recommendation for an acceptable form of naturalism. Perhaps we should be content at this juncture to record as a fact that a form of philosophical anti-naturalism which would purport to show that what is going on in cognitive science today is essentially a waste of time would meet with considerable scepticism, and that the burden of the proof would rest on the anti-naturalist.[23] He would have to show not simply that some claims are exaggerated, some phenomena likely to remain untouched, some tensions or even contradictions exist between subfields and schools, this or that strategy is bound to fail, etc.: he would have to make a convincing case that a majority of results are either unsound, or uninformative, or again require a complete reinterpretation in order to fit into some kind of conceptually accceptable picture. A tall order. Meanwhile, the realistic spirit again commends rejecting a view which would all but deny coherence and fruitfulness to this scientific and philosophical activity.

But there *is* a negative argument which the anti-naturalist could use, and in fact Hilary Putnam has developed it at length.[24] Putnam, as I said, rightly takes seriously only *anchored* forms of naturalism, and, regarding the mind, he knows of one proposal, based on the functionalist scheme which he himself propounded, and which he has faulted in theoretically deep ways. Putnam seems to conclude, in the absence of a likely stand-in, that 'computational psychology', along with associated attempts to 'naturalize' the mind, is all but hopeless. I certainly don't mean to challenge Putnam's diagnosis of functionalism's 'troubles'. But I do want to question what I take to be the logic of his case against cognitive science.

Let me start with a banal consideration from the history of science. We were doing perfectly respectable chemistry before quantum mechanics came on the scene to inform us of what the chemical bond physically consists in. Pre-quantum

23 Mention should be made of two recent collections aiming at casting doubt on the viability of cognitive science: David M. Johnson & Cristina E. Erneling (Eds.), *The Future of the cognitive revolution*. New York: Oxford University Press, 1997; and Cristina E. Erneling. & David M. Johnson (Eds.), *The Mind as a scientific object. Between brain and culture*. New York: Oxford University Press, 2005. They both contain very valuable papers, and I personally sympathize with a number of critical perspectives developed there. I would however not go as far as taking them to be more than just that: critical perspectives, which leave the target alive, though bruised.

24 Hilary Putnam, *Representation and reality*. Cambridge, MA: MIT Press 1988.

chemistry was a naturalistic inquiry into kinds and combinations of stuff, an inquiry legitimate and progressive despite not being endowed with (what we would later regard as proper) naturalistic foundations. This kind of ox-before-the-cart situation is surely not unique. Evolutionary biology is another oft-mentioned case; in fact, it seems closer to being the rule than the exception in the historical development of scientific disciplines. Now, what reasons have we to think that cognitive science is, or should be, any different? What is wrong with the idea that this flurry of activity is progressive and legitimate, despite having as yet no solid foundation? Perhaps Putnam's, and other critics', reasons for thinking it *is* wrong are the following. Cognitive psychology (and thus cognitive science as a whole, insofar as it includes cognitive psychology as a core part), they believe, is *predicated* on the representationalist-computationalist scheme. If that scheme is incoherent, cognitive psychology collapses (or requires at least a complete theoretical overhaul).

Well, this line of argument seems to me perfectly sound when applied to the sort of cognitive psychology associated with such figures as Newell and Simon, and which was barely distinguishable from artificial intelligence. (A lot more would need to be said at this point to do justice to both the program, its critics, and its eventual breakdown, but this would take us too far afield.) But it doesn't apply, at least directly, to cognitive psychology and cognitive science in their contemporary form. The pull towards believing otherwise may be due to over-reliance on *philosophers'* reconstructions of the cognitive enterprise. Inconsistencies or frailties in such reconstructions may be due to the faulty modeling, not to what is modeled, *viz*. the actual science. In fact, I have long argued that a large proportion of the work done under the label of cognitive science happily lives in a no-man's land where no dues are owed to functionalism, or connectionism, or dynamical systems, or methodological solipsism, or externalism, or learning-theoretic or biosemantic notions of representation, etc. This doesn't mean, of course, that the theoretical bases of the work shouldn't be actively sought by scientists and philosophers. But there is no ground for imputing to it, by default, a provably faulty foundation. The work should be critically examined piecemeal and directly, not through the lenses of a rationalizing philosopher.

d. Upgrading MENA to MINA

I am now in a position to strengthen my initial proposal for a form of well-tempered naturalism compatible with the constraints discussed up to this point. What MENA lacks is, I suggested, a modicum of anchoring. Yet someone who remains unconvinced by any of the current proposals will be loath to anchor it in a particular naturalistic strategy. How then can I provide my favored form of naturalism with some anchor without anchoring it entirely?

To solve this little riddle, it suffices to go back to AI. What is wrong with MENA is shown by the fact that, at the time, it would not have had the tools, nor the mission, to critically analyze AI (by which I mean 'good old-fashioned artificial intelligence' or GOFAI, in John Haugeland's terminology[25]): at the time, it seemed that scientific naturalism's best guess as to what line of inquiry to pursue regarding the mind was AI, and so MENA had no choice then but to follow suit. At the same moment, avowed anti-naturalists such as Hubert Dreyfus, and later Haugeland, as well as philosophers such as Robert Cummins and Putnam himself[26] did have the means, and the courage to do it, and did produce a convincing and enlightening critique of AI.

Now of course the cure cannot consist in amending our definition to read:

> *Engage in whatever inquiry, at any given stage of the scientific problem situation, is recommended by scientific naturalism with the aim of securing a positive result,* except if this inquiry happens to be GOFAI, *but refrain from any commitment, explicit or implicit, regarding the outcome of the inquiry.*

Not only would it be useless and unprincipled, but it would also not prevent another possible disaster, the uncritical acceptance of the next fad in cognitive science. For example, even with the proposed amendment, MENA would have nothing to say about the exaggerated claims made on behalf of fMRI-inspired research. Our responsible naturalist cannot be content with letting cognitive neuroscience based primarily on fMRI and other brain-imaging techniques simply take over cognitive science. Minimal naturalism, or MINA, in contrast with mere MENA, is not a laissez (science)-faire atttitude. Perhaps one could characterize minimal naturalism as *critical* methodological naturalism. One last constraint might put this demand in sharper focus. What experience seems to suggest is that the mind tempts those that study it, again and again, whether they be philosophers, computer scientists or neuroscientists, to treat the empirical evidence with less than the care it deserves, and altogether disregard large chunks of it. So perhaps minimal naturalism should be phrased thus:

> *Engage in whatever inquiry scientific naturalism recommends with the aim of securing a positive result,* without foregoing a critical examination of the recommendation, and with due regard to the entire empirical evidence, whether available through commonsense, phenomenology, non-naturalistic or pre-naturalistic science, or again scientific experimentation in the style of natural

---

25 John Haugeland, *Artificial Intelligence: The very idea.* Cambridge, MA: MIT Press 1989.

26 Hubert L. Dreyfus, *What Computers Can't Do,* New York: Harper & Row 1972; augm. edition: *What Computers* Still *Can't Do*, Cambridge, MA: MIT Press 1993. John Haugeland, 1978. "The nature and plausibility of cognitivism", in: *Behavioral and Brain Sciences,* 1, 1978, pp. 215-226; repr. in John Haugeland (Ed.), *Mind Design.* Cambridge, MA: MIT Press 1981. Robert Cummins, *Meaning and mental representations.* Cambridge, MA: MIT Press 1989. Putnam, *op. cit.* and "Much Ado about Not Very Much." *Daedalus* 117.1 (Winter 1988): 269-281. Repr. as "Artificial Intelligence: Much Ado about Not Very Much" in *Words and Life* (1994), pp. 391-402. See also the editor's introductory and concluding chapters in Daniel Andler (Ed.), *Introduction aux sciences cognitives.* 2nd éd., Paris: Gallimard 2004.

science. *And refrain from any commitment, explicit or implicit, regarding the outcome of the inquiry.*

In a nutshell then, minimal naturalism (which is not limited in its application to the study of the mind, though I have used that important special case as a guide) is methodological naturalism with philosophically wide open eyes, a working philosophy based on a close interaction with scientists, one which precludes neither the collaborative production of results, nor criticism and the possibility of renouncing or reorienting the collaboration.

4. Minimal naturalism in the sciences of man

The sciences of man include psychology, of course, and cognitive science can be regarded, very roughly, as psychology pursued by novel means. To that extent, the considerations developed above are ipso facto applicable to the sciences of man. Yet even in this case some interesting issues arise. One might imagine that scientific psychology *is* psychology naturalized, a branch of cognitive science. But although this is not quite false, nor is it quite true. Not only are there branches of psychology which think of themselves as both scientific and distinct from cognitive psychology (social and personality psychology come to mind), but some follow a strictly natural scientific methodology, and some even share their subject matter with cognitive psychology. Why, for example, is there a field called 'mathematical psychology', which does intersect with cognitive psychology but has not to this day been absorbed by it, and whose founding fathers are not among the heroes of cognitive science? The reasons could in part be historical, but I believe there are other, deeper, conceptually more significant reasons, having to do precisely with divergences concerning the naturalistic character of the methodologies.[27]

Linguistics presents another case, in which not only can cognitive science claim, as of now, no more than a fraction of the leading research programs, but even those were not until fairly recently uniformly 'naturalistic' in the sense in use in cognitive science today. (To dot the -i-s, these branches of linguistics had impeccable natural-scientific credentials without being fully naturalistic at least in some sense: they were stated in a rigorous, formal language allowing for cumulative knowledge and hypothetico-deductive reasoning, yet were not concerned with providing causal explanatory accounts of language, whether proximal or distal). Here again there is some degree of arbitrariness and historical contingency in the cartography of the field, but some conceptual issues are also involved.

27 The historical and conceptual origins of the split are intertwined. In fact, the historical record needs to be rectified in order for these connections to come to light. See Gary Hatfield's revisionary studies in the philosophical history of psychology, *e.g.* Gary Hatfield, "Remaking the science of the mind. Psychology as natural science", in : C. Fox/R. Porter/ R. Wokler (Eds.), *Inventing Human Science. Eighteeth-Century Domains*, Berkeley & Los Angeles: University of California Press 1995.

The suggestion is that naturalism includes, and sometimes confuses, two independent demands. The first is formal: the theoretical study of an area should be conducted in ways which are formally in accord with those of (established) natural sciences such as physics, and lead to a body of knowledge whose form is comparable to the knowledge produced by those disciplines. The second is causal and genetic: the theoretical study of an area should aim at bringing to light the causal structure of the phenomena and (in particular) show how they have come into existence.

The second demand can be met without it being clear that the first is as well; hence the debate, now essentially over, about the scientific status of biology. But many areas within the human and social sciences, including cognitive science, can boast a high grade on the second demand and fall exceedingly short of satisfying the first.

The reverse is not only also true, but it raises theoretical issues which appear now to be of central importance. In the words of a contemporary defender "of a thorough and strict naturalism", Mark Bickhard, "[i]t is distressingly easy to espouse naturalism, but nevertheless to fail in a project of naturalism. [...] Many models fail to be consistent with naturalism in spite of the best intentions of their authors. [...] An essential characteristic of any naturalistic model of any phenomena, therefore, is that it be consistent with the *natural emergence* of those phenomena."[28] On that count, it now appears as if few among the scientific branches of the human and social sciences are naturalistic.

Indeed, regarding the social sciences proper (sociology, economics, geography, anthropology, demography ...), the quantitative and formal subdisciplines which are not naturalistic (in the second sense) vastly outnumber the naturalistic programs and results. This holds not only under Bickhard's strong definition of 'naturalistic', but even in a less stringent sense, counting as naturalistic an account which provides causal mechanisms, at least in outline, and foregoing the demand for a phylogenetic story (an account of how the phenomena under scrutiny came into being in the first place, after the Big Bang and out of conditions then prevailing).

How are we to make sense then of the existence of natural sciences of man which are not 'naturalistic' in a sense acceptable to today's scientific naturalist? This is a difficult question which cannot be fully explored here. Still, we can make use of our general distinctions. There exist fully naturalistic, partly naturalistic, and anti-naturalistic answers.

The committed, up-to-date naturalist can go one of two ways. The most inclusive goal he might espouse is a full reconciliation of the formal-quantitative and causal-genetic accounts. Such would be the outcome once we find out how certain formal structures and quantitative relations can come into being through

28 Bickhard, M. H., "Critical Principles: On the Negative Side of Rationality", in: W. Herfel/C.A. Hooker (Eds.), *Beyond Ruling Reason: Non-formal Approaches to Rationality*. Forthcoming. (My italics).

the development of increasingly complex systems bound by general laws governing their component mechanisms. This ambitious naturalist wants everything, the formal relations and the causal story, and he regards classical cognitive science as an admittedly unfinished yet promising quest for such unification, which will in due course provide the expected articulation, modeled in a very general sense after the Simon-Marr-Fodor scheme of physical realization of formal processes, although possibly at variance with it.

The naturalist might also be less sanguine about the formal dimension, and be more physicalistically or biologically inclined. To her, what really counts are the causal-emergent facts. The formal descriptions might not become available, and if they do they might not play more than an instrumental role. Such would be the inclination of quite a number of cognitive neuroscientists, for example, but more generally, a scientific naturalist of that ilk would hold on to the causal-emergent goal, and renounce the formal goal if it turned out that it were unreachable within the natural-scientific perspective.

The anti-naturalist might accept the formal-quantitative accounts, and even grant them elevated status: he might for example (like most structuralists in Sartre's time) regard those accounts as expressing some invariants produced or necessitated by cultural systems set up by mankind (Levi-Strauss was unique among this group to see the anthropological invariants as a distant reflection of the natural structure of the human mind). On the other hand, the anti-naturalist will refuse to give pride of place to the causal-emergent accounts, which he would see at best as providing limiting conditions on the creative powers of historical humans.

So finally, how should a methodological naturalist, and more particularly, a minimal naturalist, view the situation? First, he would not take for granted the eventual fusion of the domain of formal-quantitative and causal-emergent accounts. Second, he would not take for granted the eventual regimentation of the entire field of human and social science (or of any of its branches) under the formal-quantitative banner. Third, he would no more take for granted the eventual triumph of the causal-emergent approach. In fact, any of the three possibilities will appear to him as unlikely, in the light of past experience, present achievements, and prospects. Like Otto Neurath,[29] he would call for an 'orchestration' of these different approaches, and with that best-possible outcome in mind, he would support, examine and possibly contribute to whatever research program is recommended by scientific naturalism. This would be his way of recognizing naturalism as today's unsurpassable philosophy for the sciences of man.

Université de Paris – Sorbonne
Ecole normale supérieure
Institut universitaire de France
daniel.andler@paris-sorbonne.fr

29 Otto Neurath, *Philosophical Papers 1913–1946*. Dordrecht: Reidel 1983.

António Zilhão

What Does it Mean to Be a Naturalist in the Human and Social Sciences? A Comment on Daniel Andler's "Is Naturalism the Unsurpassable Philosophy for the Sciences of Man in the Twenty-first Century?"

In the version of the paper "Is Naturalism the Unsurpassable Philosophy for the Sciences of Man in the Twenty-first Century?" Daniel Andler sent me a few days ago, he puts forth a position in the sciences of Man he starts calling *liberalized naturalism*. In the course of the paper's development, however, Andler's own brand of liberalized naturalism is further clarified as *minimal naturalism*. Further on, he characterizes *minimal naturalism* as *methodological naturalism with philosophically wide open eyes*. This is the complex term he ends up selecting as the designator of the position he wants to mark out. What is then *methodological naturalism with philosophically wide open eyes*?

Andler presents us his position in terms of a contrast with three other positions, namely, the positions he calls 'anti-naturalistic', 'partly naturalistic', and 'fully naturalistic'. He tells us then that what distinguishes these positions in the sciences of Man is the kind of approach they favour to the object these sciences study: the anti-naturalistic position ascribes a privileged status to a formal approach, the partly naturalistic position ascribes a privileged status to a causal approach, and the fully naturalistic position wants to reconcile the formal with the causal approaches. Methodological naturalism with philosophically wide open eyes is then characterized by Andler as a position that deems the unilateral success of any of these three positions to be highly unlikely and that therefore calls for a combination of all of them in order to obtain results that might contribute to strengthen the research program of ontological naturalism. The term *ontological naturalism* is, in turn, defined as expressing a form of commitment towards the naturalist stance, namely, full acceptance. Finally, the naturalist stance is defined both in terms of the injunction *Take natural science with the utmost seriousness* and in terms of the commitment to some form of reduction of the realm of the non-natural into the realm of the natural.

Now, I suppose that, in the context of this colloquium, the question I am required to answer is the following: do I agree with Andler's brand of 'liberalized naturalism'? Well, it is difficult to give a 'yes' or 'no' answer right away. I feel that some ground needs to be clarified first. Thus, I will postpone my answer to the closing part of my comment.

F. Stadler (ed.), *The Present Situation in the Philosophy of Science*,
The Philosophy of Science in a European Perspective, DOI 10.1007/978-90-481-9115-4_22,

Let me begin by identifying two aspects in Andler's paper with which I experience major difficulties.

In the first place, it is difficult for me to me see how blending naturalistic with anti-naturalistic views on the sciences of Man, as Andler encourages us to do, might contribute to define a philosophically coherent or stable position. This might not be a problem if Andler were just putting forth a pragmatic approach to the field of research and, consequently, refusing to take sides in the dispute between different philosophical ways of making sense of the first order knowledge actually produced by the sciences of Man; but his version of 'liberalized naturalism', no matter how liberal, is supposed to be a form of philosophical naturalism, namely, a form which implies full acceptance of the naturalist stance, and not a form of suspension of philosophical belief. I find this perplexing. However, I will not pursue this issue here, since I find the use of the labels 'fully naturalistic', 'partly naturalistic' and 'anti-naturalistic' in this context highly confusing.

Secondly, it is not at all clear to me that the strengthening of the research program of ontological naturalism in the sciences of Man, as defined by Andler, might possibly be achieved; thus, it is difficult for me to see how could one possibly contribute to such a strengthening. Let me belabour this point a little bit, as I think it is the most relevant.

As I mentioned above, liberalized naturalists of the form Andler specifies, are required to strengthen the research program of ontological naturalism.This research program is, in turn, characterized by a full acceptance of the naturalist stance. The acceptance of this stance is, in turn, characterized by the fulfilment of the two above mentioned requirements. I have troubles with the fulfilment of any of these requirements.

Here is what I find troubling with the fulfilment of the first requirement. The formulation of the injunction *Take natural science with the utmost seriousness* seems to me to imply that there is or that there should be a single field of research called 'natural science', the method of which is or should be unified and transparent. But I am not sure that there is or that there should be a method of natural science over and above the methods of the different natural sciences.

There are, of course, some general standards concerning, for instance, severity of testing, ways of making sure that the evidence is dealt with impartially, or definition of constraints on what may count as good evidence, which are common to all natural sciences. But these seem to me to be fairly general standards. In particular, I do not find it at all self-evident that many anthropologists, historians or linguists that consider themselves to be siding with interpretivism rather than with scientific explanation might not accept these general methodological standards as their own. As a matter of fact, lots of them do. However, if, in order to exclude them from the set we want to define, we try to make the methodological characterization of general natural science more specific, we will be bound to realize that natural sciences are less unified than we tended to think. We will be bound to realize, for instance, that some perfectly acceptable natural sciences do not live up to the standards of what is usually considered to be the role model of natural sciences, namely, physics.

Once we realize and accept that such a state of affairs is indeed the case, it will be highly problematic to justify a discrimination against some exceptions but not against others. And if we do not discriminate against any of the natural scientific exceptions, we will have a hard time justifying our discrimination against those non-natural sciences that abide by the very general methodological scientific standards mentioned above but do not partake of some of the more specific methodological principles followed by, e.g., physicists. This, I think, is a point Jerry Fodor made a long time ago.

Let me now deal with the second requirement. The commitment to ontological reduction seems to me to imply that it always makes sense to try, at least, to reduce the entities and properties the sciences of Man talk about to entities and properties natural sciences talk about. However, as it stands, this thesis seems to me not to be quite true. Let me introduce my point through the consideration of the following standard example.

Consider a prediction provided by standard economic theory according to which, in a normal market economy, lowering interest rates, under some relevant conditions, appropriately specifiable, boosts private investment and thus facilitates economic growth. Let us assume, for the sake of the argument, that predictions such as this one tend to turn out true more often than not and that we feel confident to provide an explanation for the surge of economic growth in terms of the lowering of the interest rates (*ceteris paribus,* of course). How are we to countenance a serious reduction of the entities referred to in an explanation belonging to this mode of discourse to natural scientific entities under some suitable definition of what these are?

The obvious route is to bring back the terms occurring in the above mentioned explanation to very complex descriptions of the underlying natural facts (I will not tackle here the more specific question of determining what kind of reductive relation is the term 'bringing back' supposed to indicate in this context). Needless to say, underlying any economic facts there are millions of human beings relating to each other in certain ways. These relations are in turn somehow managed by their brains; these brains are in turn continuously in the business of processing electro-chemically complex visual or auditory stimuli; the outcome of the cerebral processing is in turn somehow transduced into electrochemical impulses that make the bodies associated with these brains behave in some specific ways, and so on and so forth. Questions of practical feasibility apart, I suppose that it should, at least in principle, be possible to describe events such as the 'lowering of the interest rates' or the 'upsurge in private investment' in terms belonging exclusively to the levels of discourse within which we theorize about these underlying natural phenomena.

Now, if the naturalist's aim were simply to reject the idea that there might be some spooky supernatural entities governing the whole economic process, highlighting in this way the underlying psychobiophysical complexity hidden behind the entities and properties talked about in economics might indeed be an useful and illuminating reductive strategy. But I take it that, understood in this way, the naturalistic reduction we might achieve is of a not particularly informative kind. Actually, in his paper, Andler himself mentions this sort of reduction as being trivial. This then seems to mean that the idea associated with the ontological reduction aimed at in the strengthening of the program of ontological naturalism should be understood in some stronger and less trivial sense. What might this sense be?

I take it that, in order to be philosophically and epistemologically meaningful, any sort of reduction worth undertaking has to satisfy the *desideratum* of being inducive of *explanatory* progress. And here is where I think the problem lies. For even if we suppose that we might be able, *per impossibile*, to achieve a full description of the above mentioned processes which would refer only to some sort of fully naturalized entities and their properties, such a description would presumably leave us completely in the dark about the question of why the entire economic system behaves in the way it does. As a matter of fact, it seems to me that most of the information contained in such an overwhelmingly complex description would be explanatorily useless. At the same time, it also seems to me that a lot of useful and potentially explanatory information provided by hitherto unreduced disciplines such as, e.g., history, social psychology or anthropology would be completely left out of the picture. If I am right on this account, then it seems to follow that it is not the case that it always makes sense to try, at least, to reduce the entities and properties the sciences of Man talk about to entities and properties natural sciences talk about.

As a matter of fact, we can identify a pattern here which is actually more general than the problem of reducing the sciences of Man to the natural sciences. Consider for instance the case of evolutionary biology. John Maynard Smith and Richard Dawkins, two hard-nosed naturalists in any decent account of the term, have both defended a view of their subject according to which the right direction to follow in bringing back some of the central concepts of evolutionary and developmental biology to some other concepts simultaneously more basic and more explanatory is the direction of information and communication theory, not the direction of physics or microphysics. Dawkins talks of genes *being* 'long strings of pure digital information' and Maynard Smith talks of genes *being* 'symbols' and stresses the fact that the use of informational terms in biology is not metaphorical but literal, in that it implies intentionality, the property nineteenth century philosophers identified as being the mark of the mental as opposed to the physical.

Of course, we know that Dawkins's view of the genes as items of digital information encapsulated in wetware or Maynard Smith's view of natural selection as a provider of biological intentionality into living structures in no way conflict with the thesis of the causal closure of the physical realm. But this is not my point. My point is simply that just as the direction of explanatorily illuminating intra-natural theoretical connections is not determined beforehand by a previously given hierarchy of levels of natural scientificity starting in Microphysics, an explanatorily illuminating reduction of a human or social science does not have to be undertaken in the direction of some natural science or other. Conversely, I see no *a priori* reason why a reduction in the direction of some natural science or other, even if possible, would have to be explanatorily illuminating.

Now, Andler's non-triviality condition and his criticism of free-floating naturalism express concerns similar to those I have just mentioned. And these concerns seem to underly also the injunction with which Andler's terminates his final characterization of minimal naturalism, namely, the injunction "refrain from any commitment, explicit or implicit, regarding the outcome of the inquiry". But then it is hard to see how being so radically non-committal regarding the outcome of the enterprise of ontological reduction is actually compatible with the injunction that the results of your inquiry should contribute to strengthen the research program of ontological naturalism. How can you contribute to strengthen a research program by refraining from committing yourself to one of the two basic tenets in terms of which this research program is defined? Andler himself detects a threat of 'pragmatic incoherence' here, but he claims that his final formulation of minimal naturalism solves it. I do not see how.

Given the criticisms I voiced above, where do I stand then in the dispute that revolves around philosophical naturalism and the sciences of Man? From what I have already argued, it seems to follow that, according to my standpoint, if there is something specific characterizing a naturalistic view of the sciences of Man, that is neither the idea that the sciences of Man should ape the specific methods of any particular natural science nor the idea that the entities and properties they talk about should be somehow ontologically reducible to entities and properties some natural science or other talks about. So, where, if anywhere, lies the specificity of a naturalistic approach to these sciences?

My answer to this question is the following. I am convinced that the right way to construe the naturalism versus non-naturalism debate in the human and social sciences is in terms of a methodological disagreement. And this disagreement concerns the concept of *explanation* each of the perspectives endorses. According to my standpoint, what distinguishes naturalists in this dispute is the fact that they consider *causal explanation* to be the right sort of explanation human and social scientists should strive to provide in their work. On the other hand, non-naturalists, such as, e.g., interpretivists, associate explanation with rational reconstruction, understanding, role identification, or the establishment of synthetic *a priori* principles that are supposed to be constitutive of the domain of the human. This standpoint is, of course, not particularly new. It goes back to, at least, some of the remarks made by C. G. Hempel in his famous essay "Aspects of Scientific Explanation" and has been endorsed by a number of different authors since then.

Finally, let me now return to my original question. It should be clear by now that I sympathize with Andler's attempt to avoid committing himself to ontological naturalism. It is also clear that I disagree with the way he characterizes the general features of the theoretical landscape against the background of which the naturalism dispute in the sciences of Man takes place. Given his characterization of this landscape, his own version of liberalized naturalism sounds too close to being contradictory. However, against the background of my own characterization of the relevant theoretical landscape, the position summarized in his final definition loses its paradoxical aspect. As a matter of fact, it appears quite sensible and justified.

Bibliography

Robert Axelrod, *The Evolution of Cooperation.* New York: Basic Books 1984.

Tyler Burge, "Mind-Body Causation and Explanatory Practice", in: John Heil/Alfred Mele (Eds.), *Mental Causation.* Oxford: Clarendon Press 1993.

Paul Churchland, "The Logical Character of Action-Explanations", in: *The Philosophical Review* 79, 1970, pp. 214-36.

Paul Churchland, "Eliminative Materialism and the Propositional Attitudes", in: *The Journal of Philosophy* 78, 1981, pp. 67-90.

James Coleman, *Foundations of Social Theory.* Cambridge (MA): The Belknap Press of Harvard University Press 1990.

Donald Davidson, *Essays on Actions and Events.* Oxford: Oxford University Press 1980.

Donald Davidson, "Paradoxes of Irrationality", in: Richard Wohlheim/James Hopkins (Eds.), *Philosophical Essays on Freud.* Cambridge: Cambridge University Press 1982.

Donald Davidson, *Inquiries into Truth and Interpretation.* Oxford: Clarendon Press 1984.

Donald Davidson, "Could There Be a Science of Rationality", in: *International Journal of Philosophical Studies*, 3, 1995, pp. 1-16.

Richard Dawkins, *River Out of Eden: A Darwinian View of Life.* New York: Basic Books 1995.

Daniel Dennett, *The Intentional Stance.* Cambridge (MA): The MIT Press 1988.

Wilhelm Dilthey, *Einleitung in die Geisteswissenschaften. Versuch einer Grundlegung für das Studium der Gesellschaft und der Geschichte – Gesammelte Schriften, Band 1.* Göttingen: Vandenhoek and Ruprecht 2008.

William Dray, "Explanation in History", in: James H. Fetzer (Ed.), *Science, Explanation, and Rationality – The Philosophy of Carl G. Hempel.* Oxford: Oxford University Press 2000.

Jon Elster, *Explaining Social Behaviour: More Nuts and Bolts for the Social Sciences.* Cambridge: Cambridge University Press 2007.

Jerry Fodor, "Special Sciences (or: The Disunity of Science as a Working Hypothesis)", in: *Synthese,* 28, 1974, pp. 97-115.

Jerry Fodor, "Making Mind Matter More", in: J. Fodor, *A Theory of Content and Other Essays.* Cambridge (MA): The MIT Press 1990.

Dagfinn Føllesdal, "The Status of Rationality Assumptions in Interpretation and in the Explanation of Action", in: *Dialectica*, 36, 4, 1982, pp. 301-316.

Peter Gardiner (Ed.), *The Philosophy of History.* Oxford: Oxford University Press 1974.

Clifford Geertz, *The Interpretation of Cultures.* New York: The Basic Books 1973.

Carl Gustav Hempel, "Aspects of Scientific Explanation", in: Carl Gustav Hempel, *Aspects of Scientific Explanation and other Essays in the Philosophy of Science.* New York: The Free Press 1965.

Martin Hollis, *The Philosophy of Social Sciences: An Introduction.* Cambridge: Cambridge University Press 1995.

Philipp Kitcher, "Reasonable People", in: James H. Fetzer (Ed.), *Science, Explanation, and Rationality – The Philosophy of Carl G. Hempel.* Oxford: Oxford University Press 2000.

Michael Martin and Lee McIntyre (Eds.), *Readings in the Philosophy of Social Science.* Cambridge (MA): The MIT Press 1994.

John Maynard Smith, "The Concept of Information in Biology", in: *Philosophy of Science*, 67, 2000, pp. 177-194,.

David Papineau, *For Science in the Social Sciences.* London: MacMillan 1978.

David Papineau, *Philosophical Naturalism.* Oxford: Blackwell 1993.

Wesley Salmon, *Causality and Explanation.* Oxford: Oxford University Press 1998.

Wesley Salmon, *Four Decades of Scientific Explanation.* Pittsburgh (PA): The University of Pittsburgh Press 2006.

Georg Henrik von Wright, *Explanation and Understanding.* Ithaca (NY): Cornell University Press 1971.

Peter Winch, *The Idea of a Social Science and its Relation to Philosophy.* London: Routledge, 1958.

Departamento de Filosofia
Faculdade de Letras da Universidade de Lisboa
Alameda da Universidade
1600-214 Lisboa
Portugal
AntonioZilhao@fl.ul.pt

Part V
Philosophy of the Physical Sciences

Dennis Dieks

Reichenbach and the Conventionality of Distant Simultaneity in Perspective

Abstract

We take another look at Reichenbach's 1920 conversion to conventionalism, with a special eye to the background of his 'conventionality of distant simultaneity' thesis. We argue that elements of Reichenbach earlier neo-Kantianism can still be discerned in his later work and, related to this, that his conventionalism should be seen as situated at the level of global theory choice. This is contrary to many of Reichenbach's own statements, in which he declares that his conventionalism is a consequence of the arbitrariness of coordinative definitions.

1. Introduction

The history of the philosophy of physics has been shaped by a complicated and fascinating interplay between physics, philosophical ideas and external factors. This history is not only a intriguing subject for study in its own right: historical considerations can also shed light on the content of doctrines put forward by philosophers and are relevant for the appraisal of such doctrines. The aim of this paper is to illustrate this general point by a case study, namely the introduction by Hans Reichenbach of the notorious Conventionality Thesis regarding simultaneity in relativity theory.

There is an, admittedly old-fashioned, standard lore concerning the history of the conventionality thesis that goes more or less like this. At the end of the 19th century empiricism in the philosophy of physics had got the wind in its sails as a result of the work of Ernst Mach, and this empiricist Machian atmosphere decisively influenced Einstein's thinking. In his 1905 paper that established special relativity Einstein[1] accordingly adopted an empiricist and even operationalist stance. In particular, when Einstein discussed spatiotemporal notions he declared that in order to make such concepts physically meaningful we have to endow them with concrete physical content in terms of measuring procedures. For example, in the case of time at a particular place the sought *definition* of time (Einstein's term)

1 Albert Einstein, "Zur Elektrodynamik bewegter Körper", in: *Annalen der Physik* 17, 1905, 891-921. The standard English translation is in: H.A. Lorentz, A. Einstein, H. Minkowski and H. Weyl, *The Principle of Relativity*. London: Methuen 1923 (republished as a Dover edition, New York: Dover 1952).

F. Stadler (ed.), *The Present Situation in the Philosophy of Science*,
The Philosophy of Science in a European Perspective, DOI 10.1007/978-90-481-9115-4_23,

can be given as "the position of the hands of a clock situated at the same spot". Time thus defined is a purely local concept, however, so that we need a further definition to compare times at different places. For this reason Einstein famously asked himself how to synchronize clocks. He answered this question by asserting that simultaneity is *by definition* (italics used by Einstein) achieved when all clocks are set such that the velocity of light, measured with their help, is the same in all directions. With this, the characterization of time in a frame of reference becomes complete: given one standard clock, its time can be propagated everywhere by means of the simultaneity relation.

Einstein emphasized the words 'by definition' in his description of the synchronization procedure. Indeed, the temporal and spatial notions introduced earlier in his article do not yet fix the simultaneity relation, due to the fact that we cannot determine the speed of any signal if we are not yet able to compare times at different locations. If we *did* know the speed of some signal, that of light for example, we could simply synchronize clocks by sending a light signal from one clock to another and by taking into account that this signal takes a time L/c to reach its destination (with L the distance between the clocks and c the speed of light). The situation being what it is, however, it seems that we need to stipulate a synchronization procedure that fixes *both* simultaneity *and* the speed of light. Stipulations cannot be true or false, so simultaneity and the value of the speed of light come out of this analysis as not having a fact-like, but rather a *conventional* character. This latter statement should not be interpreted in the trivial sense that we have to choose units for time and length before we can say anything about the value of the speed of light: even after we have made a choice for such units it is still undecided what the speed of light along any given direction is. For although it is true that we can measure the *round-trip* velocity, by determining how much time it takes for the light to travel from clock A to clock B and back again, this will not tell us how much time was needed to go one way, from A to B. In particular, it is impossible to establish that the to and fro light speeds between A and B are equal.

If these things can only be stipulated, then it should also be possible to make other choices without coming into conflict with the facts already fixed by prior definitions. This point was worked out in a philosophically precise manner, the standard story continues, by Hans Reichenbach, especially in his epoch-making book *The Philosophy of Space and Time* (1928)[2]. Reichenbach there subsumed his investigation of simultaneity under a general analysis of the status of physical notions, according to which all independent concepts should be coordinated to concrete physical things and procedures by means of 'coordinative definitions'. Reichenbach emphasized, in line with general logical empiricist doctrine, that this

2 Hans Reichenbach, *The Philosophy of Space and Time*. New York: Dover 1957. Original German version: *Philosophie der Raum-Zeit-Lehre*. Berlin: Walter de Gruyter 1928.

coordinatization is fundamentally conventional in character: like all coordinative definitions, the definition of simultaneity is *arbitrary*[3].

The standard story thus tells us that there is a *Leitmotiv* of empiricist and operationalist considerations both in the development of special relativity itself and in the philosophy of space and time linking up with relativity. The most emblematic element of this story is the account it gives of the conventionality of relativistic simultaneity. This Thesis of the Conventionality of Distant Simultaneity not only encapsulates the empiricist philosophy that is intimately connected with relativity theory, it also relates directly to the drastic revision of temporal notions that is essential for the theory itself.

This standard account has not gone unchallenged. In particular after the appearance of Michael Friedman's *Reconsidering Logical Empiricism*,[4] it has become outdated to treat the early work of Reichenbach and other logical empiricists as a direct continuation of Mach-like empiricism. It is now well documented that at least Reichenbach's own version of logical empiricism originated from neo-Kantian considerations and that it was only under the influence of Moritz Schlick that Reichenbach after 1920 came to speak about his coordinative definitions (first proposed by him as neo-Kantian synthetic *a priori* principles) as arbitrary conventions.

In this paper I shall follow Friedman's lead, with a special eye to the specific case of the conventionality of distant simultaneity. The question I shall attempt to answer is whether a more detailed and more historically informed account of the development of Reichenbach's position than the one provided by the standard story can shed new light on the Conventionality Thesis. It is undeniable that Reichenbach, after his discussion with Schlick, explicitly and repeatedly claimed that distant simultaneity is conventional; but did he mean exactly the same thing as other conventionalists and later commentators, and was he fully consistent? I shall argue that on closer analysis traces of Reichenbach's earlier neo-Kantian stance become visible, and that these create an unresolved tension in Reichenbach's position. To start with I shall have a look at Einstein's own supposedly operationalist and conventionalist position in 1905 – the place where the whole conventionality debate has its origin. Partly drawing on another paper,[5] I shall argue that one

3 There is abundant textual support for this part of the standard account. See, e.g., Reichenbach, *The Philosophy of Space and Time, op. cit.*, sections 4 and 19; Hans Reichenbach, "La signification philosophique de la théorie de la relativité", in: *Revue Philosophique de la France et de l'Étranger* 94, 1922, pp. 5-61 (English translation in: Steven Gimbel/Anke Walz (Eds.), *Defending Einstein: Hans Reichenbach's Writings on Space, Time and Motion*. Cambridge: Cambridge University Press 2006, chapter 10); Hans Reichenbach, *Axiomatization of the Theory of Relativity*. Berkeley: University of California Press 1969, section 2 (German original: *Axiomatik der relativistischen Raum-Zeit-Lehre*. Braunschweig: Vieweg & Sohn 1924).

4 Michael Friedman, *Reconsidering Logical Positivism*. Cambridge: Cambridge University Press 1999.

5 Dennis Dieks, "The Adolescence of Relativity", in: Vesselin Petkov (Ed.), *Minkowski*

should *not* read an operationalism-based conventionalism into Einstein's statements of 1905 and later.

2. Einstein and the Definition of Space and Time

The emphasis laid by Einstein in his 1905 paper on the need to define our notions of space and time before we can even start to do physics certainly suggests operationalist sympathies. However, we should note that the 1905 paper is not the only place where Einstein expresses himself in this fashion: remarkably, he refers to the need for *definitions* of physical concepts even in contexts in which he explicitly opposes operationalist and logical empiricist ideas. For example, in his Autobiographical Notes we find Einstein reminiscing about the discovery of special relativity with the following words[6]:

> One had to understand clearly what the spatial co-ordinates and the temporal duration of events meant in physics. The physical interpretation of the spatial co-ordinates presupposed a fixed body of reference, which, moreover, had to be in a more or less definite state of motion (inertial system). In a given inertial system the co-ordinates meant the results of certain measurements with rigid (stationary) rods. … If, then, one tries to interpret the time of an event analogously, one needs a means for the measurement of the difference in time … A clock at rest relative to the system of inertia defines a local time. The local times of all space points taken together are the "time" which belongs to the selected system of inertia, if a means is given to "set" these clocks relative to each other.

This is an almost verbatim repetition of the relevant passages from the 1905 paper, including the use of the term 'define', and with the explanation that space and time coordinates *mean* what is indicated by rods and clocks; and all this without any accompanying comment that might indicate that Einstein in the nineteen-forties deemed some kind of qualification of his 1905 statements necessary. So we may safely assume that Einstein is here expressing the same view as the one he had in mind in his original relativity paper.

This is striking because elsewhere in these same autobiographical notes, and also in Einstein's 'Replies to Criticism' in the same volume[7], we find an explicit and strong rejection of Bridgman's operationalism and Reichenbach's empiricism as viable philosophies of science. For example, about Bridgman's operationalism Einstein protests[8]:

Spacetime: a Hundred Years Later. New York: Springer 2009.

6 P.A. Schilpp (Ed.), *Albert Einstein: Philosopher-Scientist*. La Salle: Open Court 1949, p. 55.

7 Schilpp, *ibid.*, pp. 665-688.

8 Schilpp, *ibid.*, p. 679.

In order to be able to consider a logical system as physical theory it is not necessary to demand that all of its assertions can be independently interpreted and "tested" "operationally"; *de facto* this has never been achieved by any theory and can not at all be achieved. In order to be able to consider a theory as a *physical* theory it is only necessary that it implies empirically testable assertions in general.

Einstein made the same point in greater detail in his Reply to Reichenbach. In his contribution to the Einstein Volume, Reichenbach had stated that the philosophical lesson to be learnt from relativity theory was that basic physical concepts must be given meaning by means of 'coordinative definitions': it is only the 'coordination' of a concrete physical object or process to the concepts in question that bestows physical significance on them. As Reichenbach wrote[9]:

For instance, the concept "equal length" is defined by reference to a physical object, a solid rod, whose transport lays down equal distances. The concept "simultaneous" is defined by the use of light-rays which move over equal distances. The definitions of the theory of relativity are all of this type; they are coordinative definitions.

Reichenbach continued by explaining that this definitional character of basic physical concepts implies that they are *arbitrary*:

Definitions are arbitrary; and it is a consequence of the definitional character of fundamental concepts that with the change of the definitions various descriptional systems arise. … Thus the definitional character of the fundamental concepts leads to a plurality of equivalent descriptions. … All these descriptions represent different languages saying the same thing; equivalent descriptions, therefore, express the same physical content.

In his response Einstein objected that any concrete physical object is subject to deforming forces, and can therefore not be used to *define* concepts. We need a *theory* of these deforming influences in order to be able to correct for them, and such a theory already uses a notion of length. Therefore, we must have an idea of what 'length' is prior to the use of any actual measuring rod. From this Einstein concludes that a concept like 'equality of length' cannot be defined by reference to concrete objects at all; such concepts "are only indispensable within the framework of the logical structure of the theory, and the theory validates itself only in its entirety".[10] The unit of length can only be supposed to be realized by an imaginary *ideal* rod, which can at its best be *approximated* by a concrete object – and this only on the condition that we are thinking of the concept of length in circumstances in which it makes sense to assume the existence of rods at all! Actual rods have thus to be adjusted on the basis of theory, and this means a reversal of order

9 Hans Reichenbach, "The Philosophical Significance of the Theory of Relativity", in Schilpp, *ibid.*, p. 295.
10 Schilpp, *ibid.*, p. 678.

compared to the analysis that begins with operational definitions and starts constructing a theory only afterwards.

Another though related point stressed by Einstein is that macroscopic devices like rods and clocks do not have a foundational role to play in the interpretation of fundamental physics. The reference to them only serves practical purposes: it makes contact with familiar everyday circumstances and thus directs our thoughts. Making this use of them is only a tentative manoeuvre, "with the obligation, however, of eliminating it at a later stage of the theory."[11]

So when we look at the historical evidence in a more detailed way, it becomes very plausible that by his use of the term *definition* in 1905 Einstein did not want to imply that we are dealing with arbitrary meaning stipulations that must precede theory construction.[12] The very same point can be made with respect to the notion of simultaneity. There is in fact a remarkable continuity in Einstein's utterances from the early twenties onwards, when he first explicitly addresses philosophical questions relating to space and time. In these philosophical writings Einstein consistently rejects the project of defining concepts along the lines of operationalism or logical empiricism. The striking fact that Einstein uses the term 'definition' to refer to the content of spatiotemporal notions even in this context illustrates that he did not realize the extent to which this term is able to excite philosophers and can give rise to misunderstandings. In itself this is quite understandable: Einstein's papers on special relativity are evidently physics papers, addressed to a physicist audience. Einstein was facing the task of convincing his readers that the spatiotemporal concepts of classical physics were not beyond discussion and that, indeed, certain changes in these concepts would make it possible, in a surprising manner, to consistently combine the two postulates of relativity theory. He attempted to demonstrate that actual measurements (of the usual kind, traditionally employed to determine spatiotemporal relations) did not prove the sole applicability of the classical notions. In particular, it was important for Einstein to make it clear that there was no empirical support for the *absoluteness* of simultaneity, nor for the pre-relativistic idea that the to and fro velocities of light have to differ in almost all inertial systems (namely those moving with respect to absolute space). The interdependence that Einstein notes between simultaneity and the value of the speed of light is employed by him to consistently apply the same synchronisation procedure, with the same value of the speed of light, in *all* inertial systems. This is a quite different project than arguing that these notions are arbitrary in any single frame of reference.

Einstein's special theory of relativity has served as a beacon for twentieth-century philosophy of science; but quite a few commentators have misinterpreted the philosophical implications of the theory. As Howard[13] correctly concludes, it

11 Schilpp, *ibid.*, p. 59.

12 See Dieks, "The Adolescence of Relativity", *op. cit.*, for more extensive argumentation concerning this point.

13 Don Howard, "Einstein and the Development of Twentieth-Century Philosophy of Sci-

was only with the downfall of logical empiricism, and the Quinean criticism of the analytic/synthetic distinction, that philosophy of science caught up with Einstein's thinking about the status of physical concepts. Reichenbach's discussions of relativity have certainly contributed to the misunderstandings. His analysis of the notion of simultaneity in particular has been instrumental in reinforcing the idea that special relativity should be seen as both the fruit and victory of a strictly empiricist philosophy of science.

However, Reichenbach's ideas are sophisticated and complex, and as we shall see they leave room for the supposition that different, conflicting conceptions were competing for priority in his thinking; not all of which fit in with the standard reading of his work. This complicated character of Reichenbach's ideas can be brought out by looking at their interesting history.

3. Reichenbach, Relativity Theory and the Apriori

Hans Reichenbach was one of the students attending Einstein's first relativity course at the University of Berlin in 1919; a year later his *Relativity Theory and Apriori Knowledge* (*Relativitätstheorie und Erkenntnis A Priori*)[14] appeared. As the title indicates, the problematic Reichenbach was dealing with in this work was the relation between Kantian philosophy and relativity. When one reads the book it very soon becomes evident that Reichenbach is not at all attacking Kant's epistemology from an empiricist point of view, using Einstein's theory as an ally – as one might expect on the basis of the lore that sees a direct empiricist link going from Mach via Einstein to Reichenbach. Quite on the contrary, Reichenbach sets himself the task of salvaging as much as possible of Kantian doctrine, given the problems the theory of relativity admittedly causes for it.[15] In his first chapter Reichenbach notes that according to special relativity the temporal order between two events is not unique in all cases: for events with spacelike separation this order depends on the choice of a frame of reference. Indeed, the simultaneity relation associated with a given inertial frame of reference determines which one of the two events is earlier; going from one frame to another means adopting different judgements about which events are simultaneous; in the case of spacelike events this change may reverse the temporal order of the events. This result is in complete contrast to Kant's doctrine of the *reine Anschauung*, according to which it is *a priori* certain that all possible events are embedded in one unique temporal series. But Reichenbach stresses that the existence of this conflict does not imply

ence", to appear in: *The Cambridge Companion to Einstein*. Cambridge: Cambridge University Press.

14 Hans Reichenbach, *Relativitätstheorie und Erkenntnis Apriori*. Berlin: Springer 1920. English translation: Hans Reichenbach, *The Theory of Relativity and A Priori Knowledge*. Berkeley: University of California Press 1965.

15 Cf. Friedman, *op. cit.*, chapter 3.

the downfall of the Kantian approach. For according to Reichenbach Kant was certainly right in pointing out that *a priori* elements are absolutely indispensable in any empirical investigation: we need to avail ourselves of concepts before we can even start studying nature.

What Reichenbach is thus arguing in 1920 is the inevitability of a 'constitutive *a priori*', consisting of a network of concepts and principles that make it first of all possible to get a grip on any field of research and that in this way 'constitute' the field. But this conceptual framework can and will change in the course of time: it is possible to develop and adapt our concepts in response to unexpected relations between empirical data and the emergence of new theoretical ideas. The contribution of human reason is therefore not given once and for all, as originally claimed by Kant, but consists in evolving principles by means of which we order the data of experience. Kant's doctrine of the *a priori* should accordingly be split up into a *constitutive* and an *apodictic* component: the constitutive *a priori* must be retained whereas the apodictic part, which says that the concepts furnished by intuition have permanent and absolute validity, should be rejected. Reichenbach's manoeuvre here is typical of neo-Kantianism: *Relativitätstheorie und Erkenntnis A Priori* can be considered a neo-Kantian discussion of relativity theory, close in spirit to the Marburg school (Cassirer et al.).

It is natural to ask how this evolution of our concepts should be thought of in detail, and Reichenbach pays explicit attention to this question. One might imagine, given the rejection of the apodictic significance of the *a priori*, that completely free conceptual changes can be made in the face of tensions in the network of our knowledge. But this is not the way Reichenbach deals with the issue in 1920. Instead, he argues that there is an important principle governing conceptual development, namely the principle of what he calls 'continuous extension' (*stetige Erweiterung*): our conceptual network is adapted in a piecemeal fashion, so that central, well-embedded elements remain unchanged and only the most 'marginal' ones undergo revision. Although in principle all our concepts may eventually change under the influence of new empirical findings, in practice some of them are virtually immune to such revision. Think, in particular, of the concepts used for describing daily experience: these will not change as a result of the evolution of science since they are so utterly central in our existing conceptual framework and indispensable for making contact with already existing knowledge. This principle of continuous extension makes it possible to relate new theories to older ones and, importantly, it justifies us in performing observations in which we implicitly use older theories.

So, even granted that there is no *a priori* foundation of our conceptual framework in the sense of a rock bottom with eternal validity, only those notions will actually change that are not linked up with direct 'un-theoretical' observation; the notions that change are those needed to deal with situations directly affected by the new theories. The better a concept is embedded in the network of concepts used to describe situations that are not directly touched by theoretical development, the less changeable it is.

Reichenbach gives several examples.[16] For instance, in ordinary measurements of lengths and times we do not need to take into account relativistic contractions and dilations, not even if the eventual goal is to test predictions of relativity theory itself. And if we look through a telescope, in order to test the predictions of general relativity, we may forget about the fact that according to general relativity in the telescope itself light does not propagate along Euclidean straight lines. Although general relativity theory necessitates drastic changes in the arsenal of concepts needed to describe the universe at large, and although the observations made with a telescope are of undoubted relevance for this cosmological description, the telescope itself and the findings arrived at with its help can be described with notions from classical pre-relativistic physics, which in turn coincide to a large extent with everyday concepts. Without this continuity of description and the approximate validity of older theories we would be at a loss in connecting our new theories with earlier observations.

4. The Flexibility of the Notion of Simultaneity

When we apply these rather conservative ideas about the evolution of concepts to the particular case of distant simultaneity we are led to an account along more or less the following lines. First of all, what we most directly measure are *local* quantities, since we ourselves are spatiotemporally local creatures that respond to local stimuli. The observation of the hands of a clock here and now is an example of such an immediate local measurement. Like all observations, also such direct ones must employ constitutive principles; but these are of a very robust kind, centrally embedded in our language and practice. For all practical purposes, the concepts relating to direct local observations are therefore immune to revision under theory change.

When it comes to the comparison of events that take place at a distance from each other, however, observations become less direct. It is true that classical theory provides us with standard procedures for making such comparisons, and in particular for establishing simultaneity at a distance, but it is not immediately obvious that these procedures can be consistently applied also within the context of the new theory – and to the extent that they can, it is not self-evident what the properties of the resulting simultaneity relation will be. Here we clearly find ourselves at a place in the network of concepts which is less central, less directly linked to observation, and therefore offering more room to flexibility and revision. This is good news, for it turns out that the concept of simultaneity *has* to become different from its classical counterpart if the two postulates of special relativity (the relativity postulate

16 Reichenbach, *Relativitätstheorie und Erkenntnis Apriori*, *op. cit.*, pp. 66-67.

that says that all inertial frames have equal status and the light postulate that says that the velocity of light is has a constant value that is independent of the velocity of the emitting source) are to be consistent – this follows deductively from the two postulates, as demonstrated in Einstein's 1905 paper.

In other words, in the initial stage of Reichenbach's philosophical discussion of relativity the emphasis is on the possibility of *concept change* in the transition to new theories. This process of conceptual evolution is interpreted as a continuous process of adaptation, which stays as close as possible to the already existing framework. The flexibility that is required is just the room needed to make the new theory possible at all; in the case of simultaneity this would lead to the standard relativistic simultaneity relation used by Einstein (corresponding to the synchronization procedure described in section 1), which deviates from the classical simultaneity relation in that different inertial observers will come to different conclusions about which events are simultaneous but is identical, qua synchronization procedure, to classical simultaneity in the ether frame (in relativity theory there is no such preferred frame, and in accordance with the relativity postulate the same procedure is applied in *all* inertial frames).

One may introduce the term 'conventionality' in this context: the concepts that are open to revision are those not completely fixed by the central, more robust concepts and the facts formulated with their aid. In this sense they are non-factual, conventional. We can decide to adjust them without coming into conflict with direct, local findings. But this type of conventionality only serves to create room for the new (frame-dependent) relativistic concept of simultaneity and has little to do with the much stronger conventionality thesis that surfaces less than two years later in Reichenbach's work.

5. The Reichenbach-Schlick Exchange and the Conventionality of Simultaneity

After the appearance of *Relativity Theory and Apriori Knowledge*, still in 1920, Reichenbach sent a copy of his book to Moritz Schlick. Schlick responded positively in a brief letter, after which a more extensive exchange developed. This correspondence, which has received much attention in the philosophical literature[17], apparently had a great impact on Reichenbach: after it Reichenbach rephrased his position completely.

In his letters Schlick takes Reichenbach to task for paying too much tribute to Kant. According to Schlick, it is exactly the *combination* of the apodictic and constitutive aspects that is characteristic of Kant; separating these two aspects in the

17 See, also for further references: Flavia Padovani, *Probability and Causality in the Early Works of Hans Reichenbach*. Geneva: dissertation at the University of Geneva, Faculty of Letters 2008. Also: Friedman, *op. cit.*

way Reichenbach has done leads to a distinctively non-Kantian stance that rather accords with empiricism. Schlick admonishes Reichenbach to avoid misunderstandings about his alliances and to eschew Kantian terminology: one should use the term 'convention', à la Poincaré, instead of speaking about '*a priori* constitutive principles'. As Schlick writes in his letter of 26 November 1920:[18]

> The central point of my letter is that I cannot find out what the difference really is between your *a priori* principles and conventions, so that we seem to agree on the essential issue. What has amazed me most in your manuscript is that you dispose of Poincaré's conventionality doctrine in only so few words.[19]

In his reply[20] Reichenbach objects that he term 'convention' may create the misunderstanding that there is no factual content in scientific statements. He points out that there is no straightforward arbitrariness in the *a priori* principles, because they cannot be conventionally changed individually, one by one. It is only the total system of such principles that can be said to admit empirically equivalent alternatives, and then still the set of alternatives is restricted (compare the discussion of geometry *cum* physics, G + F, below). As Reichenbach stresses, such a combination of principles represents an objective PROPERTY (Reichenbach's capitals) of reality.[21] A little bit earlier in the same letter he had expressed 'a profound distrust' (ein starkes Mißtrauen) about whether the choice between such alternative (i.e., empirically equivalent) systems can be made on the basis of considerations of simplicity (as Schlick, like Poincaré, would maintain); and he had also reported 'an instinctive aversion' (eine instinktive Abneigung) to the idea that such a choice is only a matter of pragmatics – he proposed to suspend judgment on this complicated issue.

But Schlick retorts on 11 December[22] that Poincaré was of course aware of such and similar complications, and concludes that a far-reaching agreement between himself and Reichenbach has now been reached.[23] Indeed, Reichenbach

18 Moritz Schlick to Hans Reichenbach, 26 November 1920, HR-015-63-22, reproduced by permission of the University of Pittsburgh, all rights reserved.

19 „Es ist der Kernpunkt meines Briefes, dass ich nicht herauszufinden vermag, worin sich Ihre Sätze a priori von den Konventionen eigentlich unterscheiden so dass wir also im wichtigsten Punkte einer Meinung waren. Dass Sie über die Poincaresche Konventionslehre mit so wenigen Worten hinweggehen, hat mich an Ihrer Schrift am meisten gewundert."

20 Hans Reichenbach to Moritz Schlick, 29 November 1920, HR-015-63-20, reproduced by permission of the University of Pittsburgh, all rights reserved.

21 Jedes mögliche System besagt in seiner Möglichkeit eine EIGENSCHAFT der Wirklichkeit.

22 Moritz Schlick to Hans Reichenbach, 11 December 1920, HR-015-63-19, reproduced by permission of the University of Pittsburgh, all rights reserved.

23 Actually, Schlick himself oscillated between the 'local' viewpoint that concepts should be given meaning by individual coordinative definitions and more holistic views. E.g., in his letter of 26 November 1920 Schlick stated that the 'arbitrariness' only enters at

from this point on starts declaring that scientific concepts are fixed by 'coordinative definitions', which 'like all definitions' are arbitrary. In the very beginning this is still occasionally mixed with references to the constitutive role of such definitions, but after 1922 there is no mention anymore of the *constitutive a priori*.[24] There are recurring warnings, though, also in Reichenbach's later work, against possible misunderstandings caused by the term 'conventionalism'. Concepts possess the status of a convention, because of their definitional character, but this does not mean that statements formulated with their help lack factual content; the designation 'conventionalism' is therefore unfortunate, Reichenbach repeatedly declares.[25]

In *The Philosophy of Space and Time* we find Reichenbach's full articulation of the idea that distant simultaneity in relativity theory is such an arbitrary convention. As Reichenbach explains,[26] to determine the simultaneity of distant events – by synchronizing two clocks A and B – we need to know the velocity of a signal (for example a light signal) connecting the clocks, but to measure such a velocity we require prior knowledge of the simultaneity of distant events, so that we get caught in a vicious circle. He concludes:

> The occurrence of this circularity proves that simultaneity is not a matter of knowledge, but of a coordinative definition, since the logical circle shows that a knowledge of simultaneity is impossible in principle. We also notice that the second characteristic of a coordinative definition, namely its arbitrariness, is satisfied. It is arbitrary which time we ascribe to the arrival of the light ray at B.

A little bit further on the same page (p. 127) we find the passage in which Reichenbach expresses this conventionality by means of his famous ε formula. Discussing the just-mentioned logical circle in any attempt to synchronize clocks A and B by sending light from A to B and back again, he writes:

> It is this consideration that teaches us how to understand the definition of simultaneity given by Einstein, $t_2 = t_1 + \frac{1}{2}(t_3 - t_1)$, which defines the time of arrival of the light ray at B as the mid-point between the time that the light was sent from A and the time that it returned to A. This definition is essential for the special theory of relativity, but it is not epistemologically necessary. Einstein's definition, too, is just one possible definition. If we were to follow an arbitrary rule restricted only to the form $t_2 = t_1 + \varepsilon(t_3 - t_1)$, $0 < \varepsilon < 1$, it would likewise be adequate and could not be called false.

---

the level of the total system of principles. "Es sind ja nur solche Konventionen gestattet, die sich in ein gewisses System von Prinzipien einfügen, und dies System *als Ganzes* wird durch die Erfahrung bestimmt; die Willkür kommt erst bei der Art seines Aufbau hinein und wird gelenkt durch das Prinzip der Einfachheit, der Ökonomie, oder, wie ich lieber gesagt habe, das Prinzip des Minimums der Begriffe." See for an illuminating discussion: Friedman, *op. cit.*, chapter 1.

24 Padovani, *op. cit.*, p. 177.

25 E.g., Reichenbach, *The Philosophy of Space and Time, op. cit.*, p. 36.

26 Reichenbach, *The Philosophy of Space and Time*, *op. cit.*, pp. 126-127.

So Reichenbach appears to have been completely convinced by Schlick that the concepts of theoretical physics in the final analysis have to be defined by ostention and possess the status of free stipulations (at least, this is how Reichenbach interpreted Schlick; see note 23). Reichenbach will rehearse this point of view explicitly and emphatically in several later publications, for example – as we have already seen – in his contribution to the 1949 Einstein volume.

6. Coordinative Definitions versus Constitutive Principles

The analysis in terms of conventional definitions as adopted by Reichenbach after his exchange with Schlick has the great advantage that it can easily be presented and argued: it is simple, systematic, logically neat and clear. However, in section 2 we already noted that this account does not accord with the actual practice of physics: individual theoretical concepts are not strictly *defined* by means of concrete empirical procedures – the relation between theory and experiment is of a more holistic nature. Of course, this observation has later become one of the essential points of criticism of logical empiricism in general (as we have noted, Einstein can be regarded as a precursor here). Remarkably, however, also in Reichenbach's own work there are signs that there were other conceptions in the back of his mind that struggled with his 'arbitrary coordinative definitions' account.

In section 2 of his 1924 book *Axiomatization of the Theory of Relativity*, entitled *The Logical Status of the Definitions*, Reichenbach distinguishes between two kinds of definitions, conceptual and coordinative. Conceptual definitions clarify the meaning of a concept by establishing relations with other concepts. If we only possessed such definitions, however, we could not do physics but would remain caught in an abstract conceptual network without empirical content, as in mathematics. We therefore need 'physical' definitions as well. Reichenbach formulates it like this:[27] "The physical definition takes the meaning of a concept for granted and coordinates it to a physical thing: it is a *coordinative definition*". Therefore, coordinative definitions *presuppose* conceptual definitions (as is explicitly declared by Reichenbach a little bit later on the same page). But this can only mean that physical concepts must already have meaning and be part of a network of interrelated concepts and principles before any 'coordination' can take place; and this introduces an element of theoretical holism in their content.

We can see how this works out in detail when we look at Reichenbach discussion of the conventionality of geometry in *The Philosophy of Space and Time*. He notoriously does so via the consideration of *deforming forces* (already introduced in his 1922 and 1924 publications), instead of by focussing directly on the arbitrariness of the definitions of length and congruence as one would expect. He explains that actual measuring rods cannot be used for the definition of the unit of

27 Reichenbach, *Axiomatization of the Theory of Relativity, op. cit.*, p. 8.

length: we have to count in the possible presence of forces that affect their length. Apparently, we already know a lot about the meaning of 'length', 'rod' and 'force' *before* the coordination, and this alerts us to the complication of the interrelations between these concepts; in physical terms, the possibility that distorting forces are present. Now, there is a systematic procedure for detecting the presence of some of these forces, namely the 'differential' ones. These affect different materials differently, so that we can compare rods of different chemical constitution, find regularities in their various responses to the force field, and finally make corrections for the differential effects. This, again, brings out that Reichenbach thinks of his coordinative definition for length in terms of a theoretically determined, *ideal* measuring rod that can only be *approximated* by an actual object. Now there is a further complication:[28] *universal* forces, which act on objects in the same way regardless of their chemical composition, cannot be detected via the just-mentioned procedure. Since only the combination G + F of geometry and physics leads to testable statements, this means that we can have different geometries, combined with different physical theories (differing from each other by positing different universal forces), that lead to exactly the same empirical predictions. This, then, is a case of global *theoretical underdetermination*, in which the theoretical framework (in this case the division of labour between G and F) fails to be completely determined by empirical data. The different geometries G say different things about which distances are equal to each other – but they *agree* on the conceptual meaning of length and congruence in the sense that they all say that the unit of length is everywhere realized by an ideal, undeformed measuring rod.

The important point here is that in this analysis of the non-uniqueness of the geometrical description the basic idea is *not* that we are free to stipulate whatever definition of congruence we happen to like. Rather, it is taken for granted that congruence is implemented by the transport of an ideal and undeformed measuring rod; the non-uniqueness of the geometry now derives from the underdetermination by empirical data of the physical theory that specifies the magnitude of the universal length deformations. Consequently, if there is conventionality here it is not the conventionality associated with the arbitrary character of stipulative, ostensive definitions. Instead we are dealing with a global kind of conventionality, associated with the underdetermination of the total theoretical system. In *The Philosophy of Space and Time* Reichenbach proposes to opt for the system that sets the universal forces equal to zero (F = 0), on the grounds that this choice leads to more 'descriptive simplicity' (not because it would be 'more true'). He adds:[29] "This conception of the problem of geometry is essentially the result of the work of Riemann, Helmholtz, and Poincaré and is known as *conventionalism*."

28 Reichenbach, *The Philosophy of Space and Time, op. cit.*, sections 3-8.

29 Reichenbach, *The Philosophy of Space and Time, op. cit.*, p. 35.

On closer inspection Reichenbach's conventionalism thus turns out to be more sophisticated than suggested by the motto "all definitions are arbitrary". In fact, that the unity of length is given by an undeformed measuring rod is handled by Reichenbach as something that is 'unconventionally' given *a priori*, before any empirical investigation can even begin. A global network of concepts specifying the relation between lengths and deforming causes is already in place, and this network plays a role comparable to that of the combination of constitutive *a priori* principles in Reichenbach's earlier work. Something similar can be said about Reichenbach's discussion of simultaneity. If the conventionality of simultaneity really were based on a supposed complete arbitrariness typical of all definitions, we would expect that no restriction at all would have to be imposed on the temporal coordination of events at different spatial positions (with respect to a given frame of reference). However, Reichenbach does restrict the conventionality of simultaneity by the requirement that the ε in his formula lie between 0 and 1. The background of this requirement can be found in sections 21 and 22 of *The Philosophy of Space and Time*, where Reichenbach introduces what he calls the 'topological coordinative definition of time order': "*If E_2 is the effect of E_1, then E_2 is called later than E_1.*"[30] Remarkably, here Reichenbach refrains from any mention of the arbitrariness of this definition. On the contrary, he emphasizes how close this 'criterion' (as he alternatively calls the definition) is to everyday experience and practice: "The procedure which we have described is used constantly in everyday life to establish a time order."[31]

Reichenbach now clarifies his restriction on ε: we should evidently not take two events as simultaneous if one of them is later than the other, and this implies that ε must be between 0 and 1. He adds: "*Simultaneity means the exclusion of causal connection*" and "The concept *simultaneous* is to be reduced to the concept *indeterminate as to time order.*"[32] There is then no complete arbitrariness in the meaning of simultaneity after all! Apparently there are principles that already fix the meaning of the concept, as in the case of length. This situation reminds us of the principle of continuous extension and the constitutive *a priori* in *The Theory of Relativity and Apriori Knowledge*. As in this earlier neo-Kantian work, Reichenbach assumes that there are global and general principles that regulate the use of spatial and temporal concepts.

Of course, if simultaneity indeed precisely *means* 'impossibility of causal connection', it follows from relativity theory that there is not one unique event at spatial position B that is simultaneous with a given event at position A. All events with spacelike separation from the given event qualify, so that a finite interval rather than a point-event is singled out as simultaneous with the event at A. This consequence is hailed by Reichenbach,[33] because it creates room for the *relativity*

30 Reichenbach, *The Philosophy of Space and Time, op. cit.*, p. 136.
31 Reichenbach, *The Philosophy of Space and Time, op. cit.*, p. 138.
32 Reichenbach, *The Philosophy of Space and Time, op. cit.*, p. 145.
33 Reichenbach, *The Philosophy of Space and Time, op. cit.*, p. 145.

of simultaneity (i.e., the result of relativity theory that whether or not two events are simultaneous depends on he frame of reference with respect to which the simultaneity relation is considered). As he writes:

> Our epistemological analysis thus leads to the discovery that the relativity of simultaneity is compatible with the intuitive conception which we connect with simultaneity. It is not this conception which is incorrect, but the conclusion derived from it that simultaneity must be uniquely determined.

This statement is striking for two reasons. First, Reichenbach here concedes that the concept of simultaneity is given meaning via antecedent notions ('the intuitive conception') that can be maintained to be correct. Second, it now becomes clear why it is important for Reichenbach to restrict the meaning of simultaneity to exactly 'lack of causal connectibility' and why he does not want to consider any additional meaning ingredients. As he states, if it would follow from the already given meaning of simultaneity that exactly *one* event at B – instead of a whole interval – would be simultaneous with any given event at A, it would be impossible to accommodate the special relativistic relativity of simultaneity. This argument, however, is mistaken. There is no harm in adding more intuitive notions to the meaning of simultaneity, and to fix one event at B as simultaneous with the event at A, if this is done in a frame-dependent way so that it is going to depend on the frame exactly *which* event is the uniquely chosen one. This is in fact the standard approach in relativity theory, and in the next section we shall see that this approach can be fitted into ideas that accord not only with actual physics but also with the 'holistic' strand in Reichenbach's own approach.

There is a clear parallel between the constitutive *a priori* principles of 1920 and the global elements in Reichenbach's later work. For example, in 1920 Reichenbach counts Euclidean geometry among the constitutive principles that in classical physics (but no longer in general relativity and in accelerated frames in special relativity) govern the meaning and use of the physical concept of congruence. The procedure of continuous extension weakens this to a local principle in the case of general relativity and accelerated frames; the global geometry will in these cases generally become non-Euclidean. After 1920 – an early example is Reichenbach's 1922 paper *The Philosophical Significance of the Theory of Relativity*[34] – Reichenbach still holds that the global geometry implicitly defines the meaning of congruence. In other words, the analysis is still top-down, going from the theoretical framework to individual concepts[35]. It is true that there is now no longer an *a priori* specification of the global geometry; Reichenbach now stresses that the geometry is underdetermined, because of the possible presence of undetectable universal forces. According to Reichenbach this underdetermination

34 Steven Gimbel/Anke Walz (Eds.), *op. cit.*, chapter 10.

35 Cf. Dennis Dieks, "Gravitation as a universal force", in: *Synthese* 73, 1987, pp. 381-397.

entails that the choice between different geometries is in principle arbitrary (he advises to adopt the option F = 0, but emphasizes that this is a merely pragmatic choice).

Conventionality and arbitrariness are therefore located on the level of *global theory choice*, linked to the theoretical underdetermination of the total theory. That underdetermination thus implies arbitrariness will probably not be accepted by most modern philosophers of science. But whatever position one takes on the question whether theory choice in such circumstances is conventional, arbitrary, purely pragmatic or not, once a choice has been made the theoretical framework functions as constitutive of the field of experience in the same way as the 1920 constitutive principles. There therefore still is a 'relativized *a priori*', relative to an evolving conceptual framework. Possibly we have to say that this theoretical framework contains important pragmatic elements – but whether this entails that the constitutive principles of the earlier 1920 neo-Kantian position were epistemically more robust and more 'truth-like' is not easy to answer. After all, questions about the *truth* of the categories or *a priori* principles are distinctly un-Kantian.

7. The Holistic Approach in Practice: Minkowski's Analysis of Space and Time

As it turns out, there actually exists an approach to relativity theory that is explicitly along the holistic lines we have outlined above. Although this seems not to have been noticed in the literature, Minkowski's famous 1908 lecture *Raum und Zeit*[36] incorporates several of the ideas that are central to Reichenbach's 1920 book – and of which some, as we have argued, can also be found in his later work.

Minkowski starts his construction of what he calls 'the objective world' by describing elementary, local physical phenomena in terms of arbitrarily chosen coordinates; he then performs transformations on the resulting equations to bring them in a simple and symmetrical form; and finally he introduces spatial and temporal concepts as embodied in that system of reference in which the equations take their desired global 'standard' form. The laws and equations Minkowski actually considers are those of Maxwell; but he makes it clear that he assumes that all laws of nature, including those yet to be discovered, e.g. those responsible for the stability of matter, will exhibit the same global spacetime symmetry properties as Maxwell's equations. In Minkowski's approach space and time are thus determined on the basis of considerations about the global form of the physical laws that represent the regularities and patterns in the physical phenomena, starting from a description of local events. The concepts needed to give this local description are assumed to be given beforehand, but for the global space and time determinations

36 Hermann Minkowski, „Raum und Zeit", in: *Physikalische Zeitschrift* 10, 1909, pp. 104-111. English translation in Lorentz et al., *op. cit.*

this is not the case and the task is accordingly to fix them via theoretical considerations.

In some more detail, Minkowski first introduces completely arbitrary coordinates in order to label events: *x, y, z* and *t*. The events in question are to be thought of as physical occurrences that happen locally to small material systems, specks of matter. The patterns among these events should now be captured by mathematical equations, laws, still expressed in the arbitrary coordinates *x, y, z* and *t*. The task then is to transform these laws such that they conform to certain form requirements, desiderata that are specified beforehand but that clearly are inspired by empirical knowledge as formalized in earlier theories. In particular, the equations should be brought into a standard form that displays spatial homogeneity and isotropy, the equivalence of inertial frames, etc. Minkowski himself describes his procedure as follows:[37]

> From the totality of natural phenomena it is possible, by successively enhanced approximations, to derive more and more exactly a system of reference *x, y, z, t*, space and time, by means of which these phenomena then present themselves in agreement with definite laws.

The reference to the *totality* of natural phenomena should be noted here. We are not dealing with local data that are pasted together by means of arbitrary conventions, but we are looking at global patterns in the world and use '*a priori*' principles (in a relativized sense!) to simplify their formulation. Concepts like spatial congruence and simultaneity are subsequently derived from this holistic analysis: the congruence and simultaneity relation are implicitly defined by the physical laws in their standard form.

Minkowski finds, of course, when he determines the standard inertial spatiotemporal coordinates this way, that there is no uniqueness. There are infinitely many solutions to the problem, corresponding to the infinitely many possible inertial frames of reference. In all these frames the laws take the same standard form. This means, in particular, that spatial congruence and simultaneity become frame-dependent. So there is the familiar relativity of simultaneity in the sense of frame-dependence; but conventionality in the sense of arbitrariness within any one frame does not emerge in a natural manner.

8. Conclusion: Holism and Reichenbach's Conventionalism

In Minkowski's approach the relativistic simultaneity relation thus emerges as part of a total theoretical package satisfying global symmetry requirements. As emphasized also by Reichenbach,[38] the existence itself of such an isotropic and

37 Lorentz et al., *op. cit.*, p. 79.

38 Reichenbach, *The Philosophy of Space and Time, op. cit.*, p. 166. See also note 21.

homogeneous description of nature is a contingent matter of fact. Although within the limits of applicability of special relativity we can give all physical laws a form in which they assign identical physical properties to all points in space and time (their standard $\varepsilon = \frac{1}{2}$ form), such a symmetrical description does generally not exist in more complicated spacetimes. In fact, most general-relativistic spacetimes do not allow a consistent global $\varepsilon = \frac{1}{2}$ simultaneity relation. The standard simultaneity relation therefore reflects, and is adapted to, an objective global symmetry property that is typical of special relativistic spacetime.

It can be argued nonetheless that even in special relativity it remains a matter of our choice whether or not we make use of this property, whether we choose a theoretical description that is adapted to the global symmetry. Reichenbach's conventionalism can be defended this way. Indeed, as we have seen, Reichenbach's conventionalism is different from the 'arbitrariness of definitions' conventionalism that assigns an arbitrary meaning to each individual term. The latter brand of conventionalism is a local affair; the former has a global, holistic character. Although Reichenbach in many places proclaimed that the local approach constituted the backbone of his conventionalism, in fact he turned out to be committed to holistic ideas that in several respects remind us of his 1920 position and permit only a conventionalism relating to the choice between theories.

Reichenbach's conventionalism boils down to a combination of the idea that local empirical data underdetermine global theories and the thesis that the choice between the various possible theories is a matter of our conventional decision. This kind of conventionalism is evidently a controversial issue in the present-day philosophy of science. Many would argue, for example, that the fact that different theoretical schemes are compatible with the same empirical data does not entail that they are equally *supported* by these data. Related to this, saying that making use of an objective property of nature is a matter of convention seems similar to holding that speaking about our daily environment in terms of tables, chairs, etc., is purely conventional; and this seems an unexciting position to take. Reichenbach's earlier approach, outlined in sections 3 and 4, seems more interesting; perhaps with the addition that the principles governing theory choice are themselves the results of a historical process of adaptation to empirical findings, in which the best fitting principles have been able to survive. But these are issues that transcend the scope of the present paper.

History and Foundations of Science
Utrecht University
P.O. Box 80.010
NL 3508 TA Utrecht
The Netherlands
D.G.B.J.Dieks@uu.nl

Mauro Dorato

On Various Senses of "Conventional" and Their Interrelation in the Philosophy of Physics: Simultaneity as a Case Study

My aim in this note[1] is to disambiguate various senses of 'conventional' that in the philosophy of physics have been frequently conflated. As a case study, I will refer to the well-known issue of the conventionality of simultaneity in the special theory of relativity, since it is particularly in this context that the above mentioned confusion is present.

My plan is to start by sketching Reichenbach's original treatment of the problem (section 1). In section 2, I will try to locate Reichenbach's problem within a much more general philosophical framework, essentially proposed by the American philosopher Wilfrid Sellars almost fifty years ago.[2] I regard this second section as particularly important, and not only as a general introduction to our topic: contemporary philosophy of physics is affected by a dangerous temptation of excessive specialization, and by an attitude that considers technicalities as ends in themselves. *Qua* philosophers, we ought to understand, as Sellars put it, "how things (in the widest possible sense of the word) hang together (in the widest possible sense of the word)".[3] In section 3, I will then distinguish among *five* different senses of "conventional", and will then study their logical relationship *vis à vis* the problem of establishing in which of these senses the relation of simultaneity could be regarded as conventional.

Not only will I press the point that, as noted by Dieks in his paper, much of the current philosophical debate on conventionality lacks contact with the issues Reichenbach's analysis was meant to address to start with.[4] Following Friedman's

1 This note originated as a comment to Dennis Dieks's presentation *The philosophy of physics in perspective*, held in Vienna in December 2008 for the meeting of the European Science Foundation. I would like to thank first and foremost Dennis Dieks, and then the audience, for the comments I received to my comments.

2 Wilfrid Sellars, "Philosophy and the Scientific Image of Man", in Robert Colodny (Ed.), *Frontiers of Science and Philosophy*, Pittsburgh: University of Pittsburgh Press, 1962, pp. 35-78.

3 Ibid.

4 See the beginning sentence of his contribution to this volume: "The history of the philosophy of physics has been shaped by a complicated and fascinating interplay between physics, philosophical ideas and external factors. This history is not only an intriguing subject for study in its own right: historical considerations can also shed light on the content of doctrines put forward by philosophers of physics and are relevant for the appraisal of such doctrines."

F. Stadler (ed.), *The Present Situation in the Philosophy of Science*,
The Philosophy of Science in a European Perspective, DOI 10.1007/978-90-481-9115-4_24,

outlook toward the history of logical positivism,[5] I will also argue that, without taking into due account the original problem inspired by Kant's philosophy – and consisting in trying to separate what in our scientific theories is "due to us" from what is contributed by the external world – the debate on the conventionality of simultaneity loses much of its interest. This concession, however, should not make us blind to the fact that the special theory of relativity has epistemically driven, verificationist foundations, inspired by the work of Mach and especially David Hume, a fact that has been very important for the subsequent history of both physics and philosophy. As to the more circumscribed question of the conventionality of simultaneity *per se*, we will see how, in an important sense of 'conventional', the relational character of "being simultaneous with" implies by itself that simultaneity *is* conventional, despite Malament's celebrated result,[6] while in another sense, simultaneity appears to be non-conventional.

1 Reichenbach's original formulation of the problem

After Reichenbach's groundbreaking *The Philosophy of Space and Time*,[7] the problem of the conventionality of simultaneity is traditionally regarded as the question of establishing whether simultaneity, besides being non-controversially *relatively* to a given inertial worldline, is also *conventional*, or non-unique, even after an arbitrary inertial worldline has been picked.

As is well-known, the original problem that Einstein had to solve in 1905 was that of synchronizing two distant clocks at rest in the same inertial frame. Adopting some terminology that has been invented only later, here is Einstein's solution. Emit a light signal at point e on the inertial worldline O toward another, parallel inertial worldline O': parallelism indicates, of course, that O and O' are in the same inertial frame. Suppose that the light signal is reflected by a mirror at point f on O' and is received again after some time at point r on O. The problem of synchronizing the two clocks that trace out the two inertial worldlines is equivalent to the question of determining which event t_f along O is simultaneous to the epistemically inaccessible reflection event f on O'.

In order to gain some generality in Einstein's simple thought-experiment, Reichenbach introduced a real number ε, $0 < \varepsilon < 1$, such that

$$t_f = \varepsilon(t_r - t_e)$$

5 M. Friedman, *Reconsidering Logical Positivism*, Cambridge: Cambridge University Press, 1999.

6 David Malament, "The Conventionality of Simultaneity", in *Noûs* 11, 1977, pp. 293-300.

7 Hans Reichenbach, *The Philosophy of Space and Time*, transl. by Maria Reichenbach and John Freund, New York: Dover, 1958, pp.123-129.

Now Reichenbach's question about the conventionality of simultaneity can be formulated with precision. Can we assume that the reflection event *f* on *O'* has occurred exactly at

$$t_f = (t_r - t_e)/2,$$

on *O*, as Einstein had originally assumed, and therefore fix the value $\varepsilon = ½$? Or, on the contrary, is the choice of one among the non-denumerable infinity of values of $\varepsilon \neq ½$ comprised between 0 and 1 absolutely unconstrained by any fact, and therefore *conventional*? In the former hypothesis, the speed of light on the two legs of the journey (from *O* to *O'* and back) is assumed to be the same (*isotropy* of the propagation of light). In the latter hypothesis, light has two different speeds on the two legs of the journey, in such a way, however, that the only measurable quantity on the part of *O*, the total two-way time, is always the same.

Reichenbach's original claim is that the choice of $\varepsilon = ½$ *is* unconstrained, or conventional, *due to the impossibility of measuring the one-way speed of light.* In order to measure the one-way speed of light, in fact, according to Reichenbach we would already need a synchronized clock on *O'*, but the whole "radar procedure" that Einstein proposed was meant to synchronize the two clocks to begin with. So the conventionality of simultaneity, for Reichenbach, is a consequence of the impossibility to attribute isotropy to light without falling into the *vicious circle* of using a criterion of simultaneity to determine the one-way speed of light and using this one-way speed in order to establish a criterion of simultaneity.

Reichenbach's argument has spurred various criticisms that have often relied on different senses of "conventional": in order to look back at the whole debate one more time, these senses need to be disentangled with care. Before doing so, however, some more general remarks are appropriate.

2 The Ambiguity of 'Conventional', and Sellar's Conflict between the Manifest and the Scientific Image

> The terms "convention" and "conventional" are flagrantly and intricately ambiguous. On the one hand, the conventional is the ordinary, the usual, the traditional, the orthodox as against the novel, the deviant, the unexpected, the heterodox. On the other hand, the conventional is the artificial, the invented, the optional, as against the natural, the fundamental, the mandatory.[8]

While this quotation obviously refers to the *social* and *epistemic* sense of convention, one that does not seem directly connected to our topic, it nevertheless helps me to make explicit the general philosophical perspective from which I want to tackle Reichenbach's original problem.

8 Nelson Goodman, "Just the Facts, Ma'am!", in Michael Krausz (Ed.), *Relativism: Interpretation and Confrontation*. Notre Dame: University of Notre Dame Press, p. 80.

In a first sense of 'convention', Goodman writes, 'conventional' seems to refer to what is the result of a widely shared, but not necessarily intentional, *agreement*, something that can be regarded as part and parcel of *common sense*. In this sense, we claim for instance that, before the revolutionary view proposed by Copernicus, the shared belief in a static Earth was part of orthodoxy, or simply of our 'conventional knowledge'.

On the other hand, and following the ancient sophists, Goodman contrasts 'conventional' also with what is 'natural'; in this second sense, it was also natural – i.e. non-conventional and mandatory – for us all to believe in the immobility of the Earth. Such a belief seems in fact to be entrenched in the so-called *folk* or *naïve physics*, whose universal and innate hold on our brains so many experiments in the cognitive science have now confirmed. From this viewpoint, scientific knowledge artificially, non-naturally and 'conventionally' goes against some of our 'natural', naïve physical beliefs, as when it convinces us that, for instance, the Earth moves along its orbit at the average speed of approximately 30 km/sec.

The conflict between Goodman's two senses of 'conventional' could then be reformulated by claiming that part of what is traditionally and *conventionally* believed by human beings (in Goodman's first sense of "convention") is so believed because of *natural, non-conventional "forces" that have been shaping our brains during our biological evolution*. Consequently, the whole scientific enterprise could be regarded as an invention of our culture based on our natural capacities, an invention that, however, had to go against (at least) *some* of our natural beliefs, like the centrality and speciality of human beings in nature, the existence of a cosmic now, the fact that moved bodies require movers, our pervasive tendency to project the notion of purpose onto the natural world, and so on.

It then seems natural to claim, along with Sellars, that one of the main aims of the philosophy of physics is to inquire into the *compatibility* of the physical image of the world (the physical image of time, space, matter, identity, etc.) with the "manifest image", which is the world of our experience, as it has been explicated by philosophers and phenomenologists of the past, and investigated empirically by neuropsychologists and cognitive scientists today. In a word, if I am right in claiming that conventional beliefs (in Goodman's first sense of 'convention') are part of the manifest image, and that they are naturally, non-conventionally believed by common sense (in Goodman's second sense of 'convention') because of evolutionary reasons, the difference between these two senses of 'convention' is *Sellars' conflict between the manifest and the scientific image of the world.*

The importance of Sellars' conflict in this paper is two-fold. Firstly, we will see one of its manifestations in the thicket of questions surrounding the conventionality of simultaneity, in particular *a propos* of our natural, naïve physical belief in the existence of a cosmic present, which found its expression in the Newtonian, *absolute* character of the relation of simultaneity. Secondly, focussing on Sellars' conflict is also particularly important in judging the importance of the a priori in the growth of scientific knowledge, *at least to the extent that many a priori presup-*

positions of science come from the manifest image of the world. How changeable is this a priori *vis à vis* the evolution of scientific theories?

3 Various senses of conventional in the philosophy of physics

The following list is not to be regarded as exhaustive,[9] but simply as a first guide to further, more detailed work.

1) 'Conventional' is whatever is opposed to synthetic or factual, that is, something that can be said of a proposition that is either analytically true, or simply devoid of any truth-value; *this is a semantic sense of 'conventional'*;
2) 'Conventional' can be regarded as what is *constitutive* of our theory (Reichenbach's *constitutive a priori,* to be contrasted with the *a priori* regarded as 'universally valid', or 'apodictic'); 'conventional' here corresponds to the formal element of knowledge in the kantian sense, something that in the construction of a scientific theory is exclusively 'due to us';
3) 'Conventional' can be regarded as what *is non-reducible/definable in terms of a physical-causal relation*: this sense involves the *causal theory of time*, and is the target of Malament's 1977 much discussed theorem on the non-conventionality of simultaneity;[10]
4) 'Conventional as more changeable', or less entrenched *epistemically* (opposed to 'analytic' in Quine's sense) because not present in direct experience. This sense is discussed by Reichenbach in his 1958 famous treatise on space and time and by Dennis Dieks in the paper contained in this volume (section 3); it is definitely an *epistemic* sense of conventional;
5) 'Conventional' can be regarded as what is referred to a choice that is supposed to *fix a gauge*. While I will not comment in a detailed way on this last sense, it is important to list it together with the others, as in some approaches the choice of a synchronization for non-inertial frames is equivalent to gauge fixing.[11]

9 A first attempt of separating different senses of conventional is due to my former teacher Robert Rynasiewicz: see his abstract in "Varieties of Conventionality", in Jacek Cachro and Katarzyna Kijania-Placek (Eds.), *Volume of Abstracts of the 11th International Congress of Logic, Methodology and Philosophy of Science*. Cracow: Kopiorama. 1999, p. 329. My list does not overlap much with his, however, and after his oral presentation, Rynasiewicz never wrote a paper.

10 See note 6.

11 David Alba, Luca Lusanna, "Generalized Radar 4-Coordinates and Equal-Time Cauchy Surfaces for Arbitrary Accelerated Observers", in *International Journal for Modern Physics*, D16, 2007, pp. 1149-1186. See also David Alba and Luca Lusanna, "Charged Particles and the Electro-Magnetic Field in Non-Inertial Frames of Minkowski Spacetime," *arXiv*.

3.1 The semantic sense of conventional: conventional as the non-synthetic

In one clear sense of the word, possibly the most central one, 'conventional' is opposed to whatever is factual or synthetic. There are two ways for a sentence to be non-synthetic: it can be devoid of any truth value, or it can be analytically true. Let me start by exemplifying the sense of convention that we are after in this section by beginning with the former alternative. After the special theory of relativity, and the discovery of the relativity of simultaneity, we know that there is no fact of the matter that could be invoked to answer the question:

Q = what is happening *right now* in Andromeda?

simply because we know that in the special theory of relativity *there is no cosmic present*: in this theory, the now does not extend in space at all (it is pointlike), or extends in space at most *locally*, in a sense of local that is in any case different from that ruling in quantum physics.[12]

The answer to question Q is conventional in this first, semantic sense, since it depends on an arbitrary choice of a reference frame or of an inertial worldline and, as such, it corresponds to no fact whatsoever. Consequently, the sentence "event *e* on Andromeda is simultaneous with an event here-now", in the post-Newtonian universes lacks a definite truth value; being an incomplete sentence (it lacks a relational term), *it is neither true nor false* or even *meaningless*.

As an instance of a sentence that is not synthetically true but still 'conventionally true' because analytically true, think of the conventionalist reading of the axioms of the geometry. In 1902, Poincaré thought of such axioms as *disguised definitions*: as such, the axioms of geometry could not be regarded as a reflection of empirical facts nor, given the multiplicity of geometries, as synthetic a priori judgments.

> The *axioms of geometry therefore are neither synthetic* a priori *judgments nor experimental facts*. They are *conventions*; our choice among all possible conventions is *guided* by experimental facts; but it remains *free* and is limited only by the necessity of avoiding all contradiction.[13]

According to Poincaré, axioms are "true" at best in the sense in which definitions like "bachelors are unmarried men" are (necessarily) "true". In another sense, however, *qua* definitions, they are neither true nor false, exactly like sentences that have no truth-maker coming from the world of facts. A definition can be useful and apt for our goals or not, but not really true or false.

Could we claim that after we fix an inertial worldline O passing through our "here-now", the answer to the above question Q is a matter of mere definition? This harder question will be tackled in the remainder of the paper.

---

12 Dennis, Dieks, "Becoming, relativity and locality", in Dennis Dieks (Ed.) *The ontology of spacetime*, Amsterdam: Elsevier, 2006, pp.157-176.

13 Henri, Poincaré, *Science and Hypothesis*, in *The Foundations of Science*. Trans. George Halsted. New York: The Science Press, 1902/1905, p. 65.

3.2 The conventional as the constitutive a priori

According to Einstein's original treatment, establishing whether "event *e* on Andromeda and an event here-now are simultaneous" necessarily requires a *Festsetzung* (a stipulation), as Einstein put it in 1905.[14] That is, given the epistemic inaccessibility of event *e* on Andromeda from our here-now and conversely (see the fourth sense of conventional) even within a single inertial frame, in order to answer any question about distant events relatively to a chosen inertial frame we need some operational/conceptual convention, like the radar method illustrated above.

To the extent that Einstein's radar convention transforms meaningless questions into empirical questions, the convention itself is also '*constitutive*' of the special theory of relativity, in Reichenbach's peculiar sense of the *constitutive a priori*.[15] And this explains the transition from the sense of conventional discussed in 3.1 to the currently discussed sense.

In a nutshell, the main idea of identifying the conventional element in a scientific theory with its constitutive element(s) is this: *not all scientific concepts in a theory are epistemically on a par, since conventional truths about some of them* (simultaneity in our case) *make empirical questions possible in Kant's sense.* Since without a convention of some sort (radar method), we could not even ask questions like *Q* above, the radar method and the resulting concept of simultaneity constitute or *ground* the whole theory.

Note that this also corresponds to Friedman's doctrine of the *relativized a priori*: Michael Friedman, following the early Reichenbach, and various other logical positivists, separates from the original kantian meaning of the a priori regarded as something pertaining to a judgment that is universally valid and unrevisable, the concept of an priori that is constitutive of a scientific theory. [16] As Reichenbach had it, such a "constitutive a priori" may change; it is therefore revisable across scientific changes, and is therefore *not* universally valid. Its flexibility is compatible with the fact that what is constitutive a priori for one theory can be abandoned in the later conceptualization of a new theoretical framework.

Why do I refer to such constitutive a priori elements of our scientific knowledge as conventions? Here I follow Moritz Schlick's oft-quoted letter to Reichenbach, one that was very important to convince the latter that he had to abandon his previous kantian language. Schlick writes:

> it is the main point of this letter that I cannot see what is the real difference between your a priori statements and conventions … The decisive place where you describe the character

14 Albert Einstein, "On the Electrodynamics of Moving Bodies", in Albert Einstein et al., *The Principle of Relativity*, New York: Dover, 1952, pp. 37-71.

15 Hans Reichenbach, *The Theory of Relativity and A Priori Knowledge*. Berkeley: University of California Press, 1965.

16 Michael Friedman, *Dynamics of Reason: The 1999 Kant Lectures at Stanford University*, Stanford: CLSI Publication, 2001.

of your a priori correspondence principles seem to me nothing short of accomplished definitions of the concept of convention.[17]

Even though Reichenbach in his later writings implicitly kept on believing in the importance of *some* constitutive a priori element in the foundations of scientific knowledge,[18] it is in any case highly significant that *after* this crucial letter and his exchange with Schlick, he will make reference to Kant more to criticize him than to vindicate or revise aspects of his thought. As a consequence, after this letter Reichenbach *will coherently abandon any form of kantian-sounding language.* This had a pragmatic motivation: the neopositivists had to make a carrier in the German University after the First World War, in a cultural environment that was dominated by neokantiants, by heirs of classical German idealism, and by phenomenologists like Husserl and their students like Heidegger. Before moving to the third sense of 'conventional', I would like to add four remarks.

First of all, while not all constitutive a priori elements in our scientific theories have to be regarded as conventions, in our case the availability of other methods (operational criteria) to fix the meaning of distant simultaneity makes the radar method conventional, or not dictated by facts. In this sense, Schlick was right.

Secondly, as Reichenbach himself makes clear, also the choice of a coordinate system in the theory of relativity is underdetermined by all possible facts. Accordingly, in his *Relativity Theory and Knowledge a priori*, he insists that the invariance with respect to (Lorentz) transformations represents the objective, factual content of reality, while the structure of what in 1920 he still calls 'reason' (the source of the whole a priori structure of a theory) is expressed by the arbitrariness of the admissible coordinate systems … 'the subjective form that makes our description possible'. So the choice of a coordinate system is conventional (because subjective and therefore a priori), and yet indispensible for the description of the physical world. In this related but slightly distinct sense of constitutive a priori, for Reichenbach the choice of a coordinate system is constitutive of the theory ('it makes our description of the world possible') because we are spatiotemporally located beings, so that we *must* describe the world from somewhere and somewhen. The a priori character depends entirely on the arbitrariness of the choice of a reference frame:

That the concept of object has an origin in reason can be revealed only by the fact that in it there are contained elements for which no choice is prescribed and that are independent of the nature of reality … The contribution of reason is not expressed in the fact that in the

17 Quoted from Alberto Coffa, *To the Vienna Station*, Cambridge: Cambridge University Press, 1993, pp. 201-2.

18 For an illustration of this claim, see Massimo Ferrari, *Categorie e a priori*, Bologna: Il Mulino, 2003.

coordination system there are invariant elements, but rather in the fact that in it there are arbitrary elements.[19]

Thirdly, note that what is constitutive of science need not be *also* constitutive of our experience of the world in Kant's sense (an aspect that neokantians tend to forget even today). In fact, there is a *prima facie conflict* between the special theory of relativity and our experience of time (see above): the latter strongly suggests a natural belief in a cosmic, absolute present, which is part and parcel of the manifest image of time, a belief that the former explicitly denies by insisting on the relativity of simultaneity. This remark creates some tension in Kant's philosophy, to the extent that in his thought the conditions of possibility of our experience are regarded also the conditions of possibility of scientific theories.

Finally, note how the first and the second sense of conventional are deeply related: the presence of conventional elements in science in the second, constitutive sense entails the view that scientific theories are "a free creation of the human mind" (as Einstein often put it), and that they are *not* simply "deducible from facts", but partially *depend on us*. In other words, scientific models and theories are human artefacts: those conventions that are constitutively *a priori* are, in analogy to the formal element of Kant's theory of knowledge, due to us, and therefore not extractable from the world of facts, which is the realm of those invariant transformations preserving the structure of spacetime.

It is therefore quite crucial to note that when we discuss the problem of the conventionality of the metric, or of the relation of simultaneity in a mere "technical setting", *we ought not to forget the struggles of the early neopositivists to confront themselves with Kant's thought, and in particular with the role of the constitutive a priori in science.*

3.3 Conventional as "non-definable in terms of a physical (causal) relation"

Malament's famous 1977 result concerns the unique definability of Einstein $\varepsilon=1/2$ simultaneity relation in terms of a time-symmetric relation of causal connectibility, and therefore in terms of the invariant structure of Minkowski spacetime. Suppose, along with the defenders of the causal theory of time, that a spatiotemporal relation x is conventional iff x is not definable or reducible in terms of a physical/causal relation. Then, Malament proves that *to the extent* that the causal theory of time ought to be endorsed (in 1977 he did not explicitly defend it, but presented his philosophical claim in a *conditional* form), his unique definability result – already implicitly present in a work by Alfred Robb[20] – rules out the claim that the relation of simultaneity is conventional.

19 Hans Reichenbach, *The Theory of Relativity and a priori Knowledge*, ibid, my translation, p. 138.

20 Alfred, Robb, *A Theory of Time and Space*. Cambridge: Cambridge University Press, 1914.

Note that this third sense of conventional seem to be totally alien to Reichenbach's original (kantian) philosophical worries about the "constitutive a priori". In this hypothesis, such a third sense (Malament's definability) would be totally *irrelevant* for the second sense (constitutivity), so that we should not confuse them under the heading "conventionality of simultaneity".

On the other hand, we could try to defend the view that causation is a constitutive a priori element of the special theory, in the sense that the objectivity and invariance of the partial temporal order available in the theory depends on the objectivity and invariance of the causal order. Of course, to the extent that causation can be regarded as a relation that is imposed by us onto the physical phenomena in the sense of a kantian category, we could as well consider it as an element that is "due to us". But this claim would need additional arguments that cannot be presented in the limited space of this paper.

All I can note in support of this claim here is the following: since in Robb's *axiomatization* of the special theory of relativity, the relation of 'being after' enjoys a foundational role, the reducibility of this relation to causation would give also the causal theory of time a "constitutive" role, since the causal relation would become the true and "primitive" building block of the theory.

This hypothesis, however, is highly controversial, since we still do not agree on what causation is. For instance, if we accepted a Humean, reductionist account of causation as mere regularity, regularities certainly don't depend on us, so that causation, while possibly constitutive of the theory in the axiomatic sense, could not be regarded as being *a priori*. An analogous conclusion would hold if we adopted a theory of causation involving a realist attitude toward potencies or causal powers.[21]

The relevance between the third sense and the first is not very easy to establish either. On the one hand, Malament's result seems certainly related to the first sense of conventional, at least to the extent that what is non-conventional *qua* definable in terms of a causal/physical relation also appears to be non-conventional *qua* factual. After all, a physical/causal relation between events represents a physical fact, albeit relational. On the other hand, if one insisted in holding that the only "facts" in SR are the invariant (worldline-independent) facts, then the relativity of simultaneity would automatically imply its conventionality.

Therefore, due to the ambiguity about the notion of fact in the special theory of relativity, and the related difficulty of establishing whether relational facts also count as fact in that theory, the thesis that Malament's result implies that simultaneity is non-conventional in the first sense remains controversial, even *after* an inertial worldline is given.

Here is another way of looking at this question, and appreciate its complexity. Consider an event a on an inertial worldline O: is it still meaningless to ask

21 For a defence of the importance of causal powers in making laws true, I refer to M. Dorato *The Software of the Universe*, Ashgate, 2005, and A. Bird, *Nature's Metaphysics*, Oxford University Press, Oxford, 2007.

whether an event e in Andromeda is occurring simultaneously with a on O, *relative* to O?

As is well-known, this question has been subject to various discussions. On the one hand, Reichenbach and Grünbaum insisted on the claim that, since we cannot measure the one-way velocity of light, we cannot deem the answer to the question as being based on matters of fact. We still need the conventional assumption that light has the same speed in all directions. So, even *after* we assign a worldline O, relatively to O simultaneity is conventional.

On the other hand, many philosophers after 1977 have taken Malament's result as having solved this problem once and for all. The relation of "being orthogonal to O", said of a straight intersecting point $a \in O$ identifies uniquely a simultaneity relation. This relation, being definable in terms of the automorphisms of the structure of Minkowski spacetime, preserves that structure. If no other simultaneity relations can preserve the structure, giving up Einstein's $\varepsilon = 1/2$ choice would amount to giving up the whole structure of Minkowski spacetime.[22] Furthermore, a relation like "is simultaneous with $a \in O$ *relative* to O" would seem to be objective and factual, *qua invariant* for all possible observers.

Debs and Redhead have recently claimed that, in a sense, *both* of these positions sketched above turn out to be correct.[23] Pick a worldline, and then ask whether two points selected by Einstein's standard synchrony are objectively simultaneous relative to that worldline: the answer must be in the positive, for the reasons just given. However, the correctness of this answer crucially depends on a previous *choice, that of eliminating Lorentz boosts from the full group of automorphisms of Minkowski spacetime.*[24] Debs and Redhead conclude that conventionalism is still with us for at least two reasons:

1) the choice of adopting a restricted set of symmetries (rather than the full set) as an invariance criterion for the objectivity of a relation is, in some sense, in itself conventional (not dictated by facts);

2) once we decide to *include* Lorentz boosts in the full group, then conventionalism seems again to be correct, because a boost will tilt the hyperplane of simultaneity orthogonal to the original worldline and will not preserve it.

Unfortunately, this irenic claim seems to forget that the choice between leaving Lorentz boost out of the automorphisms group or not is not so "free". If the question is to decide whether simultaneity if conventional even after having fixed an inertial worldline, as Reichenbach had originally posed the question, then Debs and Redhead should conclude that Malament is right, because Lorentz boost *must* be left out of the full group of automorphisms. Of course, "the choice of whether to use the line of simultaneity defined by O, O', or any one of any infinite number

22 This argument is defended in M. Friedman, *Foundations of Spacetime Theories*, Princeton University Press, 1983.

23 Talal A. Debs, Michael L.G. Redhead, *Objectivity, Invariance, and Convention*, Harvard: Harvard University Press, 2007, p.95.

24 *Ibid*, p. 97.

of inertial worldlines"[25] is fully conventional. However, this is a consequence of the relativity of simultaneity, and unless we are convinced that there is no difference between the relativity and the conventionality of simultaneity, we should stick with a distinction between the two notions. In a word, if the debate on the conventionality of simultaneity is about the uniqueness of an ε-value, *once a given worldline has been conventionally fixed*, the full Poincaré group is not a live option.

3.4 The Conventional as the epistemically "more changeable"

This epistemic sense of conventional has been clearly defined by Dieks in his paper:[26] he seems to imply that distant simultaneity is a non-local concept and that, as such, it is less firmly anchored in direct experience; therefore more open to change, or more "conventional". Concepts used in direct, local observations are in practice unrevisable "although in principle all our concepts may eventually change under the influence of new empirical findings, in practice some of them are virtually immune to such revision".[27]

Here one could raise a point that involves the meaning of *direct, local observation*. Distant simultaneity is not *directly* observable of course, if 'direct' means '*local*', but in order to decide what counts as locally and directly observable we always need a theory, namely the special theory of relativity and classical electromagnetism. *What Dieks seems to neglect is that it is always a scientific theory that decides for us what is directly, locally observable and what isn't.* Consequently, if for scientific reasons we could admit an instantaneous transmission of light, then our "direct experience of time" (in a slightly enlarged sense of "experience") would include a cosmic now. And note that as part of our manifest image of time, *we* firmly believe that there exists a cosmic present, and that simultaneity is absolute, and this seems part of our direct experience of the world. However, what seems global is instead only local, since by looking at a star in the night sky, we wrongly believe that we directly observe the light emitted by it *at present*, but we observe only light emitted light years ago. This remark is linked to the fact that, against Dieks' opinion, I think that science may dispose even of concepts that appear the result of our *direct* and most entrenched experience of the world, for the simple reason that "local observation" is a theory-laden concept. Interestingly, conventional in this fourth sense is fully synonymous with Quine's sense of synthetic, given that the latter means more revisable, because impinging on the periphery of the whole networks of beliefs in which our scientific knowledge consists.

In any case, it is because of the theory derived, non-directly accessible character of the simultaneity relation between two distant events that in order to judge

25 *Ibid.*, p. 87.

26 See the paper in this volume, section 3.

27 See Dieks's paper in this volume.

such two events as simultaneous we need some additional operational procedure. The procedure in question must translate the unobservability of the distant simultaneity of two events into a locally discernible coincidence of point-events (two distant light signals intersecting two mirrors posed in front of us). The lack of direct accessibility is the major reason for assuming that questions like Q lack factual content.

Therefore, conventional in this fourth sense, assuming positivistic theories of meaning, might be taken to explain conventional in the first sense: there is no matter of facts making a certain assertion about the simultaneity of two lightlike separated events as true or false, simply because the two events in question do not, and cannot in principle, fall within the limits of a single perception. On the other hand, it is the lack of direct epistemic access between distant events that Einstein (and Poincaré before him) has exploited to introduce some sort of a constitutive a priori convention (the second sense of the word) in the theory. As we have seen, this convention is capable of transforming a meaningless question like Q into an empirical question.

Analogously, the fact that the whole region outside the light cone centered in a point p of Minkowski spacetime is causally non-connectible with respect to p (the elsewhere region relative to p) gave philosophers additional motivations to defend the causal theory of time, already defended by Kant. The epistemic non-accessibility of distant simultaneity (its conventionality in the fourth sense) finds an explanation in the lack of a physical connection between different regions of spacetime (third sense of conventional given by the causal theory of time). If there cannot exist in principle physical signals connecting with p points in the elsewhere region relative to point p, then any sort of temporal relationship that we may fancy to introduce between spacelike related events (simultaneity included) is going to be conventional in both the third and fourth sense of the word. The only invariant temporal order is given by the causal connectibility relation, which is left invariant by the full group of automorphisms of the Minkowski spacetime. So the connection between sense three and four of conventional is certainly not unimportant.

In sum, the importance of this fourth sense of convention can hardly be exaggerated, a fact confirming that special relativity is *an epistemically based theory*. Also the point-coincidence argument, that Einstein defended later in the context of the general theory of relativity in order to avoid the dire consequences derived from the hole argument, is based on the claim that the directly observable relations are the foundational elements of any spacetime theory. This means that while I agree with Dieks that the centrality of Mach's (and Hume's) influence upon Einstein and the neopositivists needs to be re-examined with care, I think he will agree with me that it is certainly difficult to deny that the special theory of relativity has empiricist, epistemically-driven foundations.

3.5 The conventional as deriving from gauge-fixing, i.e. determining simultaneity for non inertial observers

In a recent essay, Lusanna begins by remarking that

> real observers are never inertial and for them Einstein's convention for the synchronization of clocks is not able to identify globally defined simultaneity 3- surfaces, which could also be used as Cauchy surfaces for Maxwell equations.[28]

As Lusanna clarifies, what is required in this case is

> a 3+1 splitting of Minkowski space-time, namely a foliation ... whose leaves ... [are] both a Cauchy surface for the description of physical systems and an instantaneous (in general Riemannian) 3-space of simultaneity implied by a clock synchronization convention different from Einstein's one.[29]

After a technical discussion involving the Hamiltonian constraint approach to the problem, Lusanna clarifies that all the admissible 3+1 splittings, namely all the admissible procedures for clock synchronization, and all the admissible non-inertial frames centered on time-like observers, are *gauge equivalent*.

The question that is quite interesting and novel for our problem is the following: in Lusanna's approach to establishing simultaneity for non inertial frames, the gauge fixing is linked to the *conventional* choice of an *extended physical laboratory*. The spatio-temporal phenomena as they are viewed from non-inertial frames are therefore coordinate-dependent, in the same sense in which they are coordinate dependent when we choose inertial frames. In this more general approach invoked by Lusanna, however, the inertial frames centered on inertial observers become a special case of gauge fixing. In particular:

> For each configuration of an isolated system there is a special 3+1 splitting associated to it: the foliation with space-like hyper-planes orthogonal to the conserved time-like 4-momentum of the isolated system.

Conclusion

In sum, I have tried to show that we cannot tackle the problem of the conventionality of simultaneity as if it were solvable merely with technical means: the above

28 Lusanna, Luca, "General covariance and its implications for Einstein's space-times", talk at the Meeting *La Relativita' dal 1905 al 2005: passato, presente e futuro* organized by SIGRAV and SISM, Department of Mathematics of University of Torino, June 1, 2005, p.3.

29 Ibid, p. 4.

illustrated conflation of various senses of 'conventional' can be avoided only by giving both historical and conceptual considerations their due.

On the one hand, as already pointed out by Friedman, we ought not to forget the deep involvement with Kant's philosophy both of Reichenbach and of the other members of the Vienna circle. This involvement entails that the question of the conventionality of simultaneity was for them simply a case study used to test the validity of the framework proposed by the *Critique of Pure Reason* after the new revolutionary results introduced by the two theories of relativity.

On the other hand, the historical importance of empiricist methods for the foundations and discovery of the special theory of relativity – and the relativity of simultaneity in particular – can hardly be denied. Not only is this illustrated by the fourth sense of conventional presented above, but also by important evidence provided by Einstein himself on the role of Mach and Hume's thought on the origin the 1905 theory.[30] In a letter sent by Einstein to Schlick at the end of 1915, we read:

> Your exposition is also quite right that positivism suggested relativity theory, without requiring it. Also you have correctly seen that this line of thought was of great influence on my efforts and indeed E. Mach and *still much more* Hume, whose treatise on understanding I studied with eagerness and admiration shortly before finding relativity theory. *Very possibly, I wouldn't have come to the solution without those philosophical studies*.[31] [my emphasis]

However, Dieks will hardly disagree with me on this point. It is worthwhile to recall that while Einstein later disowned it – by claiming that a good joke should not be repeated – an appeal to the verificationist/operationalist foundations of Einstein's critique of the concept of simultaneity will be of immense historical importance. This applies not only to the history of scientific philosophy (think of its influence on Wittgenstein's thought, or the Vienna circle etc.), but – considering the enormous influence that it had in Heisenberg's and Bohr's thought, and in the many physicists that still follow them – also to the interpretation of quantum mechanics.

Department of Philosophy
University of Rome 3
Viale Ostiense 234
00144, Rome
Italy
dorato@uniroma3.it

30 See J. Norton, „How Hume and Mach Helped Einstein Find Special Relativity," prepared for M. Dickson and M. Domski, eds., *Synthesis and the Growth of Knowledge: Essays at the Intersection of History, Philosophy, Science, and Mathematics*. Open Court, forthcoming.

31 A. Einstein, Letter to Schlick, December 14, 1915, *Papers*, A, Vol. 8A, Doc.165, A.

Roman Frigg and Carl Hoefer[1]

Determinism and Chance from a Humean Perspective

1. Introduction

On the face of it 'deterministic chance' is an oxymoron: either a process is chancy or deterministic, but not both. Nevertheless, the world is rife with processes that seem to be exactly that: chancy and deterministic at once. Simple gambling devices like coins and dice are cases in point.[2] On the one hand they are governed by deterministic laws – the laws of classical mechanics – and hence given the initial condition of, say, a coin it is determined whether it will land heads or tails when tossed.[3] On the other hand, we commonly assign probabilities to the different outcomes of a coin toss, and doing so has proven successful in guiding our actions. The same dilemma also emerges in less mundane contexts. Classical statistical mechanics assigns probabilities to the occurrence of certain events – for instance to the spreading of a gas that is originally confined to the left half of a container – but at the same time assumes that the relevant systems are deterministic. How can this apparent conflict be resolved?

One response to this problem would be to adopt an epistemic interpretation of probability and regard the probabilities we attach to events such as getting heads when flipping the coin or the spreading of the gas when opening the shutter as a reflection of our ignorance about the particulars of the situation rather than the physical properties of the system itself. Outcomes really are determined, but we don't know which outcome there will be and so we use probabilities to quantify our uncertainly about what will happen. There is no contradiction between determinism and probabilities thus understood.

However, this is unsatisfactory. There are fixed probabilities for certain events to occur, which are subjected to experimental test and which, in many cases, are

1 The authors are listed alphabetically; the paper is fully collaborative. To contact the authors write to r.p.frigg@lse.ac.uk and carl.hoefer@uab.es. We would like to thank Alan Hájek and Aidan Lyon for helpful comments on earlier drafts, and the audiences in Vienna, Canberra, Paris and Maryland for stimulating discussions. Frigg would like to acknowledge support from the Spanish research grant FFI2008-01580. Hoefer's research for this paper was supported by Spanish research grants HUM2005-07187-C03-02 and FFI2008-06418-C03-03/FISO.

2 Or if one insists that at bottom the world is quantum mechanical, then the problem is that probabilities like the ones we attach to coin flips don't reduce to the micro probabilities since quantum mechanics assigns values close to either 1 or 0 rather than 1/2 to events like getting heads when tossing a coin.

3 For a discussion of determinism see Earman (1986, Ch. 2).

F. Stadler (ed.), *The Present Situation in the Philosophy of Science*,
The Philosophy of Science in a European Perspective, DOI 10.1007/978-90-481-9115-4_25,

governed by probabilistic laws (such as the laws of statistical mechanics). So these probabilities seem to have nothing to do with the knowledge, or even the existence, of conscious creatures studying these systems: the chance of a coin to land heads is 0.5 and a gas is overwhelmingly likely to spread when the box is opened, no matter what anybody believes about these events. The values of these probabilities are determined by how things are, not by what we believe about them.[4] In other words, these probabilities are chances, not credences.

This is an unwelcome conclusion because chance and determinism seem to be incompatible. In this paper we argue that at least for a Humean this incompatibility is only apparent and that the problem can be resolved since Humean objective chances are compatible with there being underlying deterministic laws – Lewis' own proclamation to the contrary notwithstanding.[5]

In our discussion we focus on a simple example, a coin toss, then develop a Humean account of chance, and then show that on this account there is a non-trivial sense in which coin flips are chance events while at the same time being governed by deterministic laws. In the last section we briefly indicate that chances are introduced into statistical mechanics essentially in the same way as in the case of the coin and so the basic idea of deterministic chance developed here can be carried over to statistical mechanics without (much) further ado.

2. Flipping a Coin

Coin tossing is the most widely used example of a random process, and we are firmly convinced that the chance for getting either heads or tails is 0.5. At the same time we are also firmly convinced that coins obey the laws of mechanics and that therefore their flight as well as their landing heads or tails are determined by their initial conditions and the forces acting upon them. Can we consistently uphold both convictions?

This question has been discussed from a physics point of view by Keller (1986), and later, building on Keller's work, by Diaconis (1998) and Diaconis, Holmes and Montgomery (2007). We believe that this approach provides all the ingredients needed to explain why the chance for heads equals 0.5, and why there is no conflict between this and the fact that coins are governed by the laws of classical mechanics. However, the explanation we offer differs from Keller's and Diaconis'. We now review in some detail their arguments since they serve as the springboard for our own discussion of chance in coin flips in Section 4.

4 This point is often made in the context of statistical mechanics; see for instance Albert (2000, p. 64), Loewer (2001, p. 611) and Goldstein (2001, p. 48); see also Hoefer (2007, p. 557, pp. 563-4) and Maudlin (2007, pp. 281-2).

5 Loewer (2001; 2004) has presented a reconciliation of determinism and chance from a Humean perspective. However, we believe this reconciliation to be problematic for the reasons discussed in Frigg (2008b).

Keller introduces the following mechanical model of the coin flip. Consider a circular coin of radius r, negligible thickness, and with homogeneous mass distribution. The only force acting on the coin after being tossed is linear gravity, and the surface on which it lands is mushy so that the coin does not bounce. Furthermore the coin is flipped upwards in vertical direction with velocity v at initial height h (above the surface on which it eventually lands) so that it rotates with angular velocity ω around a horizontal axis along the diameter of the coin (i.e. we rule out precession). Solving Newton's equations for this situation (and assuming that the coin is flipped in horizontal position) yields

$$x(t)=vt-\frac{gt^2}{2}+h \tag{1}$$

$$\phi(t)=\omega t \tag{2}$$

where $x(t)$ is the coin's height at time t and $\phi(t)$ the coin's angle relative to the plane. Using the coin's radius one can then determine which point of the coin touches the surface first, and together with the assumption that the coin does not bounce this determines whether the coin lands head or tails. These calculations then allow us to determine which initial conditions result in the coin landing head and tails respectively; that is, they allow us to determine for every quadruple ($x_{0,}$ v, ϕ_0, ω) whether the coin having this initial condition lands head or tails. We have assumed that all coin tosses start at height h and that all coins leave the hand in horizontal position: $x_0=h$ and $\phi_0=0$; hence different tosses vary in their vertical velocity v and their angular velocity ω. Assuming that the coin starts heads up, initial conditions lying in the black areas of the graph shown in Figure 1 come up heads, while those lying in white areas come up tails. For this reason Keller calls the hyperbolic black and white stripes in Figure 1 the 'pre-images' of head and tails.

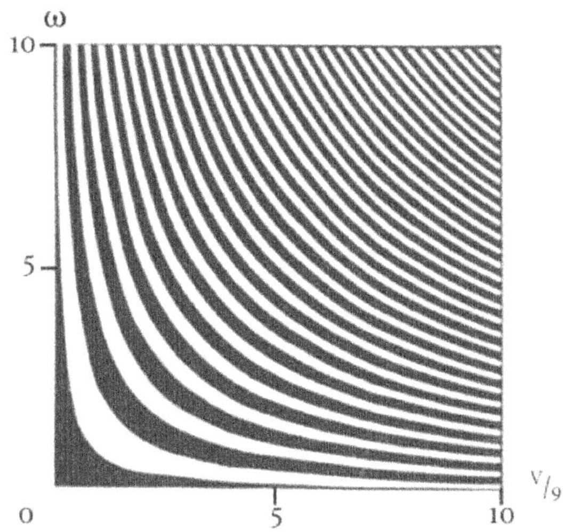

Figure 1. The pre-images of heads and tails (Diaconis 1998, p.803).

What follows from these considerations about the chance of getting heads? Keller presents an argument in two steps. The first is to regard the initial condition as a random variable with a continuous probability distribution $\rho(v,\omega)$ with support in the region shown in Figure 1 (i.e. $\omega \geq 0$ and $v \geq 0$). Then the probability for heads, $p(H)$, is given by

$$p(H) = \int_B \rho(v, \omega) dv d\omega \tag{3}$$

where B denotes the black regions in $\omega \geq 0$ and $v \geq 0$. Mutatis mutandis Equation (3) also gives the probability for tails, $p(T)$. The second step consists in showing that $p(H) = p(T) = 0.5$. To this end Keller proves a limiting theorem, basically showing that if the boundaries of the region over which the integral in Equation (3) is calculated is shifted towards infinity (i.e. if we integrate over $B' = \{(v, \omega): v \geq k, \omega \geq k\}$ and let k tend towards infinity), then $p(H) = 0.5$ *no matter what* distribution $\rho(v, \omega)$ we choose. This result becomes intuitively plausible when we realise that the stripes get thinner as the values of ω and v increase (see Figure 1), and so the integral becomes less sensitive to fluctuations in $\rho(v, \omega)$. Hence, in this limit there is a unique probability for heads.[6]

We now turn to a discussion of Humean chance and then return to the question of how to justify $p(H) = 0.5$ in Section 4. The main difference between our and Keller's approach is that we make essential use of facts about the Humean mosaic (i.e., the totality of all actual or occurrent events – see section 3.3) and thereby avoid appeal to a limiting result.

3. Humean Objective Chance

In this section we introduce the concept of Humean Objective Chance (HOC), on which our reconciliation of determinism and chance is based.[7] The views discussed here are an extension of those introduced in Hoefer (2007), but here presented in a way that pays particular attention to those features of the theory that bear on the issue of the compatibility of determinism and chance.

3.1 Defining Humean Objective Chance

The definition of chance that we present in this section differs from Lewis' canonical definition (1994, p. 480). In part this is a matter of presentation; but in part it also results from correcting certain omissions and modifying a few central

6 Diaconis *et al.* (2007) generalise this result by relaxing some of the above modelling assumption and thereby taking into account the precession of the coin. This adds interesting features to the model, but since the main features remain the same we keep using the simple model discussed in this section.

7 One might argue that 'objective chance' is a pleonasm since chances are objective by definition. True enough, but the phrase 'objective chance' has become customary in the literature and so we stick to it here.

features. Three changes are particularly crucial. First, we correct the omission of any reference to the Principal Principle (PP) in Lewis' definition. In our view PP is essential for an understanding of objective chance and therefore has to appear in one way or another in its definition. Second, our definition is of chances or chance laws alone, and is not a definition of laws of nature more generally. And finally, of course, our definition will allow for there to be genuine chances in a world that is deterministic at bottom. We return to these points in due course.

Let *e* be an event, for instance a coin coming up heads or a die landing so that it shows three spots.[8] We define chance as follows.

Definition 1 (HOC): The chance of event *e*, *ch*(*e*), is a real number in the interval [0, 1] such that:

(1) the function *ch* satisfies the axioms of probability,
(2) *ch*(*e*) is the correct plug-in for *X* in the Principal Principle, and
(3) the function *ch* supervenes on the Humean Mosaic in the right way.

Chances thus defined are Humean Objective Chances (HOC); for brevity we refer to them simply as 'chances'. We use 'THOC' to refer to the entire theory of chance presented in this section. The elements of this definition are in need of explication, and providing the needed explications is the task for this section. Let us briefly indicate what this task involves.

The first clause is straightforward, but nevertheless not entirely trivial. Lewis thought it a major problem to prove that objective chances satisfy the axioms of probability, and he argued at length that chances indeed have this property.[9] In our view there is nothing to prove here. THOC *defines* chance, and we are free to build into a definition whatever seems necessary. A function that does not satisfy the axioms of probability cannot be a chance function and so we simply require that *ch* satisfy the axioms of probability.

The second clause needs unpacking in two respects: we need to introduce PP, and we need to explicate what makes a plug-in for *X* a *correct* plug-in. Much hangs on this, and a careful exposition is imperative. For this reason we dedicate subsections to each point (Subsections 3.2 and 3.4).

The third clause is also problematic. We first have to introduce the Humean Mosaic, then say what we mean by a function supervening on the Humean Mosaic, and we then need to explicate the notion of supervening on the Humean Mosaic in the *right way*. The second clause of the definition enters here too, because an

8 Two disclaimers are in order. First, nothing in what follows depends on a more precise characterisation of events. Second, we attribute chances to events because this looks most natural in the cases we discuss. But nothing hangs on that; we could take propositions instead. In fact, as will become clear from the context, in certain formulae below letters such as *e* and *X* will stand for propositions describing events rather than directly for events. This is inconsequential for our views on chance.

9 For a discussion of Lewis' arguments see Hoefer (2007, pp. 560-62).

important part of what 'the right way' means here is: in such a way as to permit a solid argument justifying PP to be made. We turn to these issues in Subsection 3.3.

3.2 Introducing the Principal Principle

Chances, first and foremost, are guides to action. We look to chances when making decisions: if the chance for rain today is 0.95 I take my umbrella with me, but if it is 0.05 I do not. As Lewis insisted, the most central and important requirement on a theory of chance is that it make it possible to see how chances can play this action-guiding role. This aspect of chances is enshrined in PP, which establishes a connection between chances and the credences a rational agent should assign to certain events, where by 'credence' we mean an agent's subjective probability or degree of belief. The intuitive idea in PP is that a rational agent's credence for an event *e* to occur should be set equal to the chance of *e*, as long as the agent has no 'inadmissible' knowledge relating to *e*'s occurrence.

Definition 2 (Principal Principle): Let '*cr*' stand for a rational agent's credence. The Principal Principle (PP) is the rule that

$$cr(e|X\&K) = x, \tag{4}$$

where X is the proposition that the chance of e is x (i.e. X = '$ch(e) = x$'), and K is 'admissible' knowledge.

Before spelling out what we mean by admissible knowledge, let us add some clarifications about the purpose of *K*. At first sight it seems unclear why *K* should appear in Equation (4) at all, and more needs to be said the function that *K* is meant to be perform. The presence of *K* should not be interpreted as a request to gather a particular kind of knowledge before we can use PP. On the contrary, we always have knowledge about situations, and *K* simply stands for the sum of what we *de facto* happen to know. Depending on what kind of propositions *K* contains, we should or should not use Equation (4) to set our credences. The prescription is simple: if *K* contains no inadmissible knowledge then use Equation (4); if *K* does contain inadmissible knowledge then don't. In the latter case PP is silent about how to set our credences.

The question now is what counts as 'admissible' knowledge. Lewis' original characterisation is:

> Admissible propositions are the sort of information whose impact on credence about outcomes comes entirely by way of credence about the chances of those outcomes. (Lewis 1980, p. 92)

This characterisation has given rise to controversy. In fact, Lewis himself later regarded it as too imprecise and replaced it with a time-indexed version, in part in order to be able to say that all past events have chance 0 or 1. For a discussion of

Lewis' revised definition and the issue of time see Hoefer (2007, 553-5 and 558-60). We here build on this discussion and assume that these corrections are not only unnecessary, but also wrong. Chances attach to circumstances (the 'chance set-up') and not to worlds-at-specific-times. The original definition of admissibility Lewis gave was essentially right. Chance is a guide to action *when better information is not available*. So the essence of the requirement of admissibility is to exclude the agent's possession of other knowledge relevant to the occurrence of *e*, the kind of knowledge the possession of which might make it no longer sensible or desirable to set credence equal to objective chance. To use the usual (and silly) example: if you have a crystal ball that (you believe) reliably shows you future events, you may have inadmissible knowledge about a chance event such as the coin flip a week from now. If your crystal ball shows you the coin toss landing tails and you trust the ball's revelations, you would *not* be reasonable to set your credence in tails to 0.5 for that flip; you have inadmissible knowledge. This example helps make the notion of admissibility intuitively clear, and also points toward a very important fact in our world: inadmissible evidence is not something we typically have – if we did, then chances would be rather useless to have. Still, it is possible to give a slightly more precise definition of admissibility (Hoefer 2008, Ch. 2)

Definition 3 (Admissibility): A proposition *P* is admissible with respect to an outcome-specifying proposition *E* for chance set-up *S* (*E* says that event *e* occurs) iff *P* contains only the sort of information whose impact on reasonable credence about *E*, if any, comes entirely by way of impact on credence about the chances of those outcomes.

This definition makes clear that admissibility is relative to a chance set-up and its attendant possible outcome-events. It is also relative to the agent whose reasonable credence function is invoked in PP. The agent-relativity of admissibility may be more or less extreme, depending on how highly constrained a credence function must be in order to count as 'reasonable' or 'rational'. For our purposes agent-relativity is not germane, and we will assume that all reasonable agents agree about whether a proposition *P* should or should not have an impact on credence in *E*, when *P* is added to a further stock of background knowledge *K*.

3.3 Humean Supervenience

The *Humean Mosaic* (HM) is the collection of everything that actually happens; that is, all occurrent facts at all times. There is a question about what credentials something must have to be part of the mosaic. Nothing in what follows depends on how the details of this issue are resolved. What does matter is that irreducible modalities, powers, propensities, necessary connections and so forth are not part of HM. That is the 'Humean' in Humean supervenience.

The supervenience part requires that chances are entailed by the overall pattern of events and processes in HM; in other words, chances are entailed by what *actually* happens. We can make a comparison with actual frequentism, which satisfies Humean supervenience in a particularly simple way: the overall pattern of events uniquely determines the relative frequency of an event, and hence its probability. Actual frequentism has no frequency tolerance, and hence frequentist probabilities supervene on actual events. This contrasts with propensity theories, which have maximal frequency tolerance. THOC strikes a balances between these extremes by requiring that HOC's supervene on HM, but not *simply*: THOC postulates that chances are the numbers assigned to events by probability rules that are part of a *Best System* of such rules, where 'best' means that the system offers as good a combination as the actual events will allow of *simplicity, strength* and *fit*.

The idea of a Humean Best System of chances can be illustrated with a thought experiment. To this end, we introduce a fictitious creature, Lewis' Demon. In contrast to human beings who can only know a small part of the Humean mosaic, Lewis' Demon knows the entire mosaic. The demon now formulates all possible systems of probability rules concerning events in HM, i.e. rules assigning probabilities to event-types such as getting heads when tossing a coin. In the mere formulation of such rules, no interpretation of probability is assumed. The rules in these systems assign numbers to events. These numbers have to satisfy the axioms of probability – this is why they are 'probability rules' – but nothing over and above this is required at this stage. Then the demon is asked to choose the best among these systems, where the Best System (BS) is the one that strikes the best balance between simplicity, strength and fit. The probability rules of the system that comes out of this competition as the best system then, by definition, become 'chance rules', and the chance of an event *e* simply is the number that this chance rule assigns to it. That is, the chances for certain types of events to occur (given the instantiation of the setup conditions) simply are what probabilistic laws of the best system say they are.

Definition 4 (Humean *BS*-supervenience): A probability rule is Humean *BS*-supervenient on HM ('HBS-supervenes on HM', for short) iff it is part of the Best System, i.e. the system that strikes the best balance between simplicity, strength and fit given HM.

Clause (3) in Definition 1 can now be made precise: the function *ch* HBS-supervenes on HM.

Needless to say, much depends on how we understand simplicity, strength and fit. Before discussing these concepts in more detail, let us illustrate the leading idea of HBS-supervenience with an example. The question we have to ask is how certain aspects of event-patterns in HM may be captured by adding a chance rule about coin flips. Coins are fairly ubiquitous and we have the custom of flipping them to help us make choices. So the event-type we call 'a good flip of a fair coin' is widespread in HM around here. Furthermore, it is a fact, first, that in HM the relative frequency of each discernible side-type landing upward is very close to 0.5 and, second, that there are no easily discerned patterns to the flip outcomes (it is not the case, for instance, that a long sequence of outcomes consist of alternating heads and tails). THOC now asks us to consider all possible probability rules for a given class of events and then choose the one that strikes the best balance between simplicity, strength and fit. There are of course infinitely many rules. One, for instance has it that $p(H)=0.1$ and $p(T)=0.9$; another rule postulates that $p(H)=p(T)=0.5$; and yet another says that $p(H)$ is the actual frequency of heads and $p(T)$ is the actual frequency of tails. Given that the frequency of heads and tails is roughly 0.5, the first rule has bad fit; at any rate its fit is worse than the fit of the other two. But how do we adjudicate between the second and the third rule?

At this point considerations of strength come into play. In fact there may be an even better chance rule that could be part of the Best System, which would embrace coins and dice and tetrahedra and dodecahedra and other such symmetric, flippable/rollable solids. The rule would say that where such-and-such symmetry is to be found in a solid object of middling size with n possible faces that can land upward (or downward, thinking of tetrahedra), and when such objects are thrown/rolled, the chance of each distinct face being the one that lands up (or down) is exactly $1/n$. Given what we know about dice and tetrahedra and so forth, it is quite plausible that this rule belongs in the Best System; and it entails the coin-flip chances. So it enhances both simplicity and strength without much loss in fit, and hence on balance it is better than the system which sets chances equal to relative frequencies. Hence, the chance of heads on a fair flip of a coin would seem certainly to exist, and be 0.5, in a Best System for our world.

How are we to understand simplicity, strength and fit? Let us begin with simplicity. This is a notoriously difficult notion to define precisely, yet we think that there is enough one can say about it to make THOC tick. As we understand it, simplicity has two aspects, *simplicity in formulation* and *simplicity in derivation*. The former is what is usually meant when simplicity arguments are put forward: a linear relation between two variables is simpler than a polynomial of order 325, a homogenous first order differential equation is simpler than a non-linear integro-differential equation, etc. It is not easy to pin down what general rule drives these judgments, but this does not represent a serious obstacle to us because nothing in what follows depends on simplicity judgments of this kind. Another component of simplicity in formulation is how many distinct probability rules a system contains. *Ceteris paribus*, the fewer rules a system has in it, the simpler it is. The second aspect of simplicity, *simplicity in derivation*, focuses on the computational costs incurred in deriving a desired result. The question is: how many deductive steps do we have to make in order to derive the desired conclusions? Some systems allow for shorter derivations than others. It is important not to confuse simplicity in this sense with a subjective notion of a derivation being 'easy' or 'difficult'. The issue at stake here is the number of deductive steps needed to derive a conclusion, and this is a completely objective quantity, which could be quantified, for instance, by using a measure such as Kolmogorov's computational complexity (roughly, the length of the shortest programme capable of deriving the result).

Simplicity (in this latter sense) could always be improved by cutting perfectly good chance rules out of the system. However, in general improving simplicity in this way is not a good strategy because it comes at too high a cost in terms of strength. The strength of the system is measured by its scope. The wider the scope of the system, the stronger it is. In other words, the larger the part of HM that the system is able to account for, the better it fares in terms of strength. The above example illustrates the point: a system that covers only coins is weaker than a system that also covers other chance setups such as roulette wheels, dice, etc.

The Best System should not only ascribe chances to lots of event types, and do so in as simple a way as possible; it should ascribe the *right* chances! But which are the right chances? Every system assigns probabilities to possible courses of history, among them the actual course. With Lewis, we now postulate that the fit of the system is measured by the probability that it assigns to the *actual* course of history, i.e. by how likely it regards those things to happen that actually do happen. As an illustration, consider a Humean mosaic that consists of just ten outcomes of a coin flip: *HHTHTTHHTT*. It follows immediately that the first system above ($p(H)$=0.1 and $p(T)$=0.9) has worse fit than the second ($p(H)$=$p(T)$=0.5) since $0.1^5 0.9^5 < 0.5^{10}$. This example also shows that a system has better fit when it stays close to actual frequencies, as we would intuitively expect.[10]

So the ways in which we evaluate systems is objective and no appeal to 'pragmatic' or specifically 'human' values or limitations has been made. Nevertheless, we accept two (not very controversial) assumptions that assure that the Best System, whatever its concrete form, shares at least some essential characteristics with science as we, humans, know it. The first assumption is *ontological pluralism*, which denies that only basic/micro entities exist. Some hard-headed reductionists deny that anything except the basic micro entities exist. Thus, chairs, rivers, cats, trees, etc. are said not to exist. We deny this. That coins consist of atoms does not make coins unreal. Coins exist, no matter what micro physics tells us about their ultimate constitution, and so do rivers, chairs, and cats. Hence, even in a classical world, HM consists of much more than elementary particles and their trajectories.

The second assumption is *linguistic pluralism*, the posit that the language in which the Demon formulates the systems that subsequently enter into the simplicity-strength-fit competition contains terms for macroscopic kinds. That is, the language has not only the vocabulary of microphysics, but also contains terms like

10 Elga (2004) agues that this definition of fit runs into problems if there are infinitely many chancy events, and suggests a solution to based on the notion of a typical sequence. This concern is orthogonal to the problems we discuss in this paper and hence we will not pursue the matter further.

'coin' and 'river'. So we not only believe that macro objects exist, we also equip the demon with a language in which he can talk about these as *sui generis* entities.[11]

3.4 Justifying the Principal Principle

There is controversy not only over the correct formulation of PP, but also over its status. Strevens (1999) argues that it is no more possible to offer a sound argument justifying PP than it is to justify induction, and that we therefore have to accept it as something like a first principle. But not everybody shares this pessimism. In fact, we believe that the unique features of HOC permit a demonstration that it is irrational not to apply PP when its conditions are fulfilled. Space precludes a full discussion here, so we will simply present a brief version of the argument; for an in-depth discussion see (Hoefer 2008, Ch. 4).

As we have seen in the last subsection, it is a result of a careful analysis of what it means for the function *ch* to supervene on the Humean Mosaic in the right way that whenever there is a large number of instances of a chance setup, the chance of a certain outcome is close to the relative frequency of that outcome. For this reason, THOC can be understood as a (major) sophistication of finite frequentism, and understanding why PP is justified for HOC begins by recalling this affinity.[12]

There are two ways of justifying PP based on this affinity, an '*a priori*' and a 'consequentialist' one. The former is similar to the justification of PP Howson and Urbach (1993) give for von Mises frequentism. A subjective degree of belief corresponds (by definition) to the odds at which an agent feels a bet on either side of a question (*E* versus not-*E*) would be fair. An agent who has no inadmissible information pertinent to the outcomes of a *long series* of instances of chance setup *S* should have the same degree of belief in the *E*-outcome in *each* trial – having a *reason* to assign a higher or lower degree of belief to *E* on a specific trial automatically and by definition amounts to possessing inadmissible information. Hence, if an agent assigns degree of belief *p* to outcome *E* in a single trial of chance setup *S*, he should assign the same credence to an *E*-outcome in each instance of a(n indefinitely) *long series* of trials of *S;* not to do so is to take oneself to have information relevant to an *E*-outcome that does not come from *E*'s chance (which by stipulation is the same in each trial of *S*), and hence to have *inadmissible* information. Having inadmissible information makes PP inapplicable, so we may assume for

11 For further discussion of the issue of the language used in formulating laws see Lewis (1983).

12 Since clause 2 of our definition of HOC above stipulated that *ch(e)* is the correct plug-in for the Principal Principle, one might expect a quick and easy demonstration that HOC's satisfy PP: it is true by definition! Clearly, this is a bit *too* easy. The two justifications of PP for HOC offered in this section are substantial, departing from connections between HOC and frequencies of events, and are entirely non-circular.

the rest of the argument that the agent does not vary his credence in *E*-outcomes from trial to trial.

So the agent takes betting on *E* in each trial of an indefinitely long series at odds *p:(1-p)* to be fair. Assume that he bets on the same side in all trials in the sequence, i.e. either on *E* in all trials or not-*E* in all trials. Because the agent thinks the bet is fair, he must think that there is no advantage to betting on *E* rather than not-*E* (or vice versa); that is, he must be indifferent towards which side of the bet he takes. By assumption there is a chance for getting *E* on a trial, $ch(E)=q$. From the account of THOC above we know that the relative frequency of *E*'s in an indefinitely long sequence is of trials must be equal (or at least very close to) the chance of *E*. It is a simple result of probability calculus that if agents don't bet in accordance with relative frequencies, then one side of the bet is doing better. This cannot be if the agent believes the bet to be fair. It is then a simple arithmetic fact that if *q* differs non-trivially from *p*, and the agent bets on *E* at *p:(1-p)* odds throughout the long series, then the agent will certainly lose (or win) in the long run. The agent, understanding THOC and that $ch(E) = q$, must know all this too; but he cannot believe this and yet believe the long series of bets to be fair. So if $p \neq q$, the agent holds contradictory beliefs, which is irrational. So the only rational assignment of probabilities is $p = q$, as PP prescribes.

The consequentialist argument is more straightforward. It asks us, in the spirit of Humeanism, to look at HM, which not only contains outcomes of trials but also agents placing bets. If we look at all agents placing bets across the entire mosaic and check on how they are doing, we will see that those agents who set their credences equal to the chances obtain – at most places and times, at least – better results than those who adopt credences significantly different from the chances. In other words, in the wider domain just as in Las Vegas, if one has to gamble on chancy events, one does best if one knows the objective probabilities. For this reason it is rational to set one's credences to objective chances, as PP requires.[13]

3.5 The Epistemology of HOC's

Let us close this section with a brief remark about the epistemology of HOC's. At first sight, an approach to probability whose central concepts are defined in terms of everything that actually happens at any point in time and at any spatial location – the HM – and an omniscient creature – Lewis' Demon – may strike some as rather disconnected from actual human endeavours. This impression is mistaken. Needless to say, the appeal to HM and Lewis' demon are idealisations, and ones that take us rather far away from our actual epistemic situation. But this does not turn THOC into an epistemic pipe-dream. First, the limitations of actual human experience are a factor that every epistemology has to cope with. In particular, also those positions who believe in metaphysically 'thick' laws and probabilities

13 Caveats and details of the consequentialist argument are discussed in Hoefer (2007, sec. 5) and (2008, Ch. 4).

(universals, causal powers, etc.) have to base their views on the nature and character of these on actual experience and there is the possibility that future events may prove them wrong. Any view about probabilistic laws – Humean or not – has to base claims about these on our limited actual experience, and this involves an inductive leap. How to handle this leap is of course a time-honoured philosophical puzzle on which much ink has been spilled, and there is no royal road to success. The point to stress here is that the Humean is not alone with this problem. Second, the requirement that only occurrent properties be part of HM is in harmony with scientific practice as we know it, since occurrent properties are what science can observe. In this respect THOC is even closer to actual science than approaches that postulate modal entities that science can never observe. Third, the rules that are given to the Demon have an obvious 'human flavour': the omniscient Demon himself would probably not care about simplicity and strength since he knows everything anyway. These requirements are metatheoretical virtues humans value in science and hence what the Demon is asked to do is in the end 'human style' science as best as it can be done. Hence the Demon's activity is not different in kind from the endeavours of human scientists; the difference is that he can perform with perfection what we can do only inadequately.

4. Coin Flipping for Humeans

Let us now return to flipping coins. A striking feature of the discussion so far is the almost complete mismatch between how probabilities for the coin flip were treated in Sections 2 and 3 respectively. The treatment in Section 2 started with a deterministic mechanical model and sought to retrieve the 50/50 chance rule from mechanical laws plus a probability distribution over initial conditions. The approach taken in Section 3 did not mention mechanics at all and instead focussed the pattern of outcomes in HM. At least on the face of it these approaches have little in common and so the question arises whether they are compatible at all, and if so how.

In this section we argue that they are compatible, and, what is more, that they are actually complementary. In order to reach this conclusion we need to analyse the two accounts and their status *vis-a-vis* each other in greater detail. To facilitate the discussion, we set up a temporary debate between two viewpoints: 'mechanicism' *versus* 'macro-statistics', their proponents being 'mechanicists' and 'macro-statisticians'.

Mechanicists are likely to argue that their point of view is privileged since their account is based on fundamental laws: by assumption we live in a classical universe and so HM consists of trajectories of objects, among them the trajectories of coins, and classical mechanics is the fundamental theory of this universe.[14]

14 Those who also uphold micro-reductionism – the view that matter consist of atoms and

Chance rules, if there are any at all, have to be formulated in terms of the fundamental entities of HM and in the language of the fundamental theory describing them. Equation (3), supplemented with the specification of a particular distribution ρ fits the bill: it is a rule that assigns probabilities to getting heads when tossing a coin, and it does so solely in terms of basic mechanic properties. The rule is simple, has good fit, and since Keller (1986, pp. 194-6) shows that it can easily be generalised to other chance setups such as roulette wheels it also has strength. So we have good reasons to believe that within the class of all probability rules Equation (3) is the one that strikes the best balance between simplicity, strength and fit, and hence the probabilities it defines are chances in the sense of THOC.

The macro-statistician disagrees with this point of view for two reasons. The first objection is conceptual, the second technical. The conceptual objection takes issue with the mechanicist's reductionist outlook. Even if the world is classical at bottom and classical mechanics is the fundamental theory of the universe, it does not follow that everything that can be said about HM has to be said in the language of the fundamental theory. More specifically, the macro-statistician adopts a *methodological pluralism* (MP), the position that probability rules can be formulated in a macro language pertaining to a certain level of discourse, and that probabilities thus introduced are *sui generis* HOC's if the probability rules in question strike the best balance between simplicity, strength and fit relative to all other systems. To do this, they need not prove logical independence from micro-level chance rules; they need only win out in competition with alternate rules *formulated in the same language*, that of the macro-level. On this view, then, the $1/n$ rule for gambling devices is a *sui generis* chance rule because it strikes a better balance between the three basic metatheoretical virtues than any other probability rule formulated in the language of coins, wheels, throws, etc. (We come back to this principle at length below.)

The macro-statistician's technical objection to mechanicism turns on the status and mathematical form of the distribution $\rho(v, \omega)$ in Equation (3). At a general level the worry is that the mechanicist is 'cheating'. No probabilities can ever come out of a purely deterministic approach ('no probabilities in, no probabilities out'), and the mechanicist just puts them in by hand when he introduces $\rho(v, \omega)$, which is not warranted by (or even related to) any posit of mechanics. Therefore the introduction of $\rho(v, \omega)$ is an *ad hoc* manoeuvre, unmotivated from the point of view of mechanics. And, as is often the case with such manoeuvres, it may well raise more question than it answers. The first problem with $\rho(v, \omega)$ is that it is not clear what it is a distribution *for*. The most basic question we have to ask about every probability distribution is: what are these probabilities probabilities for? And it is not clear what the answer in the case of the coin is. We might take it

that the behaviour of macroscopic objects like coins eventually has to be explained in terms of the behaviour of its micro constituents – can replace the trajectory of a coin by the bundle of trajectories pertaining to the atoms making up the coin. Mutatis mutandis the arguments remain the same.

to be giving the probability of a coin flip's having initial conditions within a given range $v + dv$, $\omega + d\omega$. But nothing in mechanics can ground such a distribution.

The problem with the mathematical form of $\rho(v, \omega)$ is the following. Keller's limiting argument shows that the mathematical form of $\rho(v, \omega)$ is immaterial, and hence the question of what $\rho(v, \omega)$ to chose becomes obsolete. However, this limiting argument is of no relevance to *actual* coin tosses. Diaconis has shown in experiments that for typical coin tosses the initial upwards velocity v is about 5 mph and the frequency ω lies between 35 and 40 revolutions per second (1998, p.802). This is very far away from infinity! The problem is that once we revoke the infinite limit, it is no longer irrelevant what $\rho(v, \omega)$ one chooses. So which $\rho(v, \omega)$ is the right one to plug into Equation (3)? Intuitively one would choose a uniform distribution. For one it is simple; for another, it would give (roughly) a 0.5 probability for heads since, as becomes obvious from Figure 1, the black and the white stripes occupy approximately the same area. But nothing in the mechanical approach justifies this assumption.

Let us now step back, evaluate the arguments on either side, and explain how the two views eventually come together. Take the mechanicist's insistence on fundamental laws first. He will object to MP on the grounds that the chance for heads is not independent of the micro physics of the world. Surely, so the argument goes, there must be some dependence there! If the physics of our world was vastly different from what it is, then the chance for heads should be different too!

There is a grain of truth in this, but we must not be misled. The physics of our world might be vastly different, and yet (for *whatever* reason) the pattern of heads- and tails-outcomes in HM might be exactly the same; in which case, the chances would be the same. In most reasonably imaginable counterfactual scenarios, the physics will matter much less than the actual pattern of outcomes in HM. Macro level facts depend *ontologically* on micro-level facts – in the obvious compositional sense – but in our world they do not depend on them *constitutively* (i.e., the macro chance facts are not entailed directly or indirectly by fundamental physics; they depend on the pattern of macro events no matter what the micro physics is.). Once this is realised, the other problems have elegant solutions too. We can chop the Gordian knot in four cuts.

First, with the macro-statistician we affirm MP, from which it follows that chances for macro events like coin flips depend on the outcome pattern, not on the details of the underlying physics. (We justify MP below.)

Second, with the mechanicist we take ontological dependence seriously. The question is how to take this into account, and this is where a new element enters. We share with the mechanicists the view that Equation (3) – and similar equations – matter, but interpret them differently. This equation does not *give* us the chance for heads. We don't need to be given anything – we have the chance, and the chance is (constitutively) independent of the micro physics. Rather, we see Equation (3) both as a 'consistency check' and an explanation. Let us take these in turn. The different parts of a Best System have to be consistent with each other (which is not to say that one has to be derivable from the other). For this reason, whenever the setup conditions of a macro-level chance rule and a micro-level chance rule are the same (extensionally equivalent), then the chances they ascribe must agree or be very nearly in agreement. This, of course, does not rule out the possibility of minor adjustment. For example, assume we adopted the 50/50 rule for heads and tails. Now we know for sure that we get reductive relations right and we have the correct micro theory, and based on these we find 49/51. This is no real conflict because there is some flexibility about the macro chances and if there are very good overall reasons for making adjustments, then the Humean can make these. But there is a breaking point: if the micro theory predicts 80/20, we have to go back to the drawing board. The second element is explanation. We don't want to place too much emphasis on this, but there is the pervasive intuition that if a macro result can be derived from a more fundamental theory, there is explanation. Those who share this intuition – among them us – can see Equation (3) as providing an explanation. Those who don't can renounce explanatory goals and rest content with the consistency requirement.

The third cut is that the mechanicist has to admit that the introduction of $\rho(v, \omega)$ is a step beyond mechanics and as such $\rho(v, \omega)$ has to be justified elsewhere. But far from being a problem, this actually is an advantage. When thinking about $\rho(v, \omega)$ in the 'THOC way', we immediately have a natural interpretation of $\rho(v, \omega)$: it is the relative frequency of certain initial conditions. Of course all actual initial conditions are a collection of points in the $v-\omega$ plane, and not a continuous distribution. But arguably a continuous distribution is much simpler (in the sense of simplicity in formulation) than a huge collection of points, and so the Humean can argue convincingly that fitting a suitable continuous distribution through the points makes the system simpler and stronger. This distribution then is just an elegant summary of the actual initial conditions of all coin flips in HM.

Fourth, the common intuition that there is something epistemic about the chance of getting Heads on this flip – after all it has one and only one initial condition and given this initial condition it is determined whether it comes up heads or tails – is addressed by paying close attention to THOC's prescription about when to use chances to guide our credence. Information about the precise initial condition of a given coin flip is certainly inadmissible: such information logically implies the coin toss outcome and hence provides knowledge about the outcome of a toss that does not come by way of information about chances. The crucial point is that in typical situations in which we toss a coin, we just don't have inadmissible information, and that is why we use chances and PP to set our degrees of believe. So we use chances when we lack better knowledge.

Let us illustrate the admissibility point in some more detail. Consider the scenario described in Section 2, and an agent A who has only the usual sort of knowledge in his background K and who needs to decide how to bet on the coin flip. A should apply PP, clearly, and set his credences for heads and tails outcomes to 0.5. But now consider agent L, a Laplace-demon-in-training, who also must decide how to bet. *L* knows all that *A* knows, but – crucially – *L* also knows the exact micro-state of the world (or a big enough local region of it) just prior to the flip, and knows the laws of Newtonian mechanics. Should *L* set her credences for heads and tails outcomes equal to 0.5? Evidently not! She can calculate, on the basis of her background *K*, precisely what will happen. Let's assume she calculates that the coin will in fact land heads. *L* has inadmissible knowledge.[15] She has information relevant to whether the coin will land heads (*maximally* relevant!), and the information is *not* relevant by way of telling her about the objective chances. So *L* should not apply PP; and this is intuitively the right verdict. The conclusion is not that the objective chance of heads is 1. It is that (given the past state and the laws), the coin *will* land heads; and anyone who is aware of these facts should set their credence in heads to 1 (as the rules of subjective probability require), and not to 0.5.[16] The truth of deterministic laws entails that, given a complete-enough state of affairs at a moment of time (and perhaps boundary conditions), future events are fully determined. And this entails that *if you can get* such Laplace's-demon style information, and if you can actually calculate anything with it, then you may have better information with which to guide your credences about future events than the information HOC's give you. What is entailed, however, is not that objective chances do not exist, but rather that certain godlike beings may not have any use for them. We humans, alas, never have had nor will have either such information about initial conditions, or such demonic calculational abilities. For us, it is a good thing that objective chances exist, and that we can come to know (and use) them. With these points in mind, now we can see how determinism and non-trivial objective chances are compatible, and we also see that the admissibility clause in PP plays a crucial role in that.

We now turn to a defence of MP, which we merely stated above. Why should we subscribe to this principle? Why would a best system contain anything like chance-rules about coins and other macro objects? Let us distinguish two cases, a world in which physicalist reductionism about chance is true, and one in which it is false. Physicalist reductionism about chance is the claim that all chance-facts arise out of the laws of physics. Physicalist reductionism quite generally (not merely about chance) is popular in particular with elementary particle physicists; see for instance Weinberg (1994).

If reductionism of this kind is false, then it is obvious that the best system would contain rules about macro objects: these rules do not follow from basic laws of physics and therefore putting them into a system will greatly increase its

15 According to Lewis' official definition of admissibility, information about laws of nature and about past states of the world are fully admissible, hence *L* does not have inadmissible information. This adjudication makes it impossible for Lewis to retain non-trivial chances if the true laws of nature are deterministic. For a discussion of this point see Hoefer (2007, pp.553-555).

16 Formally, $cr(H|XK) = 1$ is required by the probability axioms, since $K \supset H$. We emphasize, it is *not correct* by contrast to say that $K \supset [ch(H) = 1.0]$.

strength. The more difficult case is if physicalist reductionism is true. If the rules about coins and wheels are but complicated applications of the laws of physics, why would we have such rules in our best system?[17] This seems to make the system less simple without adding strength. The reason to put them in nevertheless is what we above called simplicity in derivation: it is hugely costly to start from first principles every time you want to make a prediction about the behaviour of a roulette wheel. So the system becomes simpler in that sense if we write in rules about macro objects.

There is also a more intuitive argument why this independence of chances from micro physics is correct. It is the basic posit of Humeanism that the chance of a certain event HBS-supervene on the pattern of occurrence of events of the same kind in HM, and as such this chance is independent of how these events relate to other features of HM. In our concrete example this means that the chance of heads only depends on the pattern of heads in HM, or perhaps the pattern of outcomes in rolls/flips of n-sided solids with the appropriate symmetries and not on the relation that 'obtaining heads' bears to other parts of HM, in particular the basic mechanical properties of matter. As noted above, these sorts of patterns may obtain even in worlds with radically different micro-laws. Imagine a universe in which matter is a continuum and obeys something like the laws of Cartesian physics; imagine that coins exist in this universe and are tossed repeatedly. Despite the basic physics being very different, suppose it turns out that the overall pattern of outcomes of rolls/tosses of such n-sided objects in the continuum universe's HM is very similar to the pattern in our universe. What would the chance of heads be in the continuum universe? Clearly it would be given by the $1/n$ rule, since this is the best rule relative to that HM, irrespective of the micro-constitution of matter.

5. Envoy

As we have indicated in the introduction, this paper is about more than coins. In fact exactly the same considerations can be used to explain chance in statistical mechanics (SM). A full exposition of this theory is beyond the scope of this paper, but we would like to bring our discussion to a close by very briefly indicating how the insights gained with the example of the coin carry over to SM.[18] Consider

17 It may be hard to see how probability-facts could follow from fundamental physical laws, or laws plus initial conditions even, if the laws are fully deterministic. We do believe that 'no probabilities in, no probabilities out' holds here. But one might posit a fundamental-physics probability law as a supplement to the deterministic laws, precisely in order to allow derivation of probabilities for a variety of physical event types, including perhaps macro events. Loewer's version of Best System Humeanism does precisely this; see Loewer (2001).

18 For a detailed discussion of statistical mechanics see Uffink (2006) and Frigg (2008a).

a typical SM system, for instance a gas in container. The gas consists of about 10^{23} molecules. These molecules bounce around under the influence of the forces exerted onto them when they crash into the walls of the vessel and when they collide with each other. The motion of each molecule under these forces is governed by the laws of mechanics. Hence the gas is a large mechanical system: its state is fully specified by a point in its (6×10^{23}-dimensional) phase space – in this context referred to as its 'micro-state' – and its evolution over time is fully determined by the laws of mechanics.

At the same time the system is always in a certain macro-state, which is characterised by the values of macroscopic variables, in the case of a gas pressure, temperature, and volume. It is one of the fundamental posits of (Boltzmannian) SM that a system's macro-state supervenes on its micro-state, meaning that a change in the macro-state must be accompanied by a change in the micro-state. For instance, it is not possible to change the pressure of a system and at the same time keep its micro-state constant. Hence, to every given micro-state there corresponds exactly one macro-state. This determination relation, however, is not one-to-one. In fact many different micro-states can correspond to the same macro-state. We now group together all micro-states corresponding to the same macro-state, which yields a partitioning of the phase space into non-overlapping regions. We can then define an entropy function (the so-called Boltzmann entropy) that assigns a particular entropy value to every macro-state.

Systems characteristically start off in a low entropy state and then evolve into equilibrium, the macro-state with maximum entropy. The Second Law of thermodynamics tells us that this is what invariably must happen. One of the central aims of SM is to show that the Second Law – which is a purely macroscopic law – actually is a consequence of the mechanical motion of the molecules of the gas, and it does so by showing that the approach to equilibrium is overwhelmingly likely. And this is where we make contact with the coin example. In order to judge something as likely, trivially, we must introduce probabilities. SM does this by putting a uniform probability measure over the region of phase space which corresponds to the system's initial low entropy state, and then aims to show that micro conditions that lie on trajectories which eventually move towards equilibrium are overwhelmingly likely. The logic of this is like in the case of the coin, the only difference being that we sort initial conditions into ones that behave as the Second Laws requires and ones that don't, rather than into ones that yield heads and one that yield tails. Let us then mark the ones that behave as we expect white and the other ones black. We then put a measure over these all initial conditions of the same kind as ρ above. The difference just lies in the values: we now don't expect a 50/50 division between white and black, but rather something like 99.9999/0.00001 (omitting many 9s and 0s here for brevity). But the basic idea is the same: put a distribution over initial conditions and show that the outcome probabilities entailed fit well with the patterns in actual events. And indeed they do, not only the (essentially) exceptionless pattern of Second Law behaviour for macroscopic fluids, but also non-trivial probabilities for smaller collections of particles. So what we have learned from the coin also solves the problem of interpreting probabilities in SM! They can be elegantly accommodated in a Humean theory of objective chance.

References

David Albert, *Time and Chance*. Cambridge/MA and London: Harvard University Press 2000.

Persi Diaconis, "A Place for Philosophy? The Rise of Modeling in Statistical Science", in: *Quarterly of Applied Mathematics* 56, 4, 1998, pp. 797-805.

Persi Diaconis, Susan Holmes, and Richard Montgomery, "Dynamical Bias in the Coin Toss", in: *SIAM Review* 49, 2, 2007, pp. 211-235.

John Earman, *A Primer on Determinsim*. Dordrecht: Reidel 1986.

Adam Elga, "Infinitesimal Chances and the Laws of Nature", in: *Australasian Journal of Philosophy* 82, 2004, pp. 67-76.

Roman Frigg, "A Field Guide to Recent Work on the Foundations of Statistical Mechanics", in: Dean Rickles (Ed.), *The Ashgate Companion to Contemporary Philosophy of Physics*. London: Ashgate 2008a, pp. 99-196.

— "Chance in Boltzmannian Statistical Mechanics," *Philosophy of Science* (Supplement) 75, 2008b, pp. 670–681.

Sheldon Goldstein, "Boltzmann's Approach to Statistical Mechanics", in: Jean Bricmont, Detlef Dürr, Maria Carla Galavotti, Gian Carlo Ghirardi, Federico Petruccione and Nino Zanghì (Eds.), *Chance in Physics: Foundations and Perspectives*. Berlin and New York: Springer 2001, pp. 39-54.

Carl Hoefer, "The Third Way on Objctive Probability: A Sceptic's Guide to Objective Chance", in: *Mind* 116, 463, 2007, pp. 549-596.

— *Chance in the World*: Book Manuscript 2008.

Colin Howson and Peter Urbach, *Scientific Reasoning: The Bayesian Approach*. 2nd ed. Chicago: Open Court 1993.

Joseph B. Keller, "The Probability of Heads", in: *American Mathematical Monthly* 93, 3, 1986, pp. 191-197.

David Lewis, "A Subjectivist's Guide to Objective Chance", in: Richard C. Jeffrey (Ed.), *Studies in Inductive Logic and Probability. Vol. 2*. Berkeley: University of California Press 1980, reprinted in Lewis 1986, 83-132, with postscripts added.

— "New Work for a Theory of Universals", in: *Australasian Journal of Philosophy* 61, 4, 1983, pp. 343-377.

— "Humean Supervenience Debugged", in: *Mind* 103, 1994, pp.473-490.

Barry Loewer, "Determinism and Chance", in: *Studies in History and Philosophy of Modern Physics* 32, 2001, pp. 609-629.

— "David Lewis' Humean Theory of Objective Chance", in: *Philosophy of Science* 71, 2004, pp. 1115-1125.

Tim Maudlin, "What Could Be Objective About Probabilities?" in: *Studies in History and Philosophy of Modern Physics* 38, 2007, pp. 275-291.

Michael Strevens, "Objective Probability as a Guide to the World", in: *Philosophical Studies* 95, 1999, pp. 243-275.

Jos Uffink, "Compendium of the Foundations of Classical Statistical Physics", in: Jeremy Butterfield and John Earman (Eds.), *Philosophy of Physics*. Amsterdam: North Holland 2006, pp. 923-1047.

Steven Weinberg, *Dreams of a Final Theory: The Scientist's Search for the Ultimate Laws of Nature*. New York: Vintage 1994.

Roman Frigg
Department of Philosophy, Logic and Scientific Method
London School of Economics and Political Science
Houghton Street
London WC2A 2AE
England
r.p.frigg@lse.ac.uk

Carl Hoefer
Departament de Filosofia
Universitat Autònoma de Barcelona
Edifici B-7
08193 Bellaterra
Spain
carl.hoefer@uab.es

LÁSZLÓ E. SZABÓ

WHAT REMAINS OF PROBABILITY?

I

When I say "probability" I do not mean a Kolmogorovian 'probability measure' or similar *mathematical* notion. Of course, a definition – both coordinative[1] and logical – plays a constitutive role in the very concept to be defined. Nevertheless, there seems no reason to include into the *definition* of probability that it "satisfies the Kolmogorovian axioms of probability theory". For, once we know – supposedly from the rest part of the definition – what "probability" is in our world, it becomes a contingent fact of the world whether it satisfies the Kolmogorovian axioms or not; which can be known by *a posteriori* means. In other words, the aim of the so called "interpretations" of probability is not to find an interpretation of the Kolmogorovian axioms, but to give a sound meaning to scientific statements containing the term "probability".

For example, consider the following assertions of quantum mechanics or statistical mechanics:

$$p(a) = \operatorname{tr}\left(\hat{P}_a\hat{W}\right) \tag{1}$$

$$p\left(\{N_i\}_{i=1,2,\ldots}\right) = \frac{(\sum_i N_i)!}{\prod_i N_i!} \tag{2}$$

And compare them with other similar scientific assertions, like the Coulomb law,

$$\mathbf{E}(\mathbf{r}) = q\frac{\mathbf{r}-\mathbf{r}_q}{\left|\mathbf{r}-\mathbf{r}_q\right|^3} \tag{3}$$

or just a simple statement about the length of a rod:

$$l = 4\text{m} \tag{4}$$

In case (3) and (4) it is clear what the formulas assert. When we assert that the static electric field strength of a point charge is equal to $q\frac{\mathbf{r}-\mathbf{r}_q}{|\mathbf{r}-\mathbf{r}_q|^3}$, we have a previously *defined* physical quantity, electric field strength, and (3) expresses a contingent fact about this quantity.

By "definition" I mean *empirical/operational/verificationist* definition. This is not the place to argue for verificationism or operationalism. I only mention that my

1 H. Reichenbach, *The Theory of Relativity and A Priori Knowledge*. Berkeley and Los Angeles: University of California Press 1965.

F. Stadler (ed.), *The Present Situation in the Philosophy of Science*,
The Philosophy of Science in a European Perspective, DOI 10.1007/978-90-481-9115-4_26,

approach is based on a very weak operationalist/verificationist premise: scientific terms assigned to *quantities* like the ones appearing in (1)–(4) must have empirical definitions (except if the equations in question were definitions). In other words, those sentences of a scientific theory which are supposed to describe objective facts of the world must be expressible in observational/operational terms. Without this condition a scientific theory could not be empirically confirmable or disconfirmable. I believe, this view is widely accepted among physicists; although, the precise operational definition of a physical quantity can be a non-trivial issue, even in the case of basic spatiotemporal quantities.[2] However, in itself, this premise is not yet equivalent to operationalism or verificationism in general philosophical sense. It does not generally imply that a statement is necessarily meaningless if it is neither analytic nor empirically verifiable.

Now, contrary to (3) and (4), it is far from obvious what formulas (1) and (2) actually assert. What is the definition of the quantities on the left hand side of these formulas? What is the probability of an event? This is the basic question of the philosophy of probability. Strangely enough, in spite of the fact that the term "probability" is used in the everyday scientific discourse, there is a consensus in the philosophical literature[3, 4] that we have no satisfactory answer to this question.

II

The various approaches can be divided into two major groups. According to the *objectivist* school, the probability of an event is something which characterizes a feature of the *external* world; roughly speaking, it is a property of the event and the circumstances. According to the *subjectivist* approach, on the contrary, the probability of an event is something which characterizes a feature of the *internal* world; it is not a property of the event and the circumstances, but a property of a particular intentional state of mind about the event and the circumstances; a "degree" of belief. Objectivists' probability is often called "chance" (*ch*); subjectivists' probability is called "subjective probability" or "credence" (*cr*). Thus, it must be emphasized that chance and credence are not different interpretations of the same thing, but they are two different things, belonging to different types of

2 L. E. Szabó, "Empirical Foundation of Space and Time", in M. Suárez, M. Dorato and M. Rédei (eds.), *EPSA07: Launch of the European Philosophy of Science Association*. Dordrecht: Springer 2009.

3 J. Earman and W. Salmon, "The Confirmation of Scientific Hypotheses", in: M. H. Salmon, *et al.* (eds.), *Introduction to Philosophy of Science*. Englewood Cliffs, New Jersey: Prentice Hall 1992.

4 A. Hájek, "Interpretations of Probability", in E. N. Zalta (ed.), *The Stanford Encyclopedia of Philosophy* (Summer 2003 Edition), http://plato.stanford.edu/archives/sum2003/entries/probability-interpret.

phenomena. If they exist in our world, they do so independently; if they are at all connected, their connection must be a contingent fact of the world. Nevertheless, we still don't know exactly what chance is and what credence is.

III

No doubt, there is such a thing as a person's credence or belief; and, no doubt, it is meaningful to talk about the degree of a belief, as a belief can be stronger or weaker; and perhaps one can characterize it with a number between 0 and 1, just like a numeric scale from 0 to 10 can be asked to communicate the intensity of a patient's pain. And one can easily imagine some rules governing the complex mental processes that determine this number; a dynamics by which this number changes in time under various conditions.

All these things belong to the scope of ordinary empirical sciences like psychology, cognitive science, human ethology, or sociology. Strangely enough, however, in the subjectivist literature, we cannot find any reference to the results of these empirical sciences. It worth mentioning that the assertions in question are, at the same time, quite ambitious. Consider only two examples.

Lewis' Principal Principle asserts that a person's credence is strictly determined by some other mental states, namely:

$$cr(A \,|\, `ch(A) = x' \,\&\, K) = x \tag{5}$$

where '$ch(A) = x$' stands for the person's knowledge that $ch(A) = x$ and K stands for some further "admissible" (mis)information. If this is true, it is quite a strict causal/dynamical law of mental processes. Not to mention that those conditions that make proposition '$ch(A) = x$' *knowledge* (that is, *true*) and proposition K *admissible*, are conditions in the external world. So, in its stronger understanding, the principle is a statement about the relationship between the external world and one's mental states.

My second example is the Bayesian law of confirmation. It describes how the degree of a person's belief in the truth of A changes due to getting information about a new evidence E:

$$cr_{t_2}(A) = cr_{t_1}(A \,|\, E) = cr_{t_1}(A)\frac{cr_{t_1}(E \,|\, A)}{cr_{t_1}(E)} \tag{6}$$

where ch_{t_1} is the person's previous credence function based on some earlier body of believes, ch_{t_2} is the new credence function, based on the previous believes plus E. This, too, is a very strong claim about the dynamics of mental processes.

IV

How is it that the subjectivist interpretation of probability can claim so precise quantitative laws about human mind without any reference to the results of empirical sciences? There are two standard explanations of this ignorance, and both raise further problems.

The first possibility is that the whole subjectivist theory is regarded as a kind of "armchair metaphysics". Typically, in Lewis' "Subjectivist's guide" the Principal Principle is based on some "evidences" drawn from his own intuitive answers to his Questionnaire. He writes:

> I have given undefended answers to my four questions. I hope you found them obviously right, so that you will be willing to take them as evidence for what follows. If not, do please reconsider. If so, splendid – now read on.[5]

That is, we make *a priori* assertions about the real world, on the basis of our everyday pre-scientific and pre-philosophical intuitions, without any reference to the epistemic means by which the asserted facts of the world can be accessed.

However, if a statement is obviously *true* in our world, then the statement must have an obvious *meaning*; there must be an obvious way in which the statement can be *verified*, whether it is true indeed, or not. In other words, we need some empirical/coordinative definitions of the basic terms "credence" and "chance".

The second possibility is that the subjectivist theory is regarded not as a theory about real persons' believes, but about the "credences of an abstract agent". Terms like "chance", "credence" or "agent" are only mathematical terms without any reference to the real world; the whole theory is a formal/mathematical construction, just like Kolmogorov's axiomatic theory of probability, group theory, or geometry. In this mathematical construction, one can *define* a "gamble" in which the agent is making "bets" on "outcome events"; one can *define* the notion of "rational agent" as an agent whose bets and credences are in a certain relation. One may make assumptions about the "behavior of the gambling device"; for example one can assume that the "chances of the outcome events" satisfy the axioms of Kolmogorov's probability theory; and from these *premises*, for example, one can prove the Dutch book theorem.

This is, of course, possible, but I believe this is not the final aim of subjectivist interpretation of probability. At the end of the day we would like to apply these abstract constructions in the metaphysical, epistemological, and scientific discourse about the real world; therefore we need to know how to apply our theoretical terms like "credence" and "chance". We would need to understand the meaning of these terms even if the rules of subjectivist theory of probability should not be seen as a factual description of actual human reasoning, but rather as a "normative standard

5 D. Lewis, "A Subjectivist's Guide to Objective Chance", in: *Philosophical Papers Volume II*. New York: Oxford University Press 1987, p. 86.

of rationality"; since we need to understand the rules in order to follow them.

Thus, in either case, what we are missing is the empirical/operational/verificationist definitions of the basic concepts, first of all of "credence"; that is to say, a subjectivist's guide to objective chance would, first of all, require a guide to the subjectivist's credence.

V

Therefore, it could only lead to circularities if the definition of chance $ch(A)$ included any essential reference to '$ch(A) = x$' in the Principal Principle. The reason is that the alleged relation (5) holds only if proposition K is "admissible". "Admissibility" is, however, a concept the definition of which requires prior definitions of credence and chance, independently:

> A proposition P is admissible with respect to an outcome-specifying proposition E for chance set-up S (E says that event e occurs) iff P contains only the sort of information whose impact on reasonable *credence* about E, if any, comes entirely by way of impact on *credence* about the *chances* of those outcomes. [my italics][6]

VI

Let us return to the original problem of what "probability" means in the probabilistic assertions of the sciences. First, it is worth pointing out some conceptual confusion which needs to be sorted out. It is obvious that the concept of probability in science, especially in physics, is objective probability. When the behavior of a physical system is described by means of a probabilistic model, probabilities are supposed to describe some objective features of the external world; no matter if the underlying physical processes are deterministic or indeterministic. Classical statistical mechanics is a typical example. We believe that the detailed process is governed by the deterministic laws of classical mechanics; but, because of lacking the knowledge of the details, we give a less detailed probabilistic model. The probabilities in such a model are sometimes called "epistemic" probabilities. This is however a misleading terminology of the physicists, which differs from the terminology of the philosophers. For, the reason why we give a probabilistic description of the system, instead of a completely detailed deterministic one, is indeed related with a lack of knowledge, yet the probabilities in the model are objective probabilities and have nothing to do with "knowledge", "lack of information", etc. All statements of the probabilistic description of the system remain valid even if we get complete information about the details; because none of the objective features

6 R. Frigg and C. Hoefer, "Determinism and Chance from a Humean Perspective", in this volume.

of the system described by the probabilities in question changes by knowing more information about the system.

Let me give an everyday example. You are waiting for the next train in a subway station. If you knew the exact timetable, you could make predictions like "The next train will arrive in 3 minutes." If you don't know the timetable but only know that the trains come in every 5 minutes, you can make less ambitious claims. For example, you can say that "I will wait less than 5 minutes."; or you can predict the following result of a long-run experiment: "Providing that the moments at which I enter to the station are uniformly distributed in time, the long-run average of my waiting time is 2.5 minutes." Now, the validity of these claims does not change if you get know the timetable.

VII

What we observe here is nothing but a kind of Humean supervenience. Objective probabilities supervene on the collection of the particular facts of the actual history of the world, that is, all occurrent facts in all regions of spacetime; on the Humean Mosaic. And this is true, no matter if the world is deterministic or indeterministic; either in the sense that the different time slices of the actual history are not functionally related; either in the sense that there exist other possible histories of the world besides the actual one; either in the sense of a more sophisticated branching structure of possible spacetime-histories.

On the one hand this is trivially true; on the other hand, one has to put it in a more precise form: The truth or falsity of all *meaningful* statements about objective probabilities supervene on the Humean Mosaic, where "meaningful" is meant in a verificationist sense; that is, a statement is meaningful if it is expressible in terms of the Humean Mosaic.

Note, however, that this is also true for *subjective* probabilities: The truth or falsity of all *meaningful* statements about subjective probabilities supervene on the Humean Mosaic. The question is, of course: what are the meaningful statements about subjective probabilities? Another note: although the truth or falsity of all meaningful statements about both objective and subjective probabilities are determined by the actual content of Humean Mosaic, their truth or falsity can be known only by *a posteriori* means.

VIII

Nevertheless, what is chance and what is credence? We still do not have a tenable definition of probability, neither objective nor subjective. And how is it possible that physics and other empirical sciences do not notice problems arising from these unanswered fundamental questions? As to the concept of objective probability, in one of my earlier papers[7] I proposed a possible resolution which I call "No-probability Interpretation of Probability".

The key idea of my proposal is this: there is no such property of an event as its "probability". That is why the standard interpretations fail to give a sound definition of probability; and that is why empirical sciences like physics can manage without such a definition. Whenever we use the term "probability" in scientific discourse, its meaning varies from context to context: it means different dimensionless [0,1]-valued physical quantities, or more precisely, different dimensionless normalized measures composed by different physical quantities in the various specific situations. Moreover, these context-dependent meanings reduce the concept of "probability" to ordinary physical quantities of empirical meanings, like relative frequency on a finite sample, ratio of phase-space volumes, or the quantities on the right hand side of formulas (1)–(2).

Consider my example in point VI. One can calculate the average waiting time only from the fact that the trains come in every 5 minutes and that the moments of time when I enter to the station are uniformly distributed. These facts are ordinary, empirically verifiable, physical facts. The calculation requires only kinematics, without even mentioning "probabilities". If, however, someone would like to enforce a probabilistic language, the calculation of the average waiting time could be presented in the following way: I don't know when I entered to the station relative to the arrival of the train. Since all moments of time of my entering are of equal probability, I calculate with a uniform probability distribution … And the result will be the same. But, there appeared a term – probability – which has no definition. The statements containing this term, like "all moments of time of my entering to the station are of equal probability", are meaningless; it is impossible to verify whether they are true or not. As the example shows, however, the concept of "probability" is completely needless.

The research was partly supported by the OTKA Foundation, No. K 68043.

Department of Logic
Institute of Philosophy
Eötvös University
H-1518 Budapest, Pf. 32
Hungary
leszabo@phil.elte.hu

7 L. E. Szabó, "Objective probability-like things with and without objective indeterminism", *Studies in History and Philosophy of Modern Physics* 38, 2007, pp. 626-634.

Holger Lyre

Humean Perspectives on Structural Realism

Structural Realism (SR) is a moderate variant of scientific realism and can roughly be captured by the idea that we should be committed to the structural rather than object-like content of our best current scientific theories. A quick view on the list of some of the main proponents shows that SR is basically a European philosophy of science movement (and just suits our ESF Programme): John Worrall, Ioannis Votsis, Steven French, Angelo Cei, James Ladyman, Simon Saunders, Michael Esfeld, Vincent Lam, Katherine Brading, Mauro Dorato, Dean Rickles, Fred Muller, and – exceptions prove the rule – Anjan Chakravartty and John Stachel. The list is of course not exhaustive, moreover, the debate has a broad periphery. A notable example of this is Bas van Fraassen's structural empiricism.

The paper is a kind of opinionated review paper. In what follows I will pass through the most prevailing topics in recent debates over SR. My discussion will be organised, perhaps a bit unorthodoxly, in short sections, here and then I will outline my own views.

1 The notion of structure

The notion of structure is notoriously vague, and this is already one of the many problems of SR. The notion is of course not vague as far as the abstract mathematical concept of structure is concerned. Compare, for instance, Shapiro (2000):

> Define a system to be a collection of objects with certain relations among them. [...] Define a pattern or structure to be the abstract form of the system, highlighting the interrelationships among the objects, and ignoring any features of them that do not affect how they relate to other objects in the system.

The mathematical definition says that there are entities, the relata, that come equipped with a structure, but that the relata are determined by structural or relational properties only. Hence, a good working definition for SR is that structures are sets of objects, domains, with sets of relations imposed on them.

The problem is that despite the mathematical definition there exists no practical, straightforward method to extract the structural content from a given scientific theory. The problem is obvious as far as non-formalized theories in the higher special sciences are concerned, but it prevails even regarding fundamental physical theories. In this paper I do not delve into this problem, but I will mostly take the

F. Stadler (ed.), *The Present Situation in the Philosophy of Science*,
The Philosophy of Science in a European Perspective, DOI 10.1007/978-90-481-9115-4_27,

symmetry structure as the primary, genuine candidate to characterize the structural content of modern physical theories.

2 Two routes to Structural Realism

SR has a longstanding tradition in the 20th century and even earlier. There is consensus that the modern debate was initiated by John Worrall (1989). The discussion of the last two decades has actually taken two routes to SR, the Worrall-type and French-Ladyman-type route, as I prefer to call them. Worrall, Votsis (2003) and others gave arguments in favour of SR from the philosophy of science – for instance by arguing that SR's commitment to structure and not object-like content can be used as an antidote against prominent anti-realistic arguments like the pessimistic meta-induction or theory underdetermination. French-Ladyman-type authors, on the other hand, try to present arguments from the sciences directly, more precisely from the way modern science, notably physics, informs us about the ontology of the world. Meanwhile, all major fields of modern physics have been considered to strengthen arguments in favour of a structural ontology: Quantum Mechanics (French and Ladyman 2003a, Esfeld 2004), Quantum Field Theory (Cao 2003, Saunders 2003), General Relativity (Dorato 2000, Esfeld and Lam 2008, Stachel 2002), Gauge Theories (Lyre 2004a,b), Quantum Gravity (Rickles et al. 2006) or physics in general (Muller 1998, Redhead 2001). Note that the distinction between the two routes is not the same as the ESR/OSR distinction (see below). Cao (2003), for instance, proposes French-Ladyman-type ESR.

3 Ante rem versus in re structuralism

Debates on structuralism in mathematics show a similarity to structuralism in science, but must ultimately be separated from them. Shapiro (2000) is for instance known to uphold an *ante rem* structuralist position in the philosophy of mathematics, i.e. a Platonist conception of the existence of structures prior to and independent of their exemplification in the physical world. French and Ladyman (2003b) made it sufficiently clear that SR should not be confused with Platonism but is explicitly intended as a realism about structures not as abstract entities but as *in re* structures in the physical world.

4 Epistemic, Ontic and Semantic SR

As is well-known, James Ladyman (1998) first coined the distinction between Epistemic and Ontic SR. While ESR proponents believe in the structural content of theories as an epistemic constraint and, hence, uphold the view that objects may exist, but that our epistemic access is restricted to structures only, OSR proponents, according to Ladyman, take structure to be primitive and ontologically subsistent. I think, however, the distinction should be a bit more refined. In line with the usual threefold distinction between epistemic, ontic and semantic forms of scientific realism, we may accordingly distinguish between the following three options:

- Epistemic SR: science conveys true knowledge about structures,
- Semantic SR: the contents and terms of scientific theories refer to structures,
- Ontic SR: structures exist independently (from our epistemic and linguistic capacities).

5 Eliminative and non-eliminative SR

What's generally unfortunate with the above distinctions is the fact that everything still depends on our proper understanding of the term "structure". Given the mathematical definition of structure as sets of objects or relata with sets of relations imposed on them, there are, on the face of it, three possibilities:

- Epistemic SR: there are relations and relata, but that we have epistemic access to relations only,
- Non-eliminative OSR: there are relations and relata, but that there is nothing more to the relata than the relations in which they stand,
- Eliminative OSR: there are only relations and no relata.

Note that under this classification the widely debated question whether the slogan "structure is all there is" leads to the problematic position of "relations without relata" does not depend on Ladyman's ESR/OSR distinction, but rather on the distinction between non-eliminative versus eliminative versions of SR. It is perfectly possible to uphold SR as a metaphysical position about the world without being vulnerable to the "relations without relata"-problem. Well-known proponents of eliminative OSR are, or at least have initially been, Steven French and James Ladyman (French and Ladyman 2003a, French 2006, Ladyman and Ross 2007), a proponent of non-eliminative (or moderate) OSR is Michael Esfeld (2004).

6 Structurally derived intrinsic properties

I do actually believe that the above threefold distinction is still not exhaustive. General considerations about symmetry structures enforce us to assume the existence of not only relational but (in a certain sense) intrinsic properties of the relata. Technically speaking, a symmetry of a domain D is a set of one-to-one mappings of D onto itself (a.k.a. symmetry transformations), such that the structure of D is preserved. The symmetry transformations form a group and exemplify equivalence relations (i.e. a partitioning of D into equivalence classes). Naturally and necessarily, we always get certain invariants under a given symmetry. In a physical context, such invariants provide properties shared by all members of D. These properties are intrinsic properties in the sense that they belong to any member of D irrespectively of the existence of other object-like entities. On the other hand, the invariant properties do not suffice to individuate the members, since all members share the same invariant properties in a given domain. Structure invariants do not lead to individuals but to object classes only. This highlights the importance of the invariants: we use them to individuate domains, not individuals.

Now a crucial point: insofar as they are structural invariants, the intrinsic properties 'depend' (in a sense still to be determined) on the structure, we should accordingly and properly consider them as "structurally derived intrinsic properties". Nevertheless, they are intrinsic rather than relational, since they subsist irrespectively of the existence of other object-like entities.

7 Intermediate SR

We are thus left with an even more moderate non-eliminative version of SR which I shall dub "Intermediate SR" (cf. Lyre 2009). It is the view that there are relata and structurally derived properties, but that there is nothing more to the relata than the structurally derived properties, where the structurally derived properties comprise relational properties and invariants of structure as structurally derived intrinsic properties. Note further that this is still a viable SR position and does not collapse to old fashioned entity realism, since neither are we committed to essential properties nor are we committed to individuals (see below). Structurally derived properties do not individuate objects but object classes or domains of structure only.

8 An illustration: the lone electron

The following Gedankenexperiment provides an illustration of the particular nature of structurally derived intrinsic properties: Suppose a possible world with one electron only (and with relational spacetime). Does the lone electron possess an elementary charge? Under the classic view that intrinsic properties are properties an object has of itself and independently of the existence of other objects, the lone electron has certainly a charge. It seems, however, that for proponents of both eliminative and moderate OSR, who accept relational properties only, a lone electron cannot have a charge, since there are no other objects left in virtue of which the electron's charge might be considered as relational. From the point of view of Intermediate SR as another non-eliminative version including structurally derived intrinsic properties there is no problem to apply charges to lone objects. For even in the trivial case of only one member in D, the object will possess the said symmetry-invariant properties. The object has the invariance properties in virtue of the structure, the structure comes equipped with such properties. In more physical terms: even a lone electron is a proper instantiation of the *in re* U(1) gauge structure.

But couldn't we just say that the charge is relational to the structure? The problem is that in this case one cannot exclude the possibility that the structure as a relatum of the exemplification relation can exist for itself. Hence, one opens the door to unexemplified structures – a clear renunciation of *in re* structuralism and a dangerous flirt with Platonism. The idea here is that the structure we are talking about in the lone electron scenario is the U(1) structure displayed in the Maxwell equations and instantiated by that very electron. From an operationalist point of view, of course, such structure can only be observed from the behaviour of more than just one test charge. But structuralism is not *per se* committed to operationalism – both views should logically be kept disentangled.

9 A Further Argument: Gauge Invariants

The importance of structural invariants – structurally derived intrinsic properties – can most clearly be seen from the most important case of symmetry structure in modern physics, the case of gauge theoretic structures. One crucial feature of gauge symmetries is that they possess no real instantiations. Note that we must carefully distinguish between symmetries with real instantiations as opposed to symmetries without real instantiations. Examples of the former are for instance the possible space-time transformations of a physical object. Examples of the latter are scale transformations, coordinate transformations, and, in particular, gauge transformations. Therefore, a gauge theoretic characterization of a physical theory is *a fortiori* all and only a characterization by means of the symmetry invariants, since only the gauge symmetry invariants allow for a realistic interpretation. Gauge transformations possess no real instantiations (cf. Lyre 2004a,b). In the case of gauge theories, the SR commitment to structure can only be a commitment to the structure invariants. These invariants are given by the eigenvalues of the Casimir operators of the various gauge groups, which in their physical interpretation are considered to be mass, spin and the various charges. In fact, mass, spin, and charge provide paradigmatic cases of intrinsic properties of elementary particles. They are the attributes by which we classify the fundamental particle zoo. They are, in fact, the most fundamental structurally derived intrinsic properties.

10 Identity, haecceitism and metaphysical underdetermination

Another "structural attack" on traditional entity realism has to do with issues about identity and individuality in modern physics, notably quantum mechanics. The empirical indistinguishability of quantum objects has originally been regarded as a failure of Leibniz' principle of the identity of indiscernibles (PII). French (1989, 1998), however, argues that we are rather left with a kind of „metaphysical underdetermination": either quantum objects violate PII and are no individuals, or they are individuals since PII applies by reference to some kind of primitive thisness, bare particularity or haecceity (or however we may call it). The deeper lesson is that science leaves even the most profound metaphysical question about individuality underdetermined and so, following French, we better give up entity realism altogether and stick with a structural ontology. Obviously, this line of reasoning paves the way to eliminative OSR.

11 Weak discernibles

Saunders (2006) has argued that although fermions are not absolutely discernible (in terms of intrinsic monadic properties), they are nevertheless weakly discernible. Indeed, this observation can be seen as supporting structural non-eliminativism (and to give up haecceitism). To make this claim plausible consider first Black's case of two equal spheres in relational space with a distance *d*. Do such spheres violate PII? Call objects that violate PII absolutely discernible, but objects which allow for irreflexive relations weakly discernible (Quine 1976). Recall that a relation R is reflexive when for all x in the domain $R(x,x)$ holds. In the case of $\neg R(x,x)$, R is called irreflexive. For instance each Black sphere is a distance *d* apart from the other but not from itself. So the distance relation is irreflexive. The same holds in the case of fermionic particles in an entangled state for the relation of having opposite spin. Fred Muller (in print) has recently even extended this result to particles irrespective of their spin by considering the Heisenberg "commutation relation" of having complementary position and momentum. We may say that quantum objects are in fact generally weakly discernible due to the possibility of canonically conjugate variables based on the non-commutative algebra structure of quantum theory.

The case of weak discernibles accounts for the existence of relata that are weakly individuated by irreflexive relations. It runs counter to relata-eliminativism, but does at the same time not endorse full entity realism of absolute individuals. Indeed, irreflexive relations are structurally derived relations in the sense that they reflect the allowed quantum states of the non-commutative algebra structure. As in the case of structurally derived intrinsic properties, they are ontologically on a par with the structure without presupposing the independent existence of either the structure or the properties (or the relata). Rather, they are *in re* exemplifications of the structure.

12 The problem of unintended domains

Let's come to some more intricate problems of SR. Reconsider the idea of structure invariants as derived intrinsic properties. The crucial question is whether and how we will ever know about such properties as intrinsic natures of objects. Taken literally, the idea to individuate theories by means of their pure structural content (in the sense of pure mathematical structure) is far too weak. The reason lies in what one might call the "problem of unintended domains". There are in fact lots of cases where distinct physical theories show basically the same mathematical structure, hence we must qualify the structure's domain. Here are some physics examples of such "structural equivalents": (i) classical electrodynamics and hydrodynamics are based on more or less the same mathematical apparatus about unspecified 'currents' including continuity equations, theorems of Gauss and Stokes etc.; (ii) the gauge theories of strong and weak isospin are both based on SU(2); (iii) the group U(1) figures in quantum physics both as the group of temporal automorphisms and as the gauge group of QED.

Surely we've said that the domains are individuated by the structure invariants as derived intrinsic properties, but so far we did not spell out whether and how they provide an independent way to make contact with such invariant properties. Let's leave this open for the moment and discuss some further related issues first.

13 Structural underdetermination

We may exercabate the problem of unintended domains to the problem of structural underdetermination. According to the Worrall-type route to SR (as mentioned in section 2), SR can be seen as an antidote against theory underdetermination (TUD). The idea is that while TUD undermines entity content, SR seems to avoid this by not committing us to the theory's entity content but to structural content only. However, as I've argued elsewhere (Lyre, in print), there is, on the face of it, no way to make sure that the structural content of theories is not underdetermined either. On the contrary, there seem to exist cases in our best fundamental science, notably in theories of gravity, where we are directly confronted with cases of structural TUD. This means that we are confronted with structurally inequivalent but empirically equivalent theories. In such cases the structure of a theory is underdetermined by empirical evidence.

14 The Ramsey-Carnap-Lewis-account of theoretical terms

We may reiterate and generalize the two problems mentioned above. In order to do so we must reconsider the Ramsey-Carnap-Lewis-account of theoretical terms (cf. Lewis 1970). As a variant of scientific realism, SR is a realism about the unobservable. Take the classic distinction between observational and theoretical terms o_i and t_i. The Ramsey sentence of a theory T can be understood as a machinery for expressing the structural content of T. It is obtained by replacing the theoretical terms of T with bound variables: $\mathrm{T}(t_1, \ldots t_n, o_1, \ldots, o_m) \rightarrow \exists x_1, \ldots \exists x_n \, \mathrm{T}(x_1, \ldots x_n; o_1, \ldots o_m)$. Under such an account the theoretical terms are not eliminated but are expressed in terms of the structural relations between the variables x_i in T. The Ramsey sentence leaves us with a pure structural description of the theoretical knowledge about the world. The early Russell and Carnap took this as a motivation to uphold an extreme epistemic structuralism.

15 Multiple realizability, quidditism and Ramseyan Humility

Multiple realizability is in fact an immediate consequence of the Ramsey-Carnap-Lewis-account of theoretical terms. Our knowledge about the referents of the theoretical terms is just knowledge about the occupants or placeholders of descriptive causal roles. The quiddistic nature of the placeholders is indetermined, they are thus multiply realizable. A possible response is to advocate Ramseyan Humility about quiddities.

Recall that haecceitism is the view that a permutation of individuals (or tokens) makes a difference. It amounts to assume primitive thisness. We've already seen that SR, clearly in its non-eliminativist branch, dismisses haecceitism (section 10). Quidditism, on the other hand, is the view that a permutation of properties (or types) makes a difference. It amounts to assume primitive suchness. So structuralists usually reject haecceitism, but should they reject quidditism as well?

The problem not only for SR but in fact for any variant of scientific realism which commits itself to the Ramsey-Carnap-Lewis-account of theoretical terms is that quidditism amounts to making a difference without a difference. Nevertheless, David Lewis (2009) subscribes to quidditism, but at the same time advocates Ramseyan Humility, a term he has borrowed from Rae Langton's (1998) Kantian Humility. Kantian Humility, in turn, should capture Kant's view that things as we know them, phenomena, consist entirely of relations and that we have no knowledge of the intrinsic properties of things in themselves. So following Langton Kant's attitude is no idealism, but rather an epistemic humility. Accordingly, Ramseyan Humility is the view that "no amount of knowledge about what roles are occupied will tell us which properties occupy which roles" (Lewis 2009, p. 204).

A second answer to the problem of quidditism is that we might nevertheless be in contact with quiddistic natures, i.e. to advocate a more direct realism than suggested by the indirect causal and nomological knowledge provided by the Ramsey sentence (see also Schaffer 2005). And there might even be a third stance as regards quidditism, namely simply to dismiss it as an exaggerated metaphysics while at the same time claiming this to be a viable realist answer despite its apparent empiricist flavor. I will make no further attempt here to decide which way to go (in part also since, again, the problem is not special to SR but affects realism in toto).

16 The Newman problem

As is well-known, Max Newman (1928) raised a serious objection against Russell's (1927) early version of SR (see Demopoulos and Friedman (1989) for a modern resumption). The idea is that if abstract structure is all we can know from our theories about the unobservable world, then only cardinality questions are open to empirical discovery. As Newman (1928, 140) put it:

> … given any 'aggregate' A, a system of relations between its members can be found having any assigned structure compatible with the cardinal number of A.

And further:

> … the doctrine that only structure is known involves the doctrine that nothing can be known that is not logically deducible from the mere fact of existence, except ("theoretically") the number of constituting objects.

So structuralism is near-vacuous, in effect it collapses to empiricism. All we can know is just cardinality.

The point of the Newman problem is not only that relations do not suffice to pick out the intrinsic nature of the objects in the domain, but that also the nature of the relations themselves remains indetermined! According to the early Russell only *abstract* mathematical structure is known. But without further empirical qualification, any such abstract structure can be imposed on a given set (modulo cardinality constraints).

In a sense, the Newman problem is the inverse of multiple realizability. Whereas in the latter case we have multiple instantiations (collections of entities) that fit the structural description, Newman's problem amounts to saying that a given collection of entities can be endowed with any arbitrary structure, as long as the collection has the right cardinality. As van Fraassen (2008) has pointed out,

Newman's problem shows an interesting similarity to Putnam's model-theoretic problem, but we shall not delve into the details of disentangling them here.

17 Four problems revisited

We've discovered four problems in connection with SR: unintended domains (section 12), structural underdetermination (13), multirealization (15) and Newman's problem (16). They actually come in pairs. While the first pair has to do with the practical and vague notion of structure in physical theories (for instance the symmetry structure given by the symmetry groups in physics), the latter pair has to do with the precise logico-mathematical structure of a theory (cf. section 1). The difference between the two pairs is that the symmetry structure of T is most certainly not exhaustive, since the complete structure of T is almost certainly more extensive. By way of contrast, the logico-mathematical structure of the Ramsey sentence is exhaustive, insofar as the Ramsey sentence of a theory provides a complete description of T. Despite this distinction, problems 12 and 15 as well as 13 and 16 are more or less variations of the same theme – with 12 and 13 as special practical cases of the more generalized abstract cases 15 and 16. It is not at all implausible to assume that all four problems (or at least three, structural TUD is perhaps more special) are so strongly connected that they seek for a common answer. And basically, there are two routes from here, a Humean and an anti-Humean route, as I shall outline in the final sections.

18 Modal structures

Several SR proponents in recent debates have argued in favour of modal or causal structures (Chakravartty 2004, 2007; Esfeld (in print); Ladyman & Ross 2007). This means that structures are conceived as dispositional rather than categorical. The basic idea, notably in Chakravartty (2004), is to endow structures themselves with causal powers. Esfeld (in print) considers this an inevitable step in order to cope with the problem of quidditism (section 15) by assuming that the metaphysical causality behind the observable regularities has its root not in epistemically hidden quiddities but in the causal nature of the structures themselves. While Lewis believes that because of the Ramsey account of theoretical terms we have no epistemic access to quiddities (but to causal roles, i.e observable regularities only), the causal structure assumption dismisses quiddities altogether (and is, therefore, rather a dissolution to the problem).

Others even see causal structures as a possible way to overcome Newman's problem. Russell's early structuralism was about abstract structures, not about concrete *in re* structures. It was, in other words, about second and not first order relations. To overcome Newman's problem the structuralist must consider first order relations with causal powers as instantiations of abstract structures.

The causal structures strategy is perhaps a way out of the conundrum of problems 12 and 16 in particular. But, as usual, one has to pay a price. The strategy includes a double-step: first, to invoke first order relations and, second, to invoke causal powers. And the second step portrays a decisive non-Humean element, the allegedly modal or dispositional nature of structures. There are well-known difficulties connected with modal or dispositional ontologies, notably unclear identity conditions, which I shall not explore here. Rather, my project will be to outline the perspectives of SR from a strict Humean point of view.

19 A Humean response to Newman

Confronted with Newman's objection, Russell immediately realized that he must refine his position. In order to justify a particular, intended structure, we must somehow be directly acquainted with certain structural relations. Russell thus demanded "spatiotemporal copunctuality" between sense-data and physical objects as a basic relation. I cannot not discuss here whether Russell's proposal of spatiotemporal copunctuality is already the correct answer to the quest for basic relations, but I want to emphasize that his idea of knowledge about structures by acquaintance rather than mere description is, in principle, a viable solution to the notorious problems 12 and 16, perhaps even 15. It is, in fact, a solution which is also open for modern proponents of SR paving the way for a Humean conception of SR.

The essential clue is that we are not bound to relational properties only. For as we have already seen, SR must take structurally derived intrinsic properties into account (sections 6-9). We might therefore envisage direct observational acquaintance with structurally derived intrinsic properties. Whether and which placeholders of a structural description exist, i.e. whether and how a structure is instantiated, is an empirical question. And whether it is, for instance, electromagnetic or hydrodynamic current has to be distinguished on the level of observational phenomena and cannot be known from the pure theoretical and structural content alone (given the structural equivalence of the mathematical accounts). In our experimental observations we are "in contact" with the categorical, structurally derived intrinsic nature of the currents.

So the idea is basically this: Insofar as they are (structurally derived) intrinsic we need not invoke acquaintance with (causal) structures and insofar as they are categorical we need not invoke causal properties at all (be they structural or not). This paves the way for a Humean response. And finally, insofar as we assume "direct" acquaintance with them we rediscover Russell's option to circumvent Newman's problem. So we get a hybrid of a Humean and Russelian response to

Newman. Note, moreover, that weakly discernible relations are also perfectly categorical: they do not involve any quantum probabilities.

20 Humean perspectives on structural realism

A proper Humean perspective on SR is to demand categorical structures and to dismiss mysterious modalities (cf. Sparber 2009 for an account similar in spirit). Humean metaphysics, as usually construed, is based on at least three conditions:

1. a micro-physicalist supervenience base of fundamental intrinsic and categorical properties,
2. regularity (i.e. non-necessitarian) view about laws, and
3. reductionism about laws.

In an attempt to combine Humean metaphysics with SR, at least one of the three conditions must be changed. Let us consider them subsequently in the following sections.

21 Subvenient holistic structures

The first condition is best characterized in Lewis' famous conception of Humean supervenience, his view of "the world [as] a vast mosaic of local matters of particular fact" with "no difference without difference in the arrangement of qualities. All else supervenes on that" (Lewis 1986, ix-x). Meanwhile however, it is widely accepted that Humean supervenience is bound to fail. It fails according to modern science – according to the cases of quantum entanglement and gauge theoretic holism (cf. Healey 2007, chap. 4.5; Lyre 2004b; Maudlin 2007, chap. 2). Lewis even acknowledges the threat of quantum entanglement:

> maybe the lesson of Bell's theorem is exactly that there are physical entities which are unlocalized, and which might therefore make a difference between worlds … that match perfectly in their arrangements of local qualities. Maybe so. I'm ready to believe it. But I am not ready to take lessons in ontology from quantum physics as it now is. First I must see how it looks when it is purified of instrumentalist frivolity … and – most of all – … of supernatural tales about the power of the observant mind to make things jump. If, after all that, it still teaches nonlocality, I shall submit willingly to the best of authority.

But whether the quantum measurement problem has to do with frivolity or not – since the case of nonlocality can be made in gauge theories as well (a fact Lewis was obviously not aware of), it is time to realize that Humean supervenience must definitely be given up.

For proponents of Humean SR this is no bad news, since it is exactly this condition about the Lewisean Humean base which must be rejected. Instead of a mosaic of intrinsic, categorical properties, Humean SR considers *whole structures* in the supervenience base. This is a dismissal of naïve micro-physicalism, not about the categorical nature of such structures. Structures are holistic and global rather than local entities, physically exemplified and manifestly categorical. There is no need to assume causal structures, as we already saw in the discussion of Newman's problem and as we'll see now in the discussion of the second Humean condition.

22 Structural non-necessitarianism about laws

Humean SR is actually in accordance with the second condition from section 20. Structures are not arbitrary, but regular global sets of relations. Hence – and this is a quite important point – regularity, the crucial ingredient of laws, is already entailed by invoking structures. Structures are law-like. Take, for instance, the Minkowski spacetime structure of special relativity. It is a global geodesic structure exemplified by the trajectories of free falling bodies – a seemingly regular behaviour. Moreover, the behaviour of a free particle to follow geodesics is no disposition of the particle, nor is it a disposition of the geodesic structure, it is an exemplification of the manifest, categorical *in re* structure of spacetime. The same holds for other fundamental structures, for instance, the U(1)-structure of the world being exemplified by charge conservation.

Remarkably, such a structuralist regularity view about laws offers to avoid well-known problems of the orthodox regularity view. One problem is that not all regularities are law-like. Indeed, not all regularities are laws, only structures are. Under Humean SR, structures should be conceived as "world-built-in patterns" or global regularities. The holistic aspect of structures is crucial here: the particle following a geodesic is not a subsequence of disparate events which, without further explanation, show a regular behaviour. It is an exemplification of a global regularity itself – the geodesic structure.

There is, again, no reason to assume that there are "empty" laws. *In re* structuralism considers only exemplified structures. Such structures aren't necessarily exemplified at any (world) time, but they are at least globally exemplified on the whole spacetime extension. This is perhaps the most straightforward way to think of exemplification in Humean structural worlds: consider a world in which only one particle at an infinitesimally small time period has travelled a likewise infinitesimal spatial path. This particle is a proper instantiation of the full spacetime structure of that possible world.

Humean SR has furthermore the resources to explain the obvious universality of structure invariants without recourse to essentialism. Because of the holistic or global nature of structures, the structural invariants behave as universally valid. But such universality does not come equipped with mysterious necessity. It supervenes on the Humean base of structures. It is a mere regularity itself that some particular structure is instantiated. No necessities are involved here.

It follows from the same logic that Humean SR can account for exceptionless laws. Any instantiation of a structure will show the same regular behaviour encoded in the structure. Exceptions must not be expected, unless, however, the whole structure itself changes. This latter possibility can of course not be ruled out. After all, structures provide the Humean base, whether a particular structure subsists or not is a matter of pure regularity itself.

23 Non-realism about laws

The idea that structures provide the Humean base guarantees that Humean SR is in accordance with the third condition from section 20. Laws are reduced to structures, laws supervene on the structural Humean base. Some might think that SR is committed to a realism about laws because of the following argument: according to SR structures are real and laws are structures, so laws must obviously be real too. But, as we've seen, Humean SR just considers structures as global regularities and items of the Humean base. So again: whether a particular structure subsists or not is a matter of pure regularity itself. Laws aren't literally structures, and structures are only law-like in the sense that laws can be reduced to global regularities (which we call structures).

24 Transfer theory of causation

How should Humean structural realists finally construe causality? They might in fact welcome a transfer theory of causation (cf. Dowe 2000). The rough idea is that a causal process is the transmission of conserved quantities with causal interactions as intersections of such processes providing an exchange of the conserved quantities. According to fundamental physics, conserved quantities are identified with structural invariants. This is due to Noether's theorem which states that to every continuous symmetry generated by local actions there corresponds a conserved quantity. Such conserved quantities and, in turn, the causal processes and interactions are exemplifications of the fundamental structures. Structures come equipped with conserved quantities.

Some might complain that the transfer theory is non-Humean. But this is at best a problem for a micro-physicalist Humean base (according to condition 1). If we consider whole structures in the Humean base then causal processes and transfer of conserved quantities supervene on that base. And this is all the Humean needs.

25 Preliminary conclusion

Sections 18 to 24 present arguments against causal structures and provide perspectives for a Humean SR. There is no need to endow structures with causal powers. What's still missing in the picture is, perhaps, how dynamics comes into the world. We've basically outlined a static picture. And this is presumably the biggest neglect so far. Non-Humean SR with causal structures, however, doesn't solve this problem either. Metaphysical causation and physical dynamics are distinct topics, proponents of causal structures have no better grip on dynamics than opponents. Here's certainly much to be done in the future.

Admittedly, this paper was largely programmatic. We could merely touch upon some few motives and perspectives on Humean SR. But the perspectives are quite promising, perhaps promising enough to pursue them in more elaborated examinations.

Acknowledgements:
Many thanks to an anonymous referee for valuable suggestions.

References

Tian Yu Cao (2003). Structural realism and the interpretation of quantum field theory. *Synthese* 136: 3–24.

Angelo Cei and Steven French (2006). Looking for structure in all the wrong places: Ramsey sentences, multiple realizability, and structure. *Studies in History and Philosophy of Science* 37: 633–655.

Anjan Chakravartty (2004). Structuralism as a form of scientific realism. International Studies in the Philosophy of Science 18 (2 & 3): 151-171.

Anjan Chakravartty (2007). *A Metaphysics for Scientific Realism: Knowing the Unobservable.* New York: Cambridge University Press.

William Demopoulos and Michael Friedman (1985). Critical notice: Bertrand Russell's The Analysis of Matter: Its historical context and contemporary interest. *Philosophy of Science* 52: 621–639.

Mauro Dorato (2000). Substantivalism, relationism and structural spacetime realism. *Foundations of Physics* 30(10): 1605-28.

Phil Dowe (2000). *Physical Causation.* New York: Cambridge University Press

Michael Esfeld (2004). Quantum entanglement and a metaphysics of relations. Studies in *History and Philosophy of Modern Physics* 35(4): 601-617.

Michael Esfeld (in print). The modal nature of structures in ontic structural realism. *International Studies in the Philosophy of Science.*

Michael Esfeld and Vincent Lam (2008). Moderate structural realism about spacetime. *Synthese* 160: 27-46.

Bas van Fraassen (2008). Scientific *Representation: Paradoxes of Perspective*. New York: Oxford University Press.

Steven French (1989). Identity and individuality in classical and quantum physics. *Australasian Journal of Philosophy* 67: 432–446.

Steven French (1998). On the withering away of physical objects. In Elena Castellani (ed.): *Interpreting Bodies: Classical and Quantum Objects in Modern Physics*, pp. 93–113. Princeton: Princeton University Press.

Steven French and James Ladyman (2003a). Remodelling structural realism: Quantum physics and the metaphysics of structure. *Synthese* 136: 31–56.

Steven French and James Ladyman (2003b). Between platonism and phenomenalism: Reply to Cao. *Synthese* 136: 73–78.

Steven French (2006). Structure as a weapon of the realist. *Proceedings of the Aristotelian Society* 106: 167–85.

Richard Healey (2007). *Gauging What's Real: The Conceptual Foundations of Contemporary Gauge Theories*. New York: Oxford University Press.

James Ladyman and Don Ross (with David Spurrett and John Collier) (2007). *Every Thing Must Go: Metaphysics Naturalised*. Oxford: Oxford University Press.

Rae Langton (1998). *Kantian Humility: Our Ignorance of Things in Themselves*. Oxford: Oxford University Press.

David Lewis (1970). How to define theoretical terms. *Journal of Philosophy* 67: 427–446.

David Lewis (1986). *Philosophical Papers Vol. II*. Oxford: Oxford University Press.

David Lewis (2009). Ramseyan Humility. In David Braddon-Mitchell and Robert Nola (eds.): *Conceptual Analysis and Philosophical Naturalism*. Cambridge, MA: MIT Press.

Holger Lyre (2004a). *Lokale Symmetrien und Wirklichkeit*. Paderborn: Mentis.

Holger Lyre (2004b). Holism and structuralism in U(1) gauge theory. *Studies in History and Philosophy of Modern Physics* 35(4): 643-670.

Holger Lyre (2009). Structural realism and abductive-transcendental arguments. In Michel Bitbol, Pierre Kerszberg and Jean Petitot (eds.): *Constituting Objectivity. Transcendental Perspectives on Modern Physics*. Berlin: Springer.

Holger Lyre (in print). Is structural underdetermination possible? *Synthese*.

Tim Maudlin (2007): *The Metaphysics Within Physics*. New York: Oxford University Press.

Fred Muller (1998). *Structures for Everyone*. Amsterdam: Gerits & Son.

Fred Muller (in print). Withering away, weakly. *Synthese*.

Max Newman (1928). Mr. Russell's "Causal Theory of Perception". *Mind* 37(146): 137-148.

Willard V. O. Quine (1976). Grades of discriminability. Journal of Philosophy 73(5): 113-116.

Michael Redhead (2001). The intelligibility of the universe. In: Anthony O'Hear (ed.): *Philosophy at the New Millennium*. Cambridge: Cambridge University Press.

Dean Rickles, Steven French and Juha Saatsi (2006). *The Structural Foundations of Quantum Gravity*. Oxford: Oxford University Press.

Bertrand Russell (1927). *The Analysis of Matter*. London: George Allen & Unwin.

Simon Saunders (2003). Structural realism again. *Synthese* 136: 127–133.

Simon Saunders (2006). Are quantum particles objects? *Analysis* 66: 52–63.

Jonathan Schaffer (2005). Quiddistic knowledge. *Philosophical Studies* 123: 1–32.

Stewart Shapiro (2000). *Thinking about Mathematics*. New York: Oxford University Press.

Georg Sparber (2009). *Unorthodox Humeanism*. Frankfurt a.M.: Ontos.

John Stachel (2002). The relations between things versus the things between relations: The deeper meaning of the hole argument. In: David Malament (ed.): *Reading Natural Philosophy: Essays in the History and Philosophy of Science and Mathematics*. La Salle, IL: Open Court.

Ioannis Votsis (2003). Is structure not enough? *Philosophy of* Science 70: 879–890.

John Worrall (1989). Structural realism: The best of both worlds? *Dialectica* 43: 99–124.

Philosophy Department
University of Bielefeld
Germany
lyre@uni-bielefeld.de

F. A. Muller

The Characterisation of Structure: Definition versus Axiomatisation

Abstract

Crucial to structural realism is the Central Claim that *entity* B *is or has structure* $\mathfrak{S}$. We argue that neither the set-theoretical nor the category-theoretical conceptions of structure clarify the Claim in a way that serves the needs of structural realism. One of these needs is to have a viable account of reference, which almost any variety of realism needs. There is also a view of structure that can adopt *both* set-theoretical and category-theoretical conceptions of structure; this is the view that adopts B.C. van Fraassen's extension of Nelson Goodman's concept of *representation-as* from art to science. Yet the ensuing fountain of perspectives is a move away from realism, structural realism included. We then suggest that a new theory of structure is needed, one that takes the word 'structure' to express a primitive fundamental concept; the concept of structure should be axiomatised rather than defined in terms of *other* concepts. We sketch how such a theory can clarify the Central Claim in a manner that serves a descriptivist account of reference, and thereby structural realism.

1. Preamble

After having discarded a number of characterisations of *scientific realism*, i.e. realism in the philosophy of science, in his classic *The Scientific Image*, Van Fraassen provided the following minimal characterisation:

> *Science aims to give us, in its theories, a literally true story of what the world is like; and acceptance of a scientific theory involves the belief that it is true.* This is the correct statement of scientific realism.[1]

The first and foremost distinction in the variety of scientific realism called *Structural Realism* (StrR[2]) is the one between *epistemic* StrR (all that science provides is knowledge of the structure of the physical world) and *ontic* StrR (all there

1 Fraassen [1980], p. 8.

2 We shall ambiguously use abbreviation 'StrR' also for 'a structural realist'.

F. Stadler (ed.), *The Present Situation in the Philosophy of Science*,
The Philosophy of Science in a European Perspective, DOI 10.1007/978-90-481-9115-4_28,

is in the physical world is structure).[3] Yet when we take heed of the fact that knowledge implies truth, and take truth to imply ontological adequacy, the gap between epistemic and ontic StrR narrows quickly.[4] Whether epistemic or ontic, StrR needs the concept of structure. For StR, a literally true story of what the world is like will involve *structures* only, and therefore we need to know what *structures* are.

What makes StrR stand apart from other varieties of realism is that it is supposed to be more cautious, more modest, in its realist claims, in order not to fall prey to the pessimistic meta-induction over the history of science. The premise of this inductive argument is a sequence of past scientific theories that now have all been rejected. They are plausibly false. So if you think that our *currently* accepted scientific theories are not false but true, then what are you, stupid? But StrR must remain sufficiently substantive to provide a basis for the no-miracle argument: the only explanation for the fact that in sequences of successively accepted theories each theory is empirically and technologically at least as, and generically more, successful than its predecessor is that they latch on to the structure of the physical world better and better.

We take a closer look at the set-theoretical (Section 2) and the category-theoretical (Section 3) conceptions of structure and we find them inadequate to serve the needs of StrR, specifically the need to have a literal description of the referents to which terms in scientific theories refer. Then we explore the possibility of retaining both conceptions of structure by adopting Van Fraassen's concept of *representation-as* – as opposed to *representation-of* –, which can be marshalled to evade the objections leveled against set-theoretical and category-theoretical conceptions of structure when interpreted literally. The price to pay seems however too high for realism, because this adoption introduces a perspective-dependency that stands opposed to the very idea of realism (Section 4). Since by then all available options seem exhausted, we argue for the case that StrR needs a new theory of structure, that takes the concept of structure as fundamental, that is, as a primitive concept that ought to be axiomatised rather than defined (Section 5). Such a theory will serve the needs of StrR, or so we argue.

Throughout this paper we take 'structure' to mean *mathematical structure*, because in science these structures are used in science to model, to represent, to describe, to explain, to understand, etc. the world.

2. Set-Theoretical Characterisation

Although the rigorous set-theoretical characterisation of structure is well-known and widespread, easy to explain and easy to illustrate, its rigorous definition in the formal language $\mathcal{L}_\in$ of pure set-theory (ZFC, say) is rather commanding. Bourbaki was the first to provide a definition of structure of extreme generality, within the

3 Due to J. Ladyman [1998]. H. Lyre's [2009] review in this volume and Ladyman's encyclopedia article [2009] draw more distinctions within StrR.

4 As I have argued elsewhere; see Muller [2009], Section 1.

framework of his own set-theory, in Chapter IV of his *Theory of Sets* (1949).[5] We call the sets in the domain of discourse $\mathbf{V}$ of ZFC – which harbours pure sets and nothing but pure sets – answering to this definition *set-structures*. Bourbaki propounded the view that *mathematics is the study of structure* and set out, by means of set-structures, to create Law & Order in the exuberant proliferation and progressive splintering of 20th-century mathematics, which had turned the discipline into a Tower of Babel. Patrick Suppes famously came to promulgate the use of set-structures in the philosophy of science, notably to characterise scientific theories. Suppes' Slogan: to axiomatise a theory is to define a set-theoretical predicate.[6]

Informally, a set-structure is a polytuple of the following form:

$$\left\langle \textit{base sets, subset families, relations, functions, operations, constants} \right\rangle , \qquad (1)$$

or more precisely:

$$\left\langle B_1,\ \ldots,\ B_b,\ \mathfrak{F}_1,\ \ldots,\ \mathfrak{F}_s,\ R_1,\ \ldots,\ R_r,\ F_1,\ \ldots, F_f,\ O_1,\ \ldots,\ O_o,\ C \right\rangle , \qquad (2)$$

where $b \in \mathbb{N}^+$ (positive natural number) and $r, s, f, o \in \mathbb{N}$ (natural numbers). Set-structure (2) has: b base sets; s subset families, each one of any of the base sets; r relations between the members of base sets or those of the subset families; f functions each of whose domain and co-domain is one of the base sets or subset families, or is some Cartesian product-set of these sets; o operations each of which has as a domain in a Cartesian product-set of generally one of the base sets and the same base set as its co-domain; and a set C (of Constants) which contains members of the base sets or subset families that play a special rôle (as zero, one, top, bottom, , singularity, etc.). Set C is however often omitted for the sake of brevity.

Let us consider an example from physics, that we shall take as the leading example in this paper: a Helium atom (He) in a uniform magnetic field ($\mathbf{B}_0 : \mathbb{R}^3 \to \mathbb{R}^3$, $\langle x, y, z\rangle \mapsto \langle 0, 0, B_0\rangle$). A quantum-mechanical structure used to describe this composite physical system (which we shall henceforth abbreviate by HeB) is of the following type:

$$\mathfrak{S}(\mathsf{HeB}) \equiv \left\langle L^2(\mathbb{R}^3),\ H(\mathbf{B}_0),\ \psi,\ \mathrm{Pr}_t \right\rangle , \qquad (3)$$

where the four occupants are as follows.[7] The base set is the Hilbert-space of square-integrable complex functions of three real variables ('complex wave func-

5 Bourbaki [1968]. For an attempt at an accessible exposition of Bourbaki's definition, as well as a brief description of Bourbaki's programme, see Muller [1998], pp. 106–115; for a more smooth and accessible definition of structure, see Da Costa & Chuaqui [1988], who speak of 'Suppes-predicates', in honor of Patrick Suppes [1960], [1967].

6 See Suppes [1960], [2002]; see Da Costa & French [2000] for a review of developments in this area over the past 30 years. Lyre [2009] also employs set-structures.

7 The *occupants* of a polytuple like (3) are the four items in there: $L^2(\mathbb{R}^3)$ is the *first occupant*, etc. They are not the *members* of the set (3), but iterated members. See Muller [1998], p. 24, for a rigorous definition of occupant.

tions').[8] Linear function $H(\mathbf{B}_0) : \mathcal{D} \rightarrow L^2(\mathbb{R}^3)$ is the Hamiltonian, the operator that represents the physical magnitude of energy; its domain $\mathcal{D}$ lies dense in $L^2(\mathbb{R}^3)$. (The magnetic field $\mathbf{B}_0$ and other relevant physical magnitudes, such as the linear momentum of the He-atom, are present in $H(\mathbf{B}_0)$ but suppressed notation-wise.) Function

$$\psi : \mathbb{R} \rightarrow L^2(\mathbb{R}^3),\ t \mapsto \psi(t) \tag{4}$$

is the solution of the Schrödinger-equation; it is continuous iff no measurements are performed. Finally, function $\Pr_t : \Delta \mapsto \Pr_t(\Delta)$ is the Born probability measure, one for every $t \in \mathbb{R}$, that gives the probability of finding a value in $\Delta \subset \mathbb{R}$ for the energy of the He-atom when measured and when the state of the He-atom is $\psi(t)$:

$$\Pr_t(\Delta) = \langle \psi(t)|P^H(\Delta)|\psi(t)\rangle\ , \tag{5}$$

where $P^H(\Delta)$ is the relevant member of the spectral family of projectors of $H(\mathbf{B}_0)$. Mathematically there is more going on, which we have suppressed in the deceptively simple notation (3).

First of all, Hilbert-space $L^2(\mathbb{R}^3)$ is itself a structure:

$$\big\langle\, L^2(\mathbb{R}^3),\ +,\ \cdot,\ \langle\cdot|\cdot\rangle,\ \|\cdot\|,\ \mathit{0}\,\big\rangle\ , \tag{6}$$

where, first,

$$+ : L^2(\mathbb{R}^3) \times L^2(\mathbb{R}^3) \rightarrow L^2(\mathbb{R}^3) \tag{7}$$

is the operation of addition on the complex wave functions, leading to an Abelian additive group; secondly,

$$\cdot : \mathbb{C} \times L^2(\mathbb{R}^3) \rightarrow L^2(\mathbb{R}^3) \tag{8}$$

is the scalar multiplication of wave functions, which interacts distributively with addition, leading to a complex vector space; thirdly, mapping

$$\langle\cdot|\cdot\rangle : L^2(\mathbb{R}^3) \times L^2(\mathbb{R}^3) \rightarrow \mathbb{C} \tag{9}$$

is the inner-product; fourthly,

$$\|\cdot\| : L^2(\mathbb{R}^3) \rightarrow \mathbb{R}^+ \tag{10}$$

is the norm, generated by the inner-product, leading both to metrical and topological structure; and sixthly, *0* is the zero-function, the neutral element of the additive group. In turn, the reals ($\mathbb{R}$) also form some algebraic structure:

$$\big\langle\, \mathbb{R},\ <,\ \ +,\ \times,\ \{0,1\}\,\big\rangle\ , \tag{11}$$

8 More rigorously one has to identify members of $L^2(\mathbb{R}^3)$ which are equal almost everywhere, thus giving rise to a set of Lebesgue-equivalence classes of complex wave functions, denoted as $\mathcal{L}^2(\mathbb{R}^3)$.

and the complex numbers ($\mathbb{C}$) too. The natural numbers ($\mathbb{N}$) are always needed and they also form a particular structure:

$$\left\langle \mathbb{N},\ S,\ 0 \right\rangle , \tag{12}$$

where $S : \mathbb{N} \longrightarrow \mathbb{N}$ is the successor-function (all arithmetical operations can be defined inductively in terms of S).

Structure $\mathfrak{S}(\mathsf{HeB})$ (3) also harbours a Kolmogorovian probability structure:

$$\left\langle B(\mathbb{R}),\ [0,1],\ \mathrm{Pr}_t \right\rangle , \tag{13}$$

where the probability function $\mathrm{Pr}_t : B(\mathbb{R}) \longrightarrow [0,1]$ (5) is a normed measure on the Borel sets $B(\mathbb{R})$, which in turn is also a structure, a Boolean σ-lattice:

$$\left\langle B(\mathbb{R}),\ \subseteq,\ \cup,\ \cap,\ \backslash,\ \varnothing \right\rangle . \tag{14}$$

Thus the wave-mechanical structure $\mathfrak{S}(\mathsf{HeB})$ is in full splendour (permuting the order of the occupants):

$$\Big\langle \mathbb{N},\ S,\ 0;\ \ \mathbb{R},\ <,\ \ +,\ \times,\ \{0,1\};\ \ \mathbb{C},\ +,\ \times,\ \{0,1\};$$
$$L^2(\mathbb{R}^3),\ +,\ \cdot,\ \langle\cdot|\cdot\rangle,\ \|\cdot\|,\ \mathit{0}\ ;\ H(\mathbf{B}_0);\ \ \psi;\ \ B(\mathbb{R}),\ \subseteq,\ \cup,\ \cap,\ \backslash,\ \varnothing,\ [0,1],\ \mathrm{Pr}\Big\rangle \tag{15}$$

Since operations are a particular kind of functions and functions are a particular kind of relations, and relations between members of two arbitrary sets, D and R say, are subsets of their Cartesian product-set $D \times R$, and thus members of the power-set of $D \times R$, and since the Cartesian product-set $D \times R$ is a member of the 3-times iterated power-set of the union-set $D \cup R$:

$$D \times R\ \in\ \wp^3(D \cup R)\ , \tag{16}$$

one sees that starting from the infinite number sets $\mathbb{N}$, $\mathbb{R}$ and $\mathbb{C}$, the structure $\mathfrak{S}(\mathsf{HeB})$ (3) lives at a level in the cumulative hierarchy of sets $\mathbf{V}$ that is a considerable number of applications of the power-set operation higher than were $\mathbb{N}$, $\mathbb{R}$ and $\mathbb{C}$ live. We call to mind Cantor's Power Theorem, according to which the power-set $\wp(D)$ is strictly larger in cardinality than set D, to see that structure $\mathfrak{S}(\mathsf{HeB})$ (15) harbours various sets much larger than the cardinality of the continuum ($\mathbb{R}$).

The standard route for set-theoreticians is to take the finite von Neumann ordinals as the natural numbers ($\mathbb{N} \equiv \omega$); then there is a unique set of natural numbers. The structuralist route (Bourbaki's) is to define a 'natural number structure' by means of a set-theoretical structure-predicate (a Suppes-predicate), as a 'Peano structure' (12), or as a 'Dedekind structure', or as a 'Frege structure'; in all these cases there is no longer a unique 'natural number structure' but an absolute infinity of such structures (as many as there are sets in the domain of discourse $\mathbf{V}$ of

ZFC).[9] The same two routes are available for the other number structures (integers, $\mathbb{Z}$; rationals, $\mathbb{Q}$; reals, $\mathbb{R}$; complex numbers, $\mathbb{C}$): they can be constructed in **V** by set-theoretical means from $\mathbb{N} = \omega$ so as to end up with unique number structures (rationals as ordered pairs of integers, reals as Bolzano-Cauchy sequences of rationals or as Dedekind-cuts, complex numbers as ordered pairs of reals); or they can be defined by structure-predicates (see footnote 9). When one follows the first, constructive-like route, then

$$\mathbb{Z} \in \wp^4(\mathbb{N}), \quad \mathbb{Q} \in \wp^7(\mathbb{N}), \quad \mathbb{R} \in \wp^8(\mathbb{N}), \quad \mathbb{C} \in \wp^{10}(\mathbb{N}) \,. \tag{17}$$

Then for the set of wave functions from $\mathfrak{S}(\mathsf{HeB})$ (3) we have

$$L(\mathbb{R}^3) \;\in\; \wp^3\big(\wp^3(\wp^8\mathbb{N} \cup \wp^9\mathbb{N}) \;\cup\; \wp^9\mathbb{N}\big) \,, \tag{18}$$

and for the Hamiltonian:

$$H(\mathbf{B}_0) \;\in\; \wp^6\big(\wp^3(\wp^8\mathbb{N} \cup \wp^9\mathbb{N}) \;\cup\; \wp^9\mathbb{N}\big) \,, \tag{19}$$

and the wave function:

$$\psi \in \; \wp^3\Big(\wp^9\mathbb{N} \;\cup\; \wp^3\big(\wp^3(\wp^8\mathbb{N} \cup \wp^9\mathbb{N}) \;\cup\; \wp^9\mathbb{N}\big)\Big) \,, \tag{20}$$

and the probability measure:

$$\Pr_t \;\in\; \wp^3\big(\wp^8\mathbb{N} \cup \wp^9\mathbb{N}\big) \,. \tag{21}$$

For the ordered quadruple $\mathfrak{S}(\mathsf{HeB})$ (3) we then obtain:

$$\mathfrak{S}(\mathsf{HeB}) \;\in\; \wp^3\Big(L^2(\mathbb{R}^3) \;\cup\; \wp^3\big(H \;\cup\; \wp^3(\psi \cup \Pr_t)\big)\Big) \,. \tag{22}$$

But properly construed, as the ordered 28-tuple (15), structure $\mathfrak{S}(\mathsf{HeB})$ is a member of a far more involved set-structure. With (18), (19), (20), (21) and (22), one can work out exactly how many iterations of the power-set we are, with $\mathfrak{S}(\mathsf{HeB})$, beyond the first infinite ordinal level (ω) in the cumulative hierarchy, which we leave as an exercise for the willing readers. Presently it will become clear why we have bothered to point this all out.

Now, what does StrR claim with regard to structure $\mathfrak{S}(\mathsf{HeB})$ (3)? When we follow Patrick Suppes[10] in considering the class of structures like $\mathfrak{S}(\mathsf{HeB})$, and similar ones (with other physical magnitudes, mixed states, etc.), to constitute the theory of quantum mechanics (QM), then it trivially follows that all that QM tells us about physical reality, actually even all that QM *can* tell us about physical reality, such as about element of physical reality HeB, is that this physical system is or has structure $\mathfrak{S}(\mathsf{HeB})$. Thus John Worrall [1989] is right when he says that all science

9 See Muller [1998], pp. 56–64, where this is all spelled out.

10 Suppes [1960], [1967], [2002].

provides us with is knowledge of the structure(s) of the world, rather than of the nature(s) of the world. Epistemic StrR seems inevitable.

When knowledge implies truth, and truth implies ontological adequacy, then *knowing that* structure $\mathfrak{S}(\mathsf{HeB})$ *is* the structure HeB implies that $\mathfrak{S}(\mathsf{HeB})$ truly *is* the structure of HeB. Ontic StrR is just around the corner! Perhaps we should limit our claims to so-called *ontological substructures* of $\mathfrak{S}(\mathsf{HeB})$, but ontic StrR remains just around the corner.[11]

We get around the corner when we assume in addition that science tells us, or eventually will tell us, everything there is to tell about the physical world in general, and about He-atoms in uniform magnetic fields in particular (scientific optimism).[12] Nothing will be left unsaid. Since the physical world is built from atoms and according to ontic StrR they are structures, the physical world is composed of structures. *Ad fundum* structures determine everything there is in the physical world.

The conclusion seems to be that Suppes' structuralist view *on scientific theories* conjoined with a realist attitude yields epistemic StrR and optimistically also ontic StrR. As Worrall [2009] has recently put it: "Structural Realism is the only game in town." End of story?

Not yet. For what does it mean exactly to say that HeB *is or has structure* $\mathfrak{S}(\mathsf{HeB})$? This is an instance of the

Central Claim of StrR. *Being* B *is or has structure* $\mathfrak{S}$ *(a* being *is anything, any entity, that exists), independently of us, human beings, of our activities, attitudes and capacities, of our very existence.* (23)

This Central Claim stands in need of clarification, as will emerge below.

The 'is' obviously cannot mean the identity-relation, because $\mathfrak{S}(\mathsf{HeB})$ (22) is an *abstract mathematical entity*, to wit a complicated set-theoretical construction out of the empty set, living in the cumulative hierarchy of all and only pure sets, in the domain of discourse **V** of ZFC, while HeB is a *concrete physical entity*, 'out there' in the physical world. Certainly a He-atom in a uniform magnetic field (HeB) it is not *a set*.

Perhaps, then, 'is' means predication, as does 'has'. HeB *has* a structure, a very specific structure, namely $\mathfrak{S}(\mathsf{HeB})$, just as a tomato *has* a colour, a very specific colour, namely red. Let us see where this leads us.

We express properties in our language by means of predicates. The property red – if there are 'properties' – is expressed by the predicate 'red', and the ascription of the property red to a tomato is expressed by saying that 'This tomato is red' is true, or that this tomato falls under the predicate 'red'. The obvious candidate for the predicate that ascribes the wave-mechanical structure to HeB is the

11 The idea of considering *ontological substructures* of structures for realist claims was suggested more than ten years ago, and is closely related to M.L.G. Redhead's idea of 'surplus structure'. See in Muller [1998], p. 356 ff., and Redhead [1975], p. 88.

12 See Muller [2009], Section 1.

set-theoretical one that defines structure $\mathfrak{S}(\mathsf{HeB})$ (3), call it $\boldsymbol{\xi}(\cdot)$. The general form of this predicate is:[13]

$$\begin{aligned} \boldsymbol{\xi}\big(\mathfrak{S}(\mathsf{HeB})\big) \ \text{ iff } \ & \exists X_1,\ \exists X_2,\ \exists X_3,\ \exists X_4 : \\ & \mathfrak{S}(\mathsf{HeB}) = \big\langle X_1,\ X_2,\ X_3,\ X_4 \big\rangle \ \wedge \\ & X_1 = L^2(\mathbb{R}^3) \ \wedge \ X_2 = H(\mathbf{B}_0) \ \wedge \\ & X_3 = \psi \ \wedge \ X_4 = \mathrm{Pr}_t\,. \end{aligned} \tag{24}$$

This will not do either, because $\boldsymbol{\xi}(\cdot)$ (24) is an open sentence in the language $\mathcal{L}_\in$ of ZFC and thus only applies to inhabitants of $\mathbf{V}$. Our HeB does not inhabit $\mathbf{V}$ and therefore can never fall under $\boldsymbol{\xi}(\cdot)$: formally speaking, '$\boldsymbol{\xi}(\mathsf{HeB})$' is nonsense. Now what?

The way to go without leaving set-theory seems to enrich $\mathbf{V}$ with *physical systems*.[14] This makes $\mathcal{L}_\in$ a two-sorted language, with set-variables and *physical-system-variables*, say $\boldsymbol{a}$ and $\boldsymbol{b}$ for the new sort. Physical systems can be collected in sets, so that '$\boldsymbol{a} \in X$' etc. become well-formed atomic sentences of the enriched language, call it $\mathcal{L}_\in^*$. Expressions '$X \in \boldsymbol{a}$', '$X = \boldsymbol{a}$', '$\boldsymbol{a} \in \boldsymbol{b}$', '$X \notin \boldsymbol{a}$' etc. are forbidden in $\mathcal{L}_\in^*$ because the physical systems are not supposed to be entities that can have members, *they are not sets*. The axioms of ZFC have to be reformulated in the enriched language $\mathcal{L}_\in^*$, and one axiom has to be added declaring the existence of physical systems, but that's all. Thus one obtains ZFCU. No additional axioms are present in ZFCU to govern the physical systems. (It is possible to enrich ZFC with mereological axioms that govern the physical systems, by taking the *subsystem-relation* as a primitive dyadic predicate in the language additional to the membership-predicate; theory thus obtained is a conservative extension over ZFC and therefore consistent relative to ZFC, and therefore to ZF.[15] We shall not do this here; we have done it already somewhere else (see previous footnote).)

Our HeB will now hopefully become a value of the fresh variables, because, as we all know, to be is to be the value of a variable. The conclusion that HeB is *a set* will then have been avoided. But still, structure-predicate $\boldsymbol{\xi}(\cdot)$ (24) is such that only a particular kind of polytuple, hence *a set*, falls under it, namely a polytuple of the form $\mathfrak{S}(\mathsf{HeB})$ (3). Formally, from (24) we see immediately that '$\boldsymbol{\xi}(\boldsymbol{a})$' is nonsense because '$\boldsymbol{a} = \langle ., ., ., . \rangle$' is nonsense. Therefore we have to adjust $\boldsymbol{\xi}$ (24) of $\mathcal{L}_\in$ to some other predicate of $\mathcal{L}_\in^*$, say $\varphi(\cdot)$, such that '$\varphi(\boldsymbol{a})$' makes sense and structure $\mathfrak{S}(\mathsf{HeB})$ is somehow in there – as it must, because that is what QM provides. This can be achieved in two steps.

13 Notice that the right-hand-sides of the identity-statements in the definiens (24) are assumed to be antecedently defined singular terms in the language of ZFC; this is done for brevity, more standard is to write 'X_1 is a Hilbert-space'.

14 The technical term for *objects that are not sets* is *primordial elements*, or *Ur-elements*, from the German *Urelemente*. See Fraenkel [1973], pp. 23–25.

15 See Muller [1998], pp. 189–252, for details and proofs.

The first step is to let $\boldsymbol{a}$ occupy the structure polytuple:

$$\begin{aligned} \boldsymbol{\xi}^*\big(\boldsymbol{a}, \mathfrak{S}^*(\mathsf{HeB})\big) \text{ iff } \; & \exists X_1, \exists X_2, \exists X_3, \exists X_4 : \\ & \mathfrak{S}^*(\mathsf{HeB}) = \big\langle \boldsymbol{a}, X_1, X_2, X_3, X_4 \big\rangle \wedge \\ & X_1 = L^2(\mathbb{R}^3) \wedge X_2 = H(\mathbf{B}_0) \wedge \\ & X_3 = \psi \wedge X_4 = \mathrm{Pr}_t . \end{aligned} \tag{25}$$

The dyadic predicate $\boldsymbol{\xi}^*$ expresses a relation between structure $\mathfrak{S}^*(\mathsf{HeB})$ and physical system $\boldsymbol{a}$. Since $\boldsymbol{\xi}^*$ relates a physical system, which we hope to identify with concrete physical object HeB, to an abstract object, structure $\mathfrak{S}^*(\mathsf{HeB})$, which is at the end of the day still a set, just like $\mathfrak{S}(\mathsf{HeB})$, this is not quite what StrR needs. The second step is to turn $\boldsymbol{\xi}^*$ (25) into a monadic predicate of $\boldsymbol{a}$ by existentially quantifying $\mathfrak{S}^*(\mathsf{HeB})$ away:

$$\varphi(\boldsymbol{a}) \text{ iff } \exists \mathfrak{S}^*(\mathsf{HeB}) : \boldsymbol{\xi}^*\big(\boldsymbol{a}, \mathfrak{S}^*(\mathsf{HeB})\big) . \tag{26}$$

Formally, we seem to be going in the right direction. For let us compare things again to red tomatoes. Suppose there is a tomato on the plate in front of us. The sentence ‘Red(this-tomato)’ is true and the expression ‘this-tomato’ trivially *refers to* the tomato on the plate in front of us. Similarly we want to say that ‘$\varphi(\boldsymbol{a})$’ is true and that ‘$\boldsymbol{a}$’ refers to a He-atom in a uniform magnetic field. But ‘$\boldsymbol{a}$’ is a variable and variables do not refer. What we want to say instead is that $\boldsymbol{a}$ is a He-atom in a uniform magnetic field iff $\varphi(\boldsymbol{a})$, because $\varphi(\cdot)$ (26) is the set-theoretical translation of the characterisation of HeB that QM provides.[16] The extension of $\varphi(\cdot)$ then includes all and only actual (and perhaps possible) He-atoms in a uniform magnetic field. The symbol ‘HeB’ can then be officially inaugurated as a variable running over this extension, a so-called *Helium-atom-in-a-uniform-magnetic-field-$\mathbf{B}_0$-variable*. If the officially inaugurated variable ‘HeB’ assumes a value from this extension, we can say that He-atoms in a uniform magnetic field exist, or that the variable *plurally refers* to those physical systems; or if we can locate by laser cooling techniques a single He-atom in the laboratory, and give it a name, we can say that this name *singularly refers* to the atom, just as in the case of the red tomato on the plate in front of us.

This story has to be grounded in some account of *reference*. For such unobservable physical systems as He-atoms in magnetic fields, the only viable account of reference is a *descriptivist* one.[17] The relevant description here is the description of our HeB. Science, by means of (our set-theoretically reconstructed) QM,

16 He-atoms are usually characterised by their constitutive parts (a nucleus consisting of two protons and two neutrons, and two electrons) and their mass, charge and spin. Usually this can be read of ‘read off’ the Hamiltonian $H(\mathbf{B}_0)$ and therefore is included but is, unlike the uniform magnetic field $\mathbf{B}_0$, notation-wise suppressed.

17 For why the only available alternative, the Kripke-Putnam causal theory of reference, fails to provide a general account of reference for science, see Gauker [2006], pp. 130–132.

delivers this description: $\varphi(\boldsymbol{a})$ (26). Let us next take a closer look at this description.

Description $\varphi(\boldsymbol{a})$ (26) literally says: there is some particular polytuple, i.e. set-structure $\mathfrak{S}^*(\mathsf{HeB})$, that has $\boldsymbol{a}$ as its first occupant (25). One can easily prove that if $\mathfrak{S}(\mathsf{HeB})$ (3) exists in $\mathbf{V}$, then $\mathfrak{S}^*(\mathsf{HeB})$ (25) exists in $\mathbf{V}^*$, *for every* $\boldsymbol{a}$ *indiscriminately*:

$$\mathsf{ZFCU} \vdash \exists \mathfrak{S} : \boldsymbol{\xi}(\mathfrak{S}) \longrightarrow \forall \boldsymbol{a} : \varphi(\boldsymbol{a}) . \tag{27}$$

Since the antecedent can also be proved, so can the consequent:

$$\mathsf{ZFCU} \vdash \forall \boldsymbol{a} : \varphi(\boldsymbol{a}) . \tag{28}$$

Recall that the idea was to obtain a description – based on QM – such that those $\boldsymbol{a}$ falling under the description can be said to be HeB. But if $\varphi(\cdot)$ (26) is that description, then as a consequence every single physical system qualifies as a HeB (28). Which is absurd.[18] The situation is actually worse than absurd, because this all generalises. For *every* set-structure $\mathfrak{A}$ in $\mathbf{V}$, one can easily prove there are as many structures as there are physical systems in that for every physical system $\boldsymbol{b}$, there is a structure $\mathfrak{A}_{\boldsymbol{b}}$ that has $\boldsymbol{b}$ as its first occupant and that shares all its occupants with $\mathfrak{A}$. Thus every physical system is everything. The descriptions are therefore void. Not only is the putative description $\varphi(\cdot)$ (26) void, in spite of appearances to the contrary, but every other description, based on any other structure $\mathfrak{A}$, rather than $\mathfrak{S}(\mathsf{HeB})$ or $\mathfrak{S}^*(\mathsf{HeB})$, will also be void. No descriptivist account of reference can take off in the context of ZFCU.

Now we are done. Our provisional conclusion is that the set-theoretical road to physical reality for StrR seems a road to nowhere. Realism without reference, then? Hmmm. Smells like realism without reality. Before realists get *that* desperate, they should explore all other options. One option to clarify the Central Claim of StrR (23) is to replace set-theory with category-theory.

3. Category-Theoretical Characterisation

When it comes to deal with structures, in particular in abstract branches of mathematics – abstract in comparison to number theory, analysis and the geometry of figures, curves and planes –, such as algebraic topology, homology and homotopy theory, universal algebra, and what have you, a vast majority of mathematicians considers Category-Theory (CT) vastly superior to set-theory. CT also is the only rival to ZFC in providing a general theory of mathematical structure and in founding the whole of mathematics. The language of CT is two-sorted: it contains *object-variables* and *arrow-variables*. An arrow sends objects to objects; an

18 When we identify the 'objects' that Brading & Landry [2006], p. 572, take to be 'presented' by a structure as Ur-elements, then theorem (27) also makes trouble for them: everything can be 'presented' by every structure, so all them 'present' everything, or conversely, every structure can 'present' anything.

identity-arrow sends an object to itself. Simply put, *structures are categories*, and a *category* is something that has objects and arrows, such that the arrows can be composed so as to form a *composition monoid*, which means that: (i) every object has an identity-arrow, and (ii) arrow-composition is associative. The languages of CT ($\mathcal{L}_{\uparrow}$) and ZFC ($\mathcal{L}_{\in}$) are inter-translatable. In CT there is the specific category **Set**, whose objects can be identified with sets and whose arrows are maps. In ZFC one can identify objects with sets and arrows with ordered pair-sets of type $\langle f, C \rangle$, consisting of a mapping f and a co-domain C.[19]

In spite of the fact that some mathematical physicists have applied categories to physics, not a single structural realist on record has advocated replacing ZFC with CT. One of the very few critics of the use of set-theory for StrR (if not the only critic) is E. M. Landry [2007], who has argued that the set-theoretical framework does not always do the work it has been suggested to do; but even she does not openly advocate CT as the superior framework for StrR, although she does advocate it for mathematical structuralism.[20]

The objects of CT are more general than the Ur-elements one can introduce in ZFC, because whereas primordial elements are not sets, the objects of CT can be *anything*, arrows, sets, functors and categories included. Similar to ZFCU is that CT does not have axioms that somehow restrict the interpretation of 'object'. A CT-object is anything that can be sent around by an arrow, similar to the fact that a set-theoretical Ur-element is anything that can be put in a set. CT-objects obtain an 'identity', a 'nature', from the category they are in: different category, different identity. Outside categories, these objects lose whatever properties and relations they had in the category they came from and they become essentially indiscernible.

One great advantage of CT is that structures, i.e. categories, are not accompanied by all these sets that arise by iterated applications of the power-set and union-set operation, as we have seen in (6), (19), (20), (21) and (22). Nevertheless, the grim story we have been telling for StrR in the framework of ZFC, can be repeated in the framework of CT, of course with a few appropriate adjustments. Objects play the rôle that Ur-elements played even better: we end up with something very similar to (28), on top of saying that HeB definitely is not a composition monoid of objects and arrows. Since there is little point in re-telling the entire story, we leave it as an exercise for the sceptical reader. The end of the story is the same problem about reference and description we landed in with ZFCU.

Our conclusion is that the category-theoretical road to physical reality for StrR to walk on also seems a road to nowhere. Before we kiss ZFC and CT goodbye, we want to explore the possibility of retaining them *both*. This seemingly impossible possibility arises when we put the concept of *representation* center stage and see whether it can help us with clarifying the Central Claim of StrR (23).

19 See further Muller [1998], pp. 485–496.

20 When Landry [2007] argues against Suppes, French, etc. that a set-theoretical framework is *not necessary* to make things rigorous, she takes 'necessity' in a sense that is stronger than Suppes, French, etc. have ever meant it whenever they used it or sibling phrases.

4. Representation

Recently the concept of representation has gained momentum in the philosophy of science.[21] The simplest concept of representation conceivable is expressed by the following dyadic predicate: structure $\mathfrak{S}(\mathsf{HeB})$ represents HeB. S. French [2003] defended that to represent something in science is the same as to have a model for it, where models are set-structures; then 'representation' and 'model' become synonyms and so do 'to represent' and 'to model' (considered as a verb). Nevertheless, this simplest conception was quickly thrown overboard as too simple by amongst others R. N. Giere [2004], p. 743, who replaced this dyadic predicate with a quadratic predicate to express a more involved concept of representation:

$$\text{Scientist } S \text{ uses model } \mathfrak{S} \text{ to represent being } \mathsf{B} \text{ for purpose } P, \qquad (29)$$

where 'model' can here be identified with 'structure'. Another step was set by B. C. van Fraassen. As early as 1994, in his contribution to J. Hilgevoord's *Physics and our View of the World*, Van Fraassen [1994] brought Nelson Goodman's distinction between *representation-of* and *representation-as* – drawn in his seminal *Languages of Art* (1968) – to bear on science; he went on to argue that all representation in science is representation-as. We *represent* a Helium atom in a uniform magnetic field *as* a set-theoretical wave-mechanical structure $\mathfrak{S}(\mathsf{HeB})$ (3). In his new tome *Scientific Representation* [2008], Van Fraassen has moved essentially to a hexadic predicate to express the most fundamental and most involved concept of representation to date:

$$\text{Repr}(S,\, V,\, \mathfrak{S},\, \mathsf{B},\, F,\, P)\,, \qquad (30)$$

which reads: subject or scientist S is V-ing artefact $\mathfrak{S}$ to represent B as an F for purpose P. Example: In the 1920ies, Heisenberg (S) constructed (V) a mathematical object ($\mathfrak{S}$) to represent a Helium atom (B) as a wave-mechanical structure (F) to calculate its electro-magnetic spectrum (P). We concentrate on the following triadic predicate, which is derived from the fundamental hexadic one (30):

$$\text{ReprAs}(\mathfrak{S},\, \mathsf{B},\, F) \quad \text{iff} \quad \exists\, S,\, \exists\, V,\, \exists\, P:\ \text{Repr}(S,\, V,\, A,\, \mathsf{B},\, F,\, P)\,, \qquad (31)$$

which reads: abstract object $\mathfrak{S}$ *represents* being B *as* an F, so that $F(\mathfrak{S})$.

Brief historical interlude. Giere, Van Fraassen and contemporaries are not the first to include manifestations of human agency in their analysis of models and representation in science. Almost half a century ago, Peter Achinstein [1965], pp. 104-105, expounded the following as a characteristic of models in science:

> A theoretical model is treated as an approximation useful for certain purposes. (...)
> The value of a given model, therefore, can be judged from different though related viewpoints: how well it serves the purposes for which it is eimployed, and the completeness and accuracy of the representation it proposes. (...)
> To propose something as a *model* of X is to suggest it as way of representing X which

21 Suárez [2003], Giere [2004], Frigg [2006], Fraassen [1994], [2008].

provides at least some approximation of the actual situation; moreover, *it is to admit the possibility of alternative representations useful for different purposes*.

One year later, M. W. Wartofsky explicitly proposed, during the Annual Meeting of the American Philosophical Association, Western Division, Philadelphia, 1966, to consider a model as a genus of representation, to take in that representation involves "relevant respects for relevant for purposes", and to consider "the modelling relation triadically in this way: $M(S, x, y)$, where S takes x as a model of y".[22] Two years later, in 1968, Wartofsky wrote in his essay 'Telos and Technique: Models as Modes of Action' the following (our emphasis):

> In this sense, models are embodiments of purpose and, at the same time, instruments for carrying out such purposes. Let me attempt to clarify this idea. No entity is a model of anything simply by virtue of looking like, or being like, that thing. Anything is like anything else in an infinite number of respects and certainly in some specifiable respect; thus, *if I like, I may take anything as a model of anything else, as long as I can specify the respect in which I take it. There is no restriction on this.* Thus an array of teacups, for example, may be take as a model for the employment of infantry battalions, and matchsticks as models of mu-mesons, there being some properties that any of these things share with the others. But *when we choose something to be a model, we choose it with some end in view*, even when that end in view is simply to aid the imagination or the understanding. In the most trivial cases, then, the model is already *normative* and *telic*. It is *normative* in that is chosen to represent abstractly only certain features of the thing we model, not everything all at once, but those features we take to be important or significant or valuable. The model is *telic* in that *significance and value can exist only with respect to some end in view or purpose that the model serves.*[23]

Further, during the 1950ies and 1960ies the role of *analogies*, besides that of models, was much discussed among philosophers of science (Hesse, Achinstein, Girill, Nagel, Braithwaite, Wartofsky). We predict that several insights buried in the ensuing literature will be re-discovered by the contemporary division of representationalists. *End of brief historical interlude.*

On the basis of the general concept of representation (30), we can echo Wartofsky by asserting that almost anything can represent everything for someone for some purpose.[24] In scientific representations, *representans* and *representandum* (to introduce another pair of Latin barbarisms) will share some features, but not all features, because to represent is neither to mirror nor to copy. Realists, a-realists and anti-realists will all agree that ReprAs($\mathfrak{S}$, B, F) is true *only if* on the basis of $F(\mathfrak{S})$ one can save all phenomena that being B gives rise to, i.e. one can calculate or accommodate all measurement results obtained from observing B or experimenting with B. Whilst for *structural empiricists* like Van Fraassen this is also sufficient, for StrR it is not. StrR will want to add that structure $\mathfrak{S}$ of type F 'is

22 Collected in Wartofsky [1979], quotation on p. 6.

23 Collected in Wartofsky [1979], p. 142.

24 Almost anything, not everything: has anyone ever taken the universe as a whole to represent something?

realised', that $\mathfrak{S}$ of type F truly *is* the structure of being B or refers to B, so that also $F(\mathsf{B})$. StrR will want to order the representations of being B that scientists have constructed during the course of history as approaching the one and only true structure of B, its structure *an sich*, the Kantian regulative ideal of StrR. But this talk of truth and reference, of beings and structures *an sich*, is in dissonance with the concept of representation-as.

Some being B *can* be represented *as* many other things and all the ensuing representations are all hunky-dory if each one serves some purpose of some subject. That is the idea of (30). When the concept of representation-as is taken as pivotal to make sense of science, then the sort of 'perspectivalism' that Giere [2004] advocates is more in consonance with the ensuing view of science than realism is. Giere [2004] attempts to hammer a weak variety of realism into his 'perspectivalism': all perspectives are perspectives on *one and the same* reality and from every perspective *something* is said that can be interpreted realistically: in certain respects the representans resembles its representandum to certain degrees. A single unified picture of the world is however not to be had. Nancy Cartwright's *dappled world* seems more near to Giere's residence of patchwork realism. A unified picture of the physical world that realists dream of is completely out of the picture here. With friends like that, realism needs no enemies.

There is *prima facie* a way, however, for realists to express themselves in terms of representation, as follows. First, fix the purpose P to be: *to describe the world as it is*. When this fixed purpose leaves a variety of representations on the table, then choose the representation that is empirically superior, that is, that performs best in terms of describing the phenomena, because the phenomena are part of the world. This can be established objectively. When this still leaves more than one representation on the table, which thus save the phenomena equally well, choose the one that best *explains* the phenomena. In this context, Van Fraassen [1994] mentions the many interpretations of QM: each one constitutes a different representation of the same beings, or of only the same observable beings (phenomena), their similarities notwithstanding. Do all these interpretations provide equally good explanations? This can be established objectively too, but every judgment here will depend on which view of explanation is employed. Suppose we are left with a single structure $\mathfrak{A}$, of type G. Then we assert that '$G(\mathsf{B})$' is true. When this 'G' predicates structure to B, we still need to know what 'structure' literally means in order to know what it is that we attribute to B, of what $\mathfrak{A}$ is that B instantiates, and, even more important, we need to know this for our descriptivist account of reference, which realists need in order to be realists. Yes, we now have arrived where we were at the end of the previous two Sections. We conclude that this way for realists, to express themselves in terms of representation (as announced at the beginning of this paragraph), is a dead end. The concept of representation is not going to help them.

We conclude that applauding for a variety of different representations-as of the beings does not serve the aim of realism, StrR included. The need for substantive accounts of truth and reference fade away as soon as one adopts a view of science that takes the concept of *representation-as* as its pivotal concept. Fundamentally different kinds of mathematical structure, set-theoretical and category-theoretical, can then easily be accommodated. They are 'only representations'. That is moving away from realism, StrR included, *dissolving* rather than *solving* the problem for StrR of clarifying its Central Claim (23) – 'dissolved', because 'is or has' is replaced with 'is represented-as'. Realism wants to know what B *is*, not only how it can be represented for someone who wants to do something for some purpose. When we take it for granted that StrR needs substantive accounts of truth and reference, more specifically a descriptivist account of reference and then an account of truth by means of reference, then a characterisation of structure as directly as possible, without committing one to a profusion of abstract objects, is mandatory. This issue we address in the next and final Section.

5. Direct Characterisation

Suppose we have a zoological theory E about Elephants. The word 'elephant' is logically speaking an 'elephant-variable'. Since elephants are concrete observable animate beings easy to recognise, it serves no scientific purpose to think of postulates for E that will characterise what an elephant is. (This would be different if we were considering a theory of ants, of which there are about 12,000 species.) That is why one will look in vain in zoology for such postulates.

Suppose next we have a mathematical theory N about Natural Numbers. Unlike elephants, natural numbers are abstract and therefore unobservable objects, but like elephants they are easy to recognise. Children recognise elephants and natural numbers effortlessly. Since natural numbers are however crucially involved in the most rigorous intellectual praxis that human civilisation so far has produced, i.e. *mathematics*, wherein *theorems* are *proved* about natural numbers and *theorems* are *proved* about other abstract objects that employ natural numbers, it serves a mathematical purpose to think of axioms for N that will characterise what sort of abstract object a natural number is. We need to know exactly what holds for them in order to know what we can use and what we cannot use in proofs of theorems that are about them or involve them. That is why one will *not* look in vain in mathematics for such axioms. Gottlob Frege, Richard Dedekind, Giuseppe Peano and William Lawvere have provided such axioms.[25]

What StrR needs, we submit, is a theory S about Structures. Just as we can say with a clear philosophical conscience that ZFC implicitly defines the set-concept, we want to say that S implicitly defines the structure-concept.[26] Hence just as the language of ZFC, $\mathcal{L}_{\in}$, takes the set-concept as primitive by having set-variables,

25 See Muller [1998], pp. 56–64.

26 For how to obtain a clear philosophical conscience, see Muller [2004], [2005]. This concept of 'implicit definability' should not be confused E.W. Beth's concept that is expressed by the same expression, which by the way is better expressed by 'semantic definability'.

the language of pure structure theory, call it $\mathcal{L}_{\mathfrak{S}}$, must have structure-variables. The concept of structure ought not to be reduced to *other* concepts, such as *sets* (Section 2) or *objects-cum-arrows* (Section 3). The project to construct $\mathcal{L}_{\mathfrak{S}}$ and S will have to wait for another occasion. For now, let us suppose that we possess S. Will that be of any help in clarifying the Central Claim of StrR (23)? Will it provide a literal description of B that any substantive account of reference requires?

To begin with, just as an elephant-realist will take his elephant-variables in E running over at least all elephants on planet Earth, StrR will take the *structure-variables* of $\mathcal{L}_{\mathfrak{S}}$ to range over at least all structures in physical reality. The domain of discourse of S, call it $\mathbf{S}$, will of course harbour a plenitude of structures, just as $\mathbf{V}$ of ZFC harbours a plenitude of sets. When we take ZFC to provide the foundations of mathematics that is used in science, then only sets in the lower tip of $\mathbf{V}$, say below ordinal level $\omega + \omega$, will be employed. Similarly, only *some* of the structures in $\mathbf{S}$ of theory S will be candidates of which StrR will want to say that they are 'realised', or instantiated, in physical reality. Science, physics in particular, will tell us *which* ones are those candidates. StrR will then submit that those predicates in $\mathcal{L}_{\mathfrak{S}}$ that single out these candidates provide *literal descriptions of* those structures. Exactly here, within the confines of S, a substantive descriptivist account of reference will find its Archimedean point, and reference will lead the realist to truth.

Recall that the Central Claim of StrR, *being* B *is or has structure* $\mathfrak{S}$ (23), stood in need of clarification. The clarification we have in the offing with theory S runs, in summary fashion, as follows. First, we advise StrR to say that *being* B *is a structure* $\mathfrak{S}$ *of type* F, where 'F' is a predicate in $\mathcal{L}_{\mathfrak{S}}$ such that $F(\mathfrak{S})$, and where '$\mathfrak{S}$' now is a structure-variable of $\mathcal{L}_{\mathfrak{S}}$. Then StrR should simply stipulate, on the basis of $F(\mathfrak{S})$, that predicate F also supplies a structural type-description *of being* B, in other words, StrR should also assert *that* $F(\mathsf{B})$. This '$F(\mathsf{B})$' is the literal description of B that any descriptivist account of reference can take aboard.

But, wait a minute, before this paper ends: how about the inscrutability of reference?

What about it? That is a problem for everyone who wants to adopt a semantic theory that unrestrictedly aims to save all *and only* observable behaviouristic facts of ascent and dissent of language-users. As soon as more is required of a semantic theory, inscrutability arguments are blocked, which is not to deny the seriousness of the problem of what precisely these additional requirements are.[27] Since inscrutability is a problem for realists and anti-realists alike, rather than a problem for scientific realism let alone StrR in particular, the issue makes little difference in the realism debate and we have therefore ignored it. ■

27 See Williams [2008] for the strongest inscrutability result so far: a permutation argument for D.K. Lewis' Montague-based *general semantics*.

Acknowledgments
Many thanks to Ioannis Votsis (Heinrich Heine Universität, Düsseldorf, Germany) and Dennis Dieks (Utrecht University, The Netherlands) for comments.

REFERENCES

P. Achinstein, 'Theoretical Models', *British Journal for the Philosophy of Science* **16** (1965) 102–120.

N. Bourbaki, *The Theory of Sets*, Paris: Hermann, 1968 (translation of *Théorie des Ensembles*, Paris: Hermann, 1949).

K. Brading, E.M. Landry, 'Scientific Structuralism: Presentation and Representation', *Philosophy of Science* **73** (2006) 571–581.

N.C.A. da Costa, R. Chuaqui, 'On Suppes' Set-Theoretical Predicates', *Erkenntnis* **29** (1988) 95–112.

N.C.A. da Costa, S. French 'Models, Theories and Structures: Thirty Years On', *Philosophy of Science Association Proceedings* **67** (2000) S116–S127.

B.C. van Fraassen, *Scientific Representation: Paradoxes of Perspective*, Oxford: Clarendon Press, 2008.

Fraassen, B.C. van, 'Interpretation of science; science as interpretation', in: *Physics and our View of the World*, J. Hilgevoord (ed.), Cambridge: Cambridge University Press, 1994, pp. 169–187.

B.C. van Fraassen, *The Scientific Image*, Oxford: Clarendon Press, 1980.

S. French, 'A Model-Theoretic Account of Representation', *Philosophy of Science* (PSA Proceedings) **70** (2003) 1472–1483.

R. Frigg, 'Scientific Representation and the Semantic View of Theories', *Theoria* **55** (2006) 49–65.

G.N. Giere, 'How Models Are Used to Represent Reality', *Philosophy of Science* **71** (2004) 742–752.

C. Gauker, 'Scientific Realism as an Issue in Semantics', in: P. Greenough, M.P. Lynch (eds.), *Truth and Realism*, Oxford: Clarendon Press, 2006, pp. 123–136.

J. Ladyman, 'Structural Realism', *The Stanford Encyclopedia of Philosophy* (Summer 2009 Edition), E.N. Zalta (ed.), URL = ⟨http://plato.stanford.edu⟩.

J. Ladyman, 'What is Structural Realism?', *Studies in the History and Philosophy of Science* **29** (1998) 409–424.

E.M. Landry, 'Shared Structure need not be Set-Structure', *Synthese* **58** (2007) 1–17.

H. Lyre, 'Humean Perspectives on Structural Realism', in this book.

A.A. Fraenkel, *et al.*, *Foundations of Set Theory* (2nd Ed.), Amsterdam: North-Holland, 1973.

F.A. Muller, 'Withering Away, Weakly', to appear in *Synthese*, 2009.

F.A. Muller, 'The Implicit Definition of the Set-Concept', *Synthese* **138** (2004) 417-451.

F.A. Muller, 'Deflating Skolem', *Synthese* **143** (2005) 223-253.
F.A. Muller, *Structures for Everyone*, Amsterdam: A. Gerits & Son, 1998.
M.L.G. Redhead, 'Symmetry in Inter-theory Relations', *Synthese* **32** (1975) 77–.
M. Suárez, 'Scientific representation: against similarity and isomorphism', *International Studies in the Philosophy of Science* **17.3** (2003) 225–244.
P. Suppes, *Representation and Invariance of Scientific Structures*, Stanford: Centre for Logic, Language and Computation (distributed by Chicago University Press), 2002.
P. Suppes, 'What Is a Scientific Theory?', in: S. Morgenbesser (ed.), *Philosophy of Science Today*, New York: Basic Books, 1967: 55–67.
P. Suppes, 'A Comparison of the Meaning and Use of Models in Mathematics and the Empirical Sciences', *Synthese* **12** (1960) 287–301.
M.W. Wartofsky, *Models: Representation and Scientific Understanding*, Dordrecht, Holland: D. Reidel, 1979.
J.R.G. Williams, 'The Price of Inscrutability', *Nous* **42.4** (2008) 600–641.
J. Worrall, 'Structural Realism: the Best of Both Worlds?', *Dialectica* **43** (1989) 99–124.
J. Worrall, 'Realisms for Sale', talk at workshop 'Scientific Realism Revisited', London: London School of Economics, 28 April 2009.

Faculty of Philosophy
Institute for the History and Foundations of Science
Erasmus University Rotterdam
Burg. Oudlaan 50, H5–16
3062 PA Rotterdam
f.a.muller@fwb.eur.nl

Dept. of Physics & Astronomy
Utrecht University
Budapestlaan 6, IGG–3.08
3584 CD Utrecht
The Netherlands
f.a.muller@uu.nl

Index of Names

Not included are: Footnotes, Tables, References and Figures

F. Stadler (ed.), *The Present Situation in the Philosophy of Science*,
The Philosophy of Science in a European Perspective, DOI 10.1007/978-90-481-9115-4,

CPSIA information can be obtained at www.ICGtesting.com
Printed in the USA
LVOW05*2008021214
416759LV00011B/129/P